CHAOS AND COMPLEXITY IN NONLINEAR ELECTRONIC CIRCUITS

WORLD SCIENTIFIC SERIES ON NONLINEAR SCIENCE

Editor: Leon O. Chua
University of California, Berkeley

Series A.   MONOGRAPHS AND TREATISES*

*To view the complete list of the published volumes in the series, please visit:
http://www.worldscientific.com/series/wssnsa

WORLD SCIENTIFIC SERIES ON
NONLINEAR SCIENCE

Series A   Vol. 22

Series Editor: Leon O. Chua

CHAOS AND COMPLEXITY IN NONLINEAR ELECTRONIC CIRCUITS

Maciej J. Ogorzałek
University of Mining and Metallurgy
Poland

World Scientific
Singapore • New Jersey • London • Hong Kong

Published by

World Scientific Publishing Co. Pte. Ltd.

5 Toh Tuck Link, Singapore 596224

USA office: 27 Warren Street, Suite 401-402, Hackensack, NJ 07601

UK office: 57 Shelton Street, Covent Garden, London WC2H 9HE

British Library Cataloguing-in-Publication Data

A catalogue record for this book is available from the British Library.

World Scientific Series on Nonlinear Science Series A — Vol. 22

CHAOS AND COMPLEXITY IN NONLINEAR ELECTRONIC CIRCUITS

ISBN-13 978-981-02-2873-6

ISBN-10 981-02-2873-2

To
Anna
Krzysztof
Piotr

PREFACE

All real physical systems are nonlinear in nature. This simple observation is true also for electrical and electronic circuits even though many of them are designed to perform linear transformations on signals. Apart from systems designed for "linear signal processing", many systems have to be nonlinear by assumption, for example rectifiers, flip-flops, modulators and demodulators, memory cells. The design procedures for such circuits are in most cases linear and cannot account for nonidealities and nonlinear phenomena. In many cases the designed circuit, when implemented, performs in a very unexpected way, totally different from that for which it was designed. In most cases, engineers do not care about the origins and mechanisms of the malfunction; for them a circuit which does not perform as desired is of no use and has to be rejected or redesigned. Often behavior such as "excess noise", "latch-up", "false locking" etc. is reported without fully understanding the underlying phenomena. We hope that this book will help engineers to recognize and understand many bizarre cases which are encountered in everyday practice. This is the first goal of this book.

It should also be noted that during the last decade there has been significant development in research on the dynamics of nonlinear electrical and electronic circuits and in the study of chaos. This is due to two facts.

Firstly, electrical and electronic circuits constitute a group of real physical systems in which observations and measurements are relatively easy to make. Various types of behavior including numerous types of bifurcations and chaotic phenomena can be observed using general purpose laboratory tools such as oscilloscopes and spectrum analyzers. More sophisticated specialized tools for tracing solution curves in three dimensions and taking Poincaré sections can be used to pursue experimental analysis even further. Also, general purpose circuit simulation programs such as SPICE are available to designers and researchers. Such an "experimental comfort" enabled thorough studies confirming the existence of strange unexpected behavior in almost every type of electronic circuit — oscillators, filters, instrumentation circuits, switched capacitor circuits, digital circuits, power supplies, PLLs, electric machines, microwave circuits, electro-optic systems etc. The main problem remains in interpreting experimental data.

Secondly, electronic circuits offer an unprecedented opportunity for researchers — with the development of IC technology we can build cheap laboratory experimental setups that reflect properties of almost any proposed model and make measurements in real time in a real physical system for a wide range of system parameters. This makes possible experiments that are not available in any other domain of research such as physics, medicine, biology, economics etc.. A good example of such a "universal" research circuit is Chua's circuit which can mimic a large variety of bifurcations and chaotic phenomena when changing just one of the circuit elements; this circuit provides a useful paradigm for understanding these types of behavior. As a historical note let us mention that many scientists working in the field of electrical engineering and electronics made significant contributions to the development of the theory of nonlinear oscillations and chaotic behavior.

Unexpected behavior was probably first observed in the famous experiment of Van der Pol and Van der Mark but had not been recognized as chaos until relatively recently. Milestones in research into nonlinear oscillations were marked by C. Hayashi and his pupil Y. Ueda from Kyoto University who recognized, for the first time, chaotic oscillations in the forced Duffing system. These early experiments were followed by researchers in many universities around the world with a recent boost given by Leon O. Chua and his collaborators.

So far the available literature concerning chaos and complexity concentrated on two subjects: mathematical (analytical and numerical) methods, and the description and analysis of experimental data coming from various domains of science, primarily physics.

During the past decade many papers concerning specific examples appeared also in journals biased towards the area of circuits and systems (both theory and applications) and in the area of electrical engineering and electronics. These articles however do not present any systematic way of treating such phenomena and are of limited use for engineers. To our knowledge, there is no textbook available on the market which treats this subject from the viewpoint of electronic circuits.

This book is not meant either to repeat mathematical tools or to compete with the results of other authors. Except in very few places, we do not review results published by other authors. Providing basic knowledge of the mathematical notions which are needed to understand nonlinear phenomena, this book guides the reader through several encounters with chaos and complexity studied by the author himself during the past decade, from the analysis of an RC oscillator circuit and digital filters to applications using synchronized chaotic circuits and chaos control techniques.

We concentrate on specific methods for recognizing and analyzing complexity and chaos in electronic circuits. There are many specific experimental and simulation tools available for analyzing electronic and electrical circuits. These tools in some cases might be useful also in other disciplines.

We do not aim to present an in-depth study of mathematical methods. We intend rather to provide a guide for researchers and engineers in the domain of electronic circuits how to deal with bizarre phenomena, how to distinguish them among other types of behavior and how to identify the underlying mechanisms. For this purpose we give a brief review of basic notions used and mathematical tools available as it seems that many phenomena cannot be discovered and understood without introducing more abstract mathematical concepts.

We hope that readers looking at the thorough analysis of examples (employing various available analytical and experimental tools) will be able to apply the approach and tools presented here in their own fields of study. Also having an extensive list of references they will be able to find what is not included in the book but might be of interest.

Organization of the book

The basic mathematical apparatus and notions used throughout the book are introduced in Chapter 1. This should permit an interpretation and analysis of the experimental results. Further, we aim to give some insight into basic problems: how nonlinear circuits differ from linear ones, where the sources of nonlinearity arise in real circuit implementations. We introduce the notions of steady state (asymptotic) behavior and discuss various concepts for defining chaos and complex behavior. Finally a brief discussion of experimental results versus simulation results is given.

A large part of this book can be considered as "The electronic engineers' toolkit for nonlinear dynamics". Using two examples, namely the RC-ladder oscillator and simple digital filter structures, applications of various tools are described. In particular, we describe the interpretation of laboratory and simulation experiments, and some of the available analytical approaches for recognizing and verifying chaotic behavior.

In the last part of the book we discuss issues for avoiding and controlling complex systems. Possible applications in data and signal processing as well as noise generation are considered.

Even before writing the book, I realized that the subject and the underlying theoretical formalisms and models are quite multidisciplinary in nature. While such complexities make the task of presenting the subject rather challenging, they also make it very exciting and hopefully interesting also for specialists from other domains.

The origins of my interest in chaos date back to the period of preparing my PhD thesis, when studying design procedures for sinusoidal oscillators by chance I discovered strange behavior in the RC-ladder oscillator. It was the constant encouragement of my supervisor Prof. Wojciech Mitkowski and the friendly atmosphere of collaboration with the team of Prof. Jacek Kudrewicz at Warsaw University of Technology that resulted in many interesting results. Certainly these two people with exceptional personalities inspired me to become a researcher. The study of chaotic behavior in analog circuits later became my principal research interest for at least five years.

As a novice in the domain of nonlinear dynamics, I wrote a letter to Prof. Leon Chua from the University of California, Berkeley, asking his opinion about my discoveries. His comments and encouragement resulted in the first of my papers published in the IEEE Transactions on Circuits and Systems. Since then, Leon Chua has always been most supportive; in several moments of hesitation and disappointment with results, he was always there to give friendly encouragement and useful comments. Several works of mine would not have been possible without Leon.

During that time I also had the opportunity to visit the Center of Modeling, Nonlinear Dynamics, and Irreversible Thermodynamics (MIDIT) at the Technical University of Denmark, Lyngby, thanks to an invitation from Prof. Erik Mosekilde. During several visits to MIDIT not only did I learn a lot but I also had the chance to meet leading chaos researchers from many laboratories. It was a distinct pleasure for me to share an office with a pioneer in the chaos domain, Prof. Yoshisuke Ueda from Kyoto University. His enthusiasm deeply influenced my research.

During visits to many laboratories engaged in chaos or nonlinear dynamics research and by attending numerous conferences I had the opportunity to meet people who in many ways helped me in carrying out my own studies. My sincere thanks go to Prof. Martin Hasler, dear collaborator and friend, thanks to whom I could profit from excellent laboratories and the nicest possible work environment during several stays as a Visiting Professor at the Chair of Circuits and Systems, Swiss Federal Institute of Technology, Lausanne. To Martin also goes credit for carefully reading parts of this book.

I also have to mention the names of my friends Dr. Hervé Dedieu from the Swiss Federal Institute of Technology, Lausanne and Dr. Peter Kennedy from University College, Dublin. Their constant help and support while writing this book has been invaluable.

This book project would not have been possible without the help of my colleagues from the Department of Electrical Engineering at my native University in Krakow who were always most helpful. I would like to express my thanks to Prof. Wojciech Mitkowski for his guidance and many scientific discussions, and Prof. Stanisław Mitkowski for his encouragement and help in organizing my visits to and collaboration with several laboratories. Special thanks should be expressed to Dr. Zbigniew Galias who painstakingly helped me with the computer experiments, development and installation of the software, and the preparation of many of the figures included in this book. His help was really invaluable to me. In addition, Mr. Andrzej Dąbrowski helped me with many laboratory experiments.

Specific contributions came from Dr. Chai Wah Wu of IBM (formerly with UC Berkeley) who kindly supplied some figures and Dr. Peter Kennedy who supplied some circuit diagrams and data. I would like to thank both of them. I greatly appreciate the assistance and help received from Dr. Tom Parker whose INSITE software toolkit was invaluable in my investigations.

I must also mention all of the institutions which helped me to conduct the research which in part produced the results that are contained in this book. These include: my home University, AGH, for continuous financial support, the Polish National Committee of Scientific Research (KBN) for supporting my research under the grant 8T11D03109 , the Chair of Circuits and Systems at the Swiss Federal Institute of Technology in Lausanne, and the Electronics Research Laboratory of the University of California at Berkeley. Visit, to these laboratories permitted me to complete many experiments and to work i a very calm atmosphere.

CONTENTS

CHAOS AND COMPLEXITY IN NONLINEAR ELECTRONIC CIRCUITS

CHAPTER 1

BASIC NOTIONS AND DEFINITIONS

1.1 Signals and Systems

In everyday practice electrical and electronic engineers working on a variety of different problems are confronted with observations and measurements of different variables, parameters and signals coming from physical and natural systems. These systems are either powered by electrical energy and their principle of operation is electric in nature or electric signals and electronic equipment arise in the techniques which are used to sense and control them. Electrical signals may originate from industrial processes, household appliances, radio, TV, telecommunication equipment, biomedical test equipment or even natural phenomena like wind, weather or seismic activity. In virtually all domains of science and engineering we use electrical signals (voltages, currents, field intensity etc.) in one way or another. Typically these signals are sensed using probes and sensors connected to measurement equipment which in most modern approaches is computerized and serves as a data storage and processing unit. Electrical signals may also be produced as specific outputs during computer experiments. By means of transducers, electrical signals may be used to influence the behavior of the "object" of our interest; they can supply energy necessary for its operation or cause a specific type of behavior to occur.

From the analysis point of view we need to understand the physical meaning of these signals and consider mathematical models of the processes producing such signals. We need to put our observations into a more abstract framework. We need mathematical models which will be able to describe with a needed accuracy the behavior of a real process or object. To be able to analyze the nature of the signals we will use a number of mathematical notions which precisely describe properties that are of interest to the user.

For the purpose of this work we will assume that signals which we observe in practice are produced by systems.

Definition 1.1 (System – intuitive definition) *We will consider a system as an object, a device or a mechanism (phenomenon) performing certain operations or producing some phenomena, and having a number of connections*

1

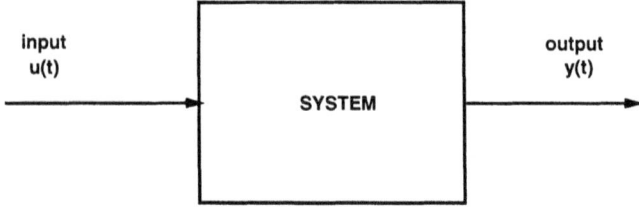

Figure 1.1: Schematic representation of the system.

with the outside world—some of these connections are classified as inputs (by means of which the outside world influences the system behavior) and some as outputs (which give the reaction of the system to the inputs). The system performs operations on the input signals to produce the output signals. It is often denoted symbolically in the form of a block diagram shown in Fig.1.

Definition 1.2 (System - proper mathematical setting) *From the mathematical point of view a system is described by a triplet (U, Y, F), where U is the set of input signals, Y is the set of output signals and the operation (law) $F : U \rightarrow Y$ describes the correspondence between given inputs and outputs (reaction of the system in terms of the outputs to given inputs).*

- **Remark:**

Usually u and y are functions of time (we call then the system *temporal system*) $u = u(t), y = y(t)$.

It should be stressed here that this definition is very general and does not impose any limitations on the classes of the input and output signals (these could be electrical, mechanical or any other variables, continuous in time or discontinuous, corrupted by noise etc.). Furthermore, the operator describing the "action" of the system can take a very general form starting from summation or multiplication, through nonlinear functions, to time averages, probabilistic measures, time-delay operations and integro-differential operators etc..

Simple electronic examples of systems could be amplifiers, filters, A/D and D/A converters, transmission channels etc.

Following the type of signals on which the system operates we can classify systems in the following way:

- **Analog** systems for which the input and output signals are analog ($u : \mathbb{R} \mapsto \mathbb{R}^n$, $y : \mathbb{R} \mapsto \mathbb{R}^n$ or $u : \mathbb{R}^+ \mapsto \mathbb{R}^n$, $y : \mathbb{R}^+ \mapsto \mathbb{R}^n$). Example: analog filter, audio amplifier etc.

- **Discrete-time systems** for which the input and output signals are discrete in time but take analog values ($u : \mathbb{Z} \mapsto \mathbb{R}^n$, $y : \mathbb{Z} \mapsto \mathbb{R}^n$ or ($u : \mathbb{Z}^+ \mapsto \mathbb{R}^n$, $y : \mathbb{Z}^+ \mapsto \mathbb{R}^n$). Example: Switched-capacitor filter.

- **Digital** systems (discrete-time and discrete-value) for which the input and output signals are defined for discrete time moments and can take values from a finite set only. Example: digital filter.

Obviously there exist systems which are mixed analog/digital such as an A/D converter for which the input signals are analog and the outputs are digital. Also one can easily identify systems that do not fall into any of the above-mentioned categories like measuring instruments where outputs might be the read-outs in alpha-numeric form and many others.

The definition of a system given above, although admitting a wide variety of possible operations and signals, suffers from one major drawback. In many cases and especially in nonlinear systems it is not possible to define properly the operation linking the system outputs with its inputs. Either the mathematical description can be given only in the form of implicit equations which are impossible to solve explicitly for the outputs, or determination of the output signals requires knowledge of some additional internal variables referred to as **states** of the system. In this case we say that the system has **memory**. To be able to deal with systems with memory we have to augment our set of definitions.

Throughout this book, we will restrict ourselves to deterministic systems and we will consider in particular **deterministic dynamical systems**. The notion of **dynamics** is closely associated with time evolution of the system variables (internal states and outputs). A **deterministic** system is defined as one for which knowledge of its initial state at some initial time t_0, equations of evolution and input signals fully determines its state and outputs for any $t > t_0$.

In strict mathematical terms we can introduce the following definition:

Definition 1.3 (Dynamical system—abstract definition) *The triplet* (X, T, π), *where X is a metric space[a] T is a group [b] and π is a continuous*

---

[a] A metric space is a pair (X, ρ) where X is a set (later called the space, and its elements called points) and ρ is a mapping $\rho : X \times X \mapsto R^+ \cup 0$ called the distance, satisfying

$$\rho(x, y) = 0 \Leftrightarrow x = y$$
$$\rho(x, y) = \rho(y, x) \text{ for all } x, y \in X$$
$$\rho(x, y) \leq \rho(x, z) + \rho(z, y) \text{ for all } x, y, z \in X$$

[b] A nonempty set T is called a semi-group if an operation $*$ on its elements can be defined

operation from $X \times T$ into X is called a dynamical system if for all $x \in X$ and all $t_1, t_2 \in T$ the following conditions are satisfied:

$$\pi(x, 0) = x \tag{1.1}$$

$$\pi(\pi(x, t_1), t_2) = \pi(x, t_1 + t_2) \tag{1.2}$$

Again, if $T = \mathbb{Z}$ we will say that we consider a discrete dynamical system, and if $T = \mathbb{R}$ we will have a continuous-time (analog) system.

In some situations we consider $T = \mathbb{Z}^+$ or $T = \mathbb{R}^+$; in this case T is a semi-group. In these cases we have a so-called **semi-dynamical system**.

• **Note:** In this definition the input signals are not explicitly mentioned. To define dynamical system in mathematical terms we just describe the time evolution of the state and put some mild restriction on the properties on the operation describing this evolution.

In engineering applications we tend to consider a less abstract definition in which the sets of inputs, states and outputs and operations linking these signals are explicitly defined. This framework enables us to consider for example general problems of finding classes of output signals for given classes of inputs or addressing the common control theory questions (like controllability) for general classes of input signals.

Definition 1.4 (Dynamical System—control engineer's definition)
The quintuplet (U_s, X, Y_s, F_1, F_2), where $U_s = \{u : T \to U\}$ is called the input space (sometimes called also the space of controls), T is a linearly ordered set interpreted as time, U is the set of momentarily values of inputs , X is the state space (often called in the literature the phase space), $Y_s = \{y : T \to Y\}$ is the output space , Y is the set of values of the outputs.

• **Note:** *U_s and Y_s are sets of functions of time.*

F_1 is the state transition operator linking the values of the state vector at two moments of time:

$$F_1 : X \times U_s \times T \times T \to X \tag{1.3}$$

F_2 is the output function:

$$F_2 : X \times U \times T \to Y \tag{1.4}$$

---

such that the following conditions are satisfied:
1. $T \times T \ni (a, b) \mapsto a * b \in T$
2. the operation is associative ie. $a * (b * c) = (a * b) * c$ for all $a, b, c \in T$,
If there exists so-called neutral element $e \in T$ satisfying $a * e = e * a = a$ for all $a \in T$ we call such a set a semi-group with a neutral element.
Further, if for every $a \in T$ there exists a $b \in T$ (called an inverse element) such that $a * b = b * a = e$ the set T is called a group.

The quintuplet is called a **dynamical system** *if the following conditions are satisfied:*

1. **Evolution property** *for the operation F_1 meaning that:*
   if for all $t_1, t_2, t_3 \in T$, $t_1 \leq t_2 \leq t_3$ and for all $u_1, u_2 \in U_s$ such that $u_1(t) = u_2(t)$ for $t \in [t_2, t_3]$ we have:

$$F_1(x(t_1), u_1, t_1, t_3) = F_1(F_1(x(t_1), u_1, t_1, t_2), u_2, t_2, t_3) \qquad (1.5)$$

   and

$$F_1(x(t_1), u_1, t_1, t_1) = x(t_1) \qquad (1.6)$$

2. **Completeness condition** *for F_1 ie.:*
   for all $t, s \in T$, $s < t$, for all $u_1, u_2 \in U_s$ such that $u_1(h) = u_2(h)$ for $h \in [s, t)$ we have

$$F_1(x(s), u_1, s, t) = F_1(x(s), u_2, s, t) \qquad (1.7)$$

   Often instead of the completeness condition we require that F_1 satisfies the **causality** *condition which is easily obtained by fixing s in the completeness condition as a minimal element of the set T (sometimes referred to as "beginning of time" eg. for $T = [t_0, \infty)$ we have to put $s = t_0$).*

3. **Separability condition** *for the inputs meaning that:*
   if u_1, $u_2 \in U_s$ then also $u_3 \in U_s$, where $u_3(t) = u_1(t)$ for $t \in [t_1, t_2)$ and $u_3(t) = u_2(t)$ for $t \in [t_2, t_3)$ where $t_1 < t_2 < t_3$, $t_1, t_2, t_3 \in T$. (Comment: This condition is clearly introduced to be able to define later various control problems and combine control signals).

Interpretation of the functions F_1 and F_2 is straightforward—if we know the state of the system $x(t_0)$ at some time moment $t_0 \in T$ and we know the values of $u(t)$ for $t > t_0$, then the state of the system at $t > t_0$ is only a function of the initial state, the initial and actual time and the input function in the interval between the initial and actual times, and can be calculated using

$$x(t) = F_1(x(t_0), u, t_0, t) \qquad (1.8)$$

The output of the system is a function of the values of the actual time and the state input functions at that moment and can be determined by

$$y(t) = F_2(x(t), u(t), t) \qquad (1.9)$$

As can be seen from the above discussion and the three problem statements there is some freedom in choosing the definition of a system — depending on purpose such definition could serve.

• **Note:** There exist also definitions of system in which initial conditions or parameter values can be considered as inputs. The latter case is often considered in bifurcation studies where influence of parameter changes on dynamic behavior is the principal problem to be solved.

Traditionally the general class of systems is divided into **autonomous** and **non-autonomous** ones. A system is said to be autonomous if it has no external inputs.

In this work we will mainly consider dynamical systems described (in mathematical terms, "generated") by ordinary differential equations of the form:

$$\frac{d\mathbf{x}(t)}{dt} = \mathbf{F}(\mathbf{x}(t), \mathbf{u}(t), t), \ \ \mathbf{x}(t_0) = \mathbf{x}_0 \tag{1.10}$$

with an output function defined as a linear combination of states and inputs

$$\mathbf{y}(t) = \mathbf{C}^T \mathbf{x}(t) + \mathbf{D}\mathbf{u}(t) \tag{1.11}$$

where $\mathbf{x}(t) \in \mathbb{R}^n$, $\mathbf{u}(t) \in \mathbb{R}^m$, $\mathbf{F} : \mathbb{R}^n \times \mathbb{R}^m \times \mathbb{R} \mapsto \mathbb{R}^n$.

This kind of description is often shortened by hiding the functions of time $\mathbf{u}(t)$ in the general dependence of the right-hand side of the equation on time.

$$\frac{d\mathbf{x}(t)}{dt} = \mathbf{F}(\mathbf{x}(t), t), \ \ \mathbf{x}(t_0) = \mathbf{x}_0 \tag{1.12}$$

Where: $\mathbf{x} \in \mathbb{R}^n$, $\mathbf{F} : \mathbb{R}^n \times \mathbb{R} \mapsto \mathbb{R}^n$.

Thus we obtain two equivalent forms corresponding to two general definitions of the dynamical system introduced before.

For such a description we will say that the system is autonomous if the right-hand side of equation (1.12) does not depend explicitly on time, ie.:

$$\frac{d\mathbf{x}(t)}{dt} = \mathbf{F}(x(t)), \ \ \mathbf{x}(t_0) = \mathbf{x}_0 \tag{1.13}$$

The function $\mathbf{F} : \mathbb{R}^n \mapsto \mathbb{R}^n$ is called a **vector field**. The solution of equation (1.13) with the initial condition $\mathbf{x}_0$ is called a **trajectory** and denoted by $\varphi(\mathbf{x}_0, t)$. We will also often denote such a solution simply by $\mathbf{x}(t)$. Trajectory represents a set of pairs "(point coordinates, time)".

The following theorem tells us exact conditions under which the description in terms of differential equations defines a dynamical system in the sense of definition (1.3):

Theorem 1.1 *Let us consider an autonomous system of differential equations (1.13) with a continuous function* $\mathbf{F} : \mathbb{R}^n \mapsto \mathbb{R}^n$. *Let us assume that for every* $\mathbf{x} \in \mathbb{R}^n$ *there exists exactly one solution* $\varphi(\mathbf{x}, t)$ *passing through* $\mathbf{x}$ *and defined on* $\mathbb{R}$ *satisfying the condition* $\varphi(\mathbf{x}, 0) = \mathbf{x}$. *Then the transformation* $\pi(\mathbf{x}, t) \overset{\Delta}{=} \varphi(\mathbf{x}, t)$ *defines a continuous-time dynamical system on* $\mathbb{R}^n$.

Theorem 1.2 *If the transformation* $\mathbf{F}$ *satisfies a global Lipschitz condition, ie.:*

$$\exists K > 0 : \|\mathbf{F}(\mathbf{x}) - \mathbf{F}(\mathbf{y})\| \le K \|\mathbf{x} - \mathbf{y}\| \ \forall \mathbf{x}, \mathbf{y} \in \mathbb{R}^n \tag{1.14}$$

then the condition on existence and uniqueness of solutions mentioned in the previous theorem is satisfied.

• **Note:** For many mathematical models of systems (eg. differential equations) the solutions exist only on subsets of $\mathbb{R}$; they do not exist for all times, eg. they do not exist for $t \to -\infty$ or can be defined only on an interval $t \in [t_0, t_f)$. From the engineer's point of view this can be viewed as a modeling error as real systems always possess solutions for all times.

In the case of discrete-time systems we have the following theorem which clarifies the properties of function f which are required to guarantee the generation of a dynamical system structure:

Theorem 1.3 *If* $f : X \to X$ *is a continuous bijective function and* $\varphi(x, n) \overset{\Delta}{=} f^n(x)$ *then* $(X, \mathbf{Z}, \varphi)$ *is a discrete-time dynamical system. If* f *is not bijective then* (X, φ) *is a semi-dynamical system.*

There exists a special class of discrete-time dynamical systems which are often encountered when analyzing continuous dynamical systems. These are so-called Poincaré maps. Let us introduce the definition of this notion next.

Definition 1.5 (Poincaré map) *Let us denote by* $\varphi_t = \varphi(., t)$ *a family of solutions of the system (1.13). Let us define locally a hyper-plane* $\Sigma \subset \mathbb{R}^n$ *of dimension* $n - 1$, *transversal^c to* φ_t *at some point* $\mathbf{x}_1$. *Let us assume further that there exists a point* $\mathbf{x}_2 = \varphi(\mathbf{x}_1, T) \in \Sigma$ *which belongs to the trajectory of* $\mathbf{x}_1$ *and at which the trajectory intersects* Σ *(ie.* Σ *is also transversal to flow at* $\mathbf{x}_2$). *We will assume that the trajectory for* $t \in (0, T)$ *does not pass through* Σ. *Let* $U \subset \Sigma$ *be some neighborhood of* $\mathbf{x}_1$. *Then the Poincaré map (return map)* $\Pi : U \to \Sigma$ *for point* $\mathbf{x}_u \in U$ *is defined by*

$$\Pi(\mathbf{x}_u) = \Pi_\Sigma(\mathbf{x}) \overset{\Delta}{=} \varphi(\mathbf{x}_u, \tau) \tag{1.15}$$

where $\tau = \tau(\mathbf{x}_u)$ *is the time after which the trajectory* $\varphi(\mathbf{x}_u, t)$ *for the first time returns (and intersects) to* Σ.

---

^{c}Transversal intersection is such that vector tangent to the flow at the intersection point together with the vectors spanning the hyper-plane span whole n-space.

In the definition given above, it is sufficient to consider a bundle of trajectories that correspond to recurrent motion i.e. after a finite time they return to a neighborhood of their initial states. In many systems, trajectories are not recurrent, at least in certain regions of their state space.

We have the following theorem which guarantees the existence of a continuous Poincaré map:

Theorem 1.4 *If the hyper-plane Σ is transversal to the flow φ_t at points $\mathbf{x}_1$ and $\mathbf{x}_2 = \varphi(\mathbf{x}_1, T)$, then there exists an open neighborhood U of $\mathbf{x}_1$ and exactly one mapping $\tau : U \mapsto \mathbb{R}$ which is C^1 and for every $\mathbf{x} \in U$, $\varphi_{\tau(\mathbf{x})}(\mathbf{x}) \in \Sigma$ and $\tau(\mathbf{x}_1) = T$.*

In this way instead of looking at the behavior of system trajectories we can analyze the properties of series of points being successive intersections of a trajectory with the Poincaré plane $\mathbf{x}_0, \mathbf{x}_1, \mathbf{x}_2, \ldots$ (this is valid in regions of the phase space where the Poincaré map is well-defined by system trajectories). The series $\Pi^k(\mathbf{x}_0)$ is called the **trajectory of the Poincaré map** for the point $\mathbf{x}_0$. We will denote by

$$\Pi^n(\mathbf{x}_u) = \Pi(\Pi^{n-1}(\mathbf{x}_u)) \tag{1.16}$$

the n-times composition of the Poincaré map. Equation (1.13) has a *periodic solution* if and only if for the associated Poincaré map there exist a positive integer n and an initial point $\mathbf{x}_*$ such that $\Pi^n(\mathbf{x}_*) = \mathbf{x}_*$. The point $\mathbf{x}_*$ is a *fixed point* of the map Π and is the point of intersection of the periodic *orbit* γ of (1.13) with the plane Σ. The stability of this fixed point reflects the stability properties of the orbit γ. Looking at the trajectory one could say that the periodic trajectory makes n loops before closing itself in the space. If $n = 1$ we have so-called period-one orbit. In more general mathematical setting we have the following result concerning the stability of fixed points of the Poincaré map:

Theorem 1.5 *Let $\mathbf{x}_1$ and $\mathbf{x}_2$ be any two points belonging to some periodic orbit γ. Let Σ_1 be a hyper-plane transversal to the flow at $\mathbf{x}_1$ and let Σ_2 be a hyper-plane transversal to the flow at $\mathbf{x}_2$. Then the Jacobian matrices $S_1 = D\Pi_{\Sigma_1}(\mathbf{x}_1)$ and $S_2 = D\Pi_{\Sigma_2}(\mathbf{x}_2)$ are similar i.e. there exists a non-singular matrix P such that $S_1 = P^{-1}S_2P$.*

A consequence of this theorem is that the eigenvalues of the Jacobian matrix of the Poincaré map generated by a flow (solutions of a continuous-time dynamical system) do not depend on the choice of the point on the periodic orbit or on the choice of the transversal plane.

1.2 Basic properties of solutions

Throughout this book we will be interested in asymptotic properties of solutions of dynamical systems and in particular of (1.13) i.e. we will analyze the behavior of system solutions for $t \to \infty$. [d] Asymptotic behaviors of systems can be very complicated and even bizzare [140,165,166,256]. To be able to distinguish between them we need to introduce some further notions and definitions. [39,171]

Definition 1.6 (Trajectory) *The set* $\gamma(x) = \{(\pi(x,t),t), t \in T\}$ *(set of all pairs (x,t) — "(point,time)") is called a trajectory of point x. The sets* $\gamma^+(x) = \{\pi(x,t) : t \geq 0, t \in T\}$ *and* $\gamma^-(x) = \{\pi(x,t) : t \leq 0, t \in T\}$ *are called a positive- and negative semi-trajectories of point x respectively. The transformation $\pi_x : T \to X$ by $\pi_x(t) = \pi(x,t)$ is called the* **motion** *through x.*

Definition 1.7 (Orbit) *The set all points $\Gamma(x) = \{\pi(x,t) : t \in T\}$ is called an orbit of point x.*

Orbit is a geometric object (a set of points).

Definition 1.8 (Invariant set) *The set $S \subset X$ ($S \subset \mathbb{R}^n$) such that*

$$\pi(x,t) \in S \ \forall x \in S \text{ and } \forall t \in T$$

is called an invariant set.

In terms of trajectories, the invariant set contains whole trajectories i.e. for all $x \in S$, $\gamma(x) \in S$.

Typical examples of limit sets are fixed points, closed (periodic) trajectories, integral manifolds etc. We will be interested in particular in bounded invariant sets.

Definition 1.9 (Non-decomposable set) *Set $S \subset X$ ($S \subset \mathbb{R}^n$) is called non-decomposable if it is closed and invariant and for any $x, y \in S$ and for any $\varepsilon > 0$ there exist $x = x_0, x_1, \ldots, x_n = y$ and $t_1, \ldots t_n \geq 1$ such that the distance between $\pi(x_{i-1}, t_i)$ and x_i is smaller then ε.*

Definition 1.10 (Minimal set) *Set $S \subset X$ ($S \subset \mathbb{R}^n$) is called minimal if it is non-empty, closed and invariant and no subset of it has these properties.*

Definition 1.11 (Fixed point) *The point $x \in X$ is called a fixed point if for every $t \in T$ we have $\pi(x,t) = x$.*

Definition 1.12 (Periodic point) *The point $x \in X$ is called a* **periodic point** *if x is not a fixed point and there exists $\tau > 0$ such that $\pi(x,\tau) = x$. The number $\tau_0 = \inf\{t > 0 : \pi(x,t) = x\}$ is called the period of the periodic point x. If the point x is periodic then the trajectory $\gamma(x)$ is also called periodic.*

---

[d]In engineering practice such a solution, if it exists, is called a steady state.

Definition 1.13 (Quasi-periodic motion) *The motion π_x is called* **quasi-periodic** *if $\forall \varepsilon > 0$ there exists a relatively dense set^e of numbers τ_n, $n \in \mathbf{Z}$ such that*

$$\rho(\pi(x,t), \pi(x,t+\tau_n)) < \varepsilon \tag{1.17}$$

for every $t \in T$ and every integer n, where ρ denotes the distance between points.

Definition 1.14 (Limit sets) *The set $\omega(x)$ of all points p, such that*

$$\exists \{t_i\} : \lim_{t_i \to \infty} \pi(x, t_i) = p$$

is called the positive limit set of point x.

The set $\alpha(x)$ of all points p, such that

$$\exists \{t_i\} : \lim_{t_i \to -\infty} \pi(x, t_i) = p$$

is called the negative limit set of point x.

Definition 1.15 (Positive (negative) recursive set) *Set $S \subset X$ ($S \subset \mathbb{R}^n$) is called positively (resp. negatively) recursive with respect to the set $B \subset X$ if for every $x = x(t) \in B$, (for some $t \in T$) there exists $s > t$ (resp. $s < t$) such that $\pi(x, s) \in S$.*

1.3   Stability notions

We will use several notions of stability to help us establish the asymptotic properties of solutions.

Definition 1.16 (Stability in the sense of Lyapunov)
The trajectory $\gamma(x_0) = \{\pi(x_0, t) : t \in T\}$ is stable in the sense of Lyapunov if for every $\varepsilon > 0$ exists $\delta > 0$ such that:

$$||x - x_0|| < \delta \Rightarrow sup_{t \geq 0} ||\pi(x, t) - \pi(x_0, t)|| < \varepsilon \tag{1.18}$$

If, in addition, $\lim_{t \to \infty} ||\pi(x,t) - \pi(x_0,t)|| = 0$ then the solution $\pi(x_0, t)$ is called **asymptotically stable** *in the sense of Lyapunov.*

Stability in the sense of Lyapunov is a *local* property valid in the vicinity of a chosen solution.

Comments:

---

^{e}The set D of real numbers is called *relatively dense* if there exists $t_0 > 0$ such that $D \cap (t - t_0, t + t_0) \neq \emptyset$ for all $t \in \mathbb{R}$, i.e. D has at least one *accumulation point*.

- For systems described by equations of the type (1.12) the analysis of the stability of a chosen solution $x_0(t) = x(x_0, t)$ can be transformed into analysis of the stability of the zero solution of an associated system. Substituting $x(t) = x_0(t) + z(t)$ we obtain for the new variable $z(t)$ the equation:

$$\frac{dz(t)}{dt} = G(z(t), t) \tag{1.19}$$

where $G(z, t) = F(x_0(t) + z(t), t) - F(x_0(t), t)$, $G(0, t) = 0$. Following the definition of stability in the Lyapunov sense, stability of $x_0(t)$ (trajectory through x_0) is equivalent to stability of the zero solution of (1.19).

- **Linearization**

Let us assume that the function $G(z, t)$ is C^1 with respect to z. Then there exists a matrix $A(t)$ such that for small $||z||$ we have:

$$G(z, t) = A(t)z + g(z, t) \text{ where } \lim_{z \to 0} \frac{||g(z, t)||}{||z||} = 0 \tag{1.20}$$

The matrix $A(t) = \frac{\partial F(x,t)}{\partial x}|_x = x_0(t)$ is called the Jacobian matrix and the equation

$$\frac{dz(t)}{dt} = A(t)z(t) \tag{1.21}$$

is called the linearized equation for the problem (1.19) .

- Let us assume that $G(z, t)$ is periodic in time or a constant matrix. Then the matrix $A(t)$ in (1.21) is also periodic or constant. We have then the following theorem due to Lyapunov:

Theorem 1.6 (Lyapunov) *If all the characteristic exponents of the linearized equation (1.21) have negative real parts (ie. if its zero-solution is asymptotically stable) then the zero solution of the equation (1.19) is also asymptotically stable. If at least one characteristic exponent of (1.21) has a positive real part then the zero solution of (1.19) is not stable in the Lyapunov sense.*

- In an autonomous system (1.13) the function F does not depend on time. Let x^0 denote a fixed point of (1.13) ie. $F(x^0) = 0$. The linearized equation around the fixed point is a differential equation with fixed coefficients;

$$\frac{dz(t)}{dt} = Az(t) \text{ where } A = \frac{\partial F(x)}{\partial x}|_{x=x^0} \tag{1.22}$$

In this case the stability properties of solutions are determined by the eigenvalues of the matrix A ie. the zeros of the characteristic polynomial $W(\lambda) = det(\lambda I - A)$. If all the zeros have negative real parts then the fixed point is asymptotically stable in the sense of Lyapunov. A polynomial $W(\lambda)$, all of whose roots have negative real parts, is called *Hurwitz*.

- *Eigenvalues and stability of equilibrium points*

Let x^* be an *equilibrium point* ie. a solution of the equation:

$$\frac{d\mathbf{x}(t)}{dt} = \mathbf{F}(x(t)) = 0 \qquad (1.23)$$

A complete description of stability of this equilibrium is contained in the linearization of equation 1.13 about this equilibrium point. Eigenvalues of the matrix $\mathbf{A} = \frac{\partial F(x)}{\partial x}|_{x=x^*}$ determine the stability type of the equilibrium point.

If the real parts of all of the eigenvalues of $\mathbf{A}$ are strictly negative, then the equilibrium point x^* is *asymptotically stable* and is called a *sink* because all nearby trajectories converge towards it.

If any of the eigenvalues has a positive real part, the equilibrium point is *unstable*; if all eigenvalues have positive real parts, the equilibrium point is called a *source*. An equilibrium point which has both stable and unstable eigenvalues is called a *saddle*.

An equilibrium point is said to be *hyperbolic* if all of the eigenvalues of $\mathbf{A}$ have non-zero real parts. All hyperbolic equilibrium points are either unstable or asymptotically stable.

For discrete-time dynamical system:

$$\mathbf{x}_{k+1} = \mathbf{G}(\mathbf{x}_k)$$

stability of equilibria x^* is determined by the eigenvalues of the linearization $\mathbf{A} = \frac{\partial G(x)}{\partial x}|_{x=x^*}$ of the vector field $\mathbf{G}$, evaluated at x^*.

The equilibrium point is classified as *stable* if all of the eigenvalues of $\mathbf{A}$ are strictly less than unity in modulus, and *unstable* if any has modulus greater than unity.

- *Eigenvalues, eigenvectors, eigenspaces, stable and unstable manifolds*

Associated with each distinct eigenvalue λ of the Jacobian matrix $\mathbf{A}$ is an eigenvector $\mathbf{v}$ defined by

$$\mathbf{A}\mathbf{v} = \lambda\mathbf{v} \qquad (1.24)$$

A real eigenvalue γ has a real associated eigenvector $\vec{\eta}$. Complex eigenvalues of a real matrix occur in pairs of the form $\sigma \pm j\omega$. The real and imaginary parts of the associated eigenvectors $\vec{\eta}_r \pm j\vec{\eta}_c$ span a plane called a *complex eigenplane*.

For linear (linearized) systems the n_s-dimensional subspace of $\mathbb{R}^n$ associated with the stable eigenvalues of the Jacobian matrix is called the *stable eigenspace*, denoted $E^s(x^*)$. The n_u-dimensional subspace corresponding to the unstable eigenvalues is called the *unstable eigenspace*, denoted $E^u(x^*)$. Geometrically speaking these eigenspaces are lines, planes or hyper-planes.

The analogs of the stable and unstable eigenspaces for a general nonlinear system are called the local stable and unstable *manifolds*[f] $W^s(x^*)$ and $W^u(x^*)$.

The stable manifold $W^s(x^*)$ is defined as the set of all states from which trajectories remain in the manifold and converge under the flow to x^*. The unstable manifold $W^u(x^*)$ is defined as the set of all states from which trajectories remain in the manifold and diverge under the flow from x^*.

By definition, the stable and unstable manifolds are *invariant* under the flow (if $\mathbf{x} \in W^s$, then $\varphi(\mathbf{x}, t) \in W^s$). Furthermore, the n_s- and n_u-dimensional *tangent* spaces to W^s and W^u at x^* are E^s and E^u. In the special case of a linear or affine vector field $\mathbf{F}$, the stable and unstable manifolds are simply the eigenspaces E^s and E^u themselves[171].

• In systems with periodic driving (having a periodic input signal) the function $F(x, t)$ in (1.12) is a periodic function of t. Let $x_T(t)$ be a periodic solution of equation (1.12) with a period which is commensurate with (i.e. rationally related to) that of the driving signal. In such a case, the matrix $A(t) = \frac{\partial F(x,t)}{\partial x}\big|_{x=x_T(t)}$ is a periodic function of time. In this case, the stability analysis of solutions can be reduced the analysis of a system of differential equations with constant coefficients, due to the famous Floquet theorem:

Theorem 1.7 (Floquet) *Let $A(t)$ be a matrix of continuous periodic functions with period T_0. Let us consider a differential equation of the form:*

$$\frac{dx(t)}{dt} = A(t)x(t) \tag{1.25}$$

In this case there exists a nonsingular quadratic matrix $P(t)$ of continuous periodic functions of period T_0 such that the change of variables $x(t) = P(t)z(t)$ brings equation (1.25) into the form

$$\frac{dz(t)}{dt} = Qz(t) \tag{1.26}$$

with a constant Q matrix.

Unfortunately there is no simple way for finding $P(t)$ and Q. This theorem is however important from the qualitative point of view because the solutions will always have a form of a linear combination of solutions of the equation

---

[f]An m-dimensional manifold is a geometrical object every small section of which looks like $\mathbb{R}^m$. For example, a limit cycle of a continuous-time dynamical system is a one-dimensional manifold.

with constant coefficients and continuous periodic functions. The eigenvalues of the matrix Q are called the characteristic multipliers of the equation (1.25).

• Let us consider next the stability of periodic solutions of the equation (1.13). Let $x_T(t) \triangleq x_T(x + T)$ be a nontrivial (non-constant) periodic solution of the autonomous system (1.13). Such a solution can be stable or unstable in the sense of Lyapunov but never asymptotically stable. To prove this it is enough to see that for any t_0 $x_T(t + t_0)$ is also a solution of the equation (1.13) and if t_0 is small then $x_T(0)$ is very near $x_T(t_0)$ and the difference $x_T(t) - x_T(t + t_0)$ is a periodic function not vanishing to 0 as $t \to \infty$. For periodic solutions of autonomous systems we need to introduce a special notion of stability, namely *orbital stability*.

Definition 1.17 (Orbital stability) *Let $x_T(t)$ be a periodic solution of equation (1.13) and let $\gamma_T = \{x_T(t) : t \in \mathbb{R}\}$ be the periodic orbit (trajectory) associated with this solution. In the state space this orbit represents a closed trajectory. Let $\rho(z, \gamma_T)$ be the distance between a point z and the periodic orbit γ_T.*

The periodic solution $x_T(t)$ is called orbitally stable if for every $\varepsilon > 0$ exists $\delta > 0$ such that

$$||x_0 - x_T(0)|| < \delta \text{ implies } \rho_{t \geq 0}\{x(x_0, t), \gamma_T\} < \varepsilon \qquad (1.27)$$

If, in addition,

$$\lim_{t \to \infty} \rho\{x(x_0, t), \gamma_T\} = 0 \qquad (1.28)$$

we say that the periodic solution $x_T(t)$ is orbitally asymptotically stable.

Definition 1.18 (Stability in the sense of Poisson) *The point $x \in X$ is called positively (negatively) stable in the sense of Poisson if every neighborhood U of the point x is positively (resp. negatively) recursive in respect to the one-element set x ie. for every $t \in T$ there exists $s > t$ such that $\pi(x, s) \in U$. (An equivalent condition can be used for this definition, namely: point x is positively (negatively) stable in the Poisson sense if $x \in \omega(x)$ (resp. $x \in \alpha(x)$).*

Definition 1.19 (Non-wandering point) *Point $x \in X$ is called non-wandering if every neighborhood U of x is positively recursive with respect to itself ie. $\forall t \in T$, $\forall y \in U$, $\exists s > t$ such that $\pi(y, s) \in U$.*

Definition 1.20 (Stability in the sense of Lagrange) *For any $x \in X$ the motion π_x is stable in the sense of Lagrange (positively or negatively) if the set $\overline{\gamma(x)}$ $(\overline{\gamma^+(x)}$ or $\overline{\gamma^-(x)})$ is compact.*

In typical situations where $X = \mathbb{R}^n$, stability in the sense of Lagrange is equivalent to boundedness of the sets $\gamma(x)$, $\gamma^+(x)$, $\gamma^-(x)$.

All these notions of stability are closely related to a concept which is widely used when studying nonlinear dynamical systems and chaos, namely **sensitive dependence on initial conditions** .

Definition 1.21 (Sensitive dependence on initial condition) *The mapping $\xi : X \mapsto X$ (defined by some discrete-time dynamical system or a Poincaré map associated with solutions of a continuous-time dynamical system) is said to have sensitive dependence on initial conditions[102] if there exists a number $\tau > 0$ such that for all $x \in X$ and for any neighborhood U of x, there exist $y \in U$ and $n > 0$ such that $\rho(\xi^n(x), \xi^n(y)) > \tau$ (where $\rho(x, y)$ denotes the distance between x and y).*

It means that there exists a **separation constant** τ such that in every infinitely small neighborhood of the point $x \in S$ there is always a point y which will eventually but not necessarily permanently move away from x to a distance of at least τ.

1.3.1 Attracting sets and attractors

Definition 1.22 (Attracting set) *An invariant set $A \subset X$ ($A \subset \mathbb{R}^n$) is called* attracting *if there exists some neighborhood U of A, such that*

$$\forall x \in U : \varphi(x, t) \in U \text{ for } t \geq 0 \text{ and } \lim_{t \to \infty} \varphi(x, t) = A$$

Definition 1.23 (Attractor) *An attracting, bounded invariant set $A \subset X$ ($A \subset \mathbb{R}^n$) which contains a dense trajectory[9] is called an* attractor.

This means that an attractor has a neighborhood $U(A)$ such that for any $x_0 \in U(A)$ the positive semi-trajectory starting at x_0 remains in $U(A)$ for $t \in [0, \infty)$ and tends to A as $t \to \infty$.

Definition 1.24 (Domain of attraction) *The set $A_\Omega = \bigcup_{t \leq 0} \varphi(U, t)$ is called the* domain of attraction *of the attractor A (It is a union of pre-images of points contained in the neighborhood U mentioned in the previous definition).*

For systems described by differential equations (1.12) this definition can be reformulated for domain of attraction of an asymptotically stable solution $x^*(t)$. Namely:

Definition 1.25 (Domain of attraction for solutions of ODE) *The set of all initial points x_0 for which the solutions $x(x_0, t)$ satisfy:*

$$\lim_{t \to \infty} \|x(x_0, t) - x^*(t)\| = 0 \tag{1.29}$$

---

[9] The condition on existence of a dense trajectory means that the attractor is a minimal set in the sense that it does not contain any other attractor, i.e. it is non-decomposable.

is called the domain of attraction of the solution $x^*(t)$.

1.4   Classification of attracting limit sets

Birkhoff[30] proposed a basic classification of trajectories and their corresponding limit sets which was further refined by Andronov. They have proposed the following classification of asymptotic behavior of system trajectories and their associated attracting limit sets:

- trajectories constant in time (limit set – a point),

- Periodic trajectories (limit set – closed curve),

- Quasi-periodic trajectories (limit set - torus),

- Chaotic trajectories. Recurrent, bounded in space trajectories not belonging to any of the above-mentioned classes, (stable in the Poisson sense) ; limit set of the Cantor type often referred to as "strange attractor"[h]

Trajectories belonging to this last class will be our principal interest throughout this book. We will try to refine their description to be able to understand the underlying mechanisms and analyze them.

- **Notes:**

There exist also *special trajectories* which are doubly (in positive and negative time) to equilibrium points or periodic trajectrories. Such trajectories are called *homoclinic* if they constitute a self-link between an equilibrium or periodic orbit and *heteroclinic* if they link two different equilibria or periodic orbits. Homoclinic and heteroclinic trajectories will be discussed also in Chapter 6.

From the electronic engineer's point of view limit sets correspond to so-called steady state behavior ie. after all the transients die out.

1.5   Structural stability and bifurcations

To discuss the notion of *structural stability* we have to consider that the system behavior depends on a set of parameters μ. For example the dynamics of the system is described by a parameterized differential equation of the form:

$$\frac{d\mathbf{x}}{dt} = \mathbf{F}(\mathbf{x}, \mu), \quad \mathbf{x}(0) = \mathbf{x}_0 \tag{1.30}$$

---

[h]A Cantor set is a closed, non-empty set not possessing either internal points or isolated points ie. in every neighborhood of any point belonging to this set there are always points belonging and not belonging to it.

Structural stability refers to sensitivity of the dynamic behavior observed in the system to changes in the parameters. The vector field **F** will be called structurally stable if sufficiently close vector fields **F'** have equivalent dynamics (in terms of existence of a continuous invertible function which transforms **F** into **F'**. The term *conjugate* is sometimes used as a synonym for *equivalent*.

The dynamic behavior observed in the system may vary qualitatively when changing its parameters. For example tuning an oscillator one can clearly identify parameter values for which we observe onset of oscillations - thus we can observe stable equilibrium (no oscillations) or limit cycle (oscillations) depending on the chosen parameter value in the same circuit.

Qualitative changes in the observed dynamic behavior when changing system parameters are called *bifurcations*. The chosen variable parameter for the variation of which we observe changes in dynamic behavior is called the *bifurcation parameter*.

The value of bifurcation parameter at which a qualitative change in observed dynamic behavior is called a *bifurcation point*.

1.5.1 Basic types of bifurcations

In this section we will describe the most common types of bifurcations only. For high-dimensional systems very complex types of bifurcations not considered here can be encountered[2,14,15]. For understanding the phenomena analyzed later in this book we will consider only local bifurcations: the *Hopf bifurcation*, the *saddle-node* or *fold* bifurcation, the *symmetry-breaking* and the *period-doubling* or *flip* bifurcation. These bifurcations are called *local* because they may be understood by linearizing the system close to an equilibrium point or limit cycle.

• **Hopf bifurcation**

The Hopf bifurcation occurs when an equilibrium point changes stability from stable to unstable and a stable limit cycle is born. Looking at the eigenvalues of the linearized system we find that Hopf bifurcation occurs when a pair of complex eigenvalues moves out from the left-hand half of the complex plane to the right-hand half-plane passing the imaginary axis (ie. the sign of the real parts of a pair of complex eigenvalues changes from negative to positive). In electronic circuits this type of bifurcation is easily observable in all generators — at the onset of oscillations the Hopf bifurcation occurs. Hopf bifurcation also occurs for periodic orbits — in such a case a torus is born.

• **Saddle-node bifurcation (fold)**

The simplest case of saddle-node bifurcation is the case in which in the system we have two orbits — a stable one and an unstable one which when changing the bifurcation parameter, approach each-other and disappear at the bifurcation point.

This type of bifurcation is also typical in systems having multiple attractors. In a saddle-node bifurcation, one of two attractors loses its stability and "jumps" to the other. In the simplest case we observe jumps between two equilibrium points. Switching of states in a flip-flop or low-high transition in a memory cell corresponds in terms of dynamics to a saddle-node bifurcation. A common example of this in electronic circuits is also a Schmitt trigger. At the threshold for switching, a stable equilibrium point corresponding to the "high" state merges with the high-gain region's unstable saddle-type equilibrium point and disappears. After a switching transient, the trajectory settles to the other stable equilibrium point, which corresponds to the "low" state.

A saddle-node bifurcation may also manifest itself as a switch between periodic attractors of different size, between a periodic attractor and a chaotic attractor, or between a limit cycle at one frequency and a limit cycle at another frequency.

• **Symmetry-breaking bifurcation**

This is a specific type of bifurcation of periodic orbits (equilibria for maps) in which when changing the value of the bifurcation parameter a single stable periodic orbit of period T splits into two stable orbits of the same period co-existing at the same time. This kind of bifurcation often occurs in systems with symmetric nonlinearities (an example of such behavior will be shown when analyzing bifurcations in the RC-ladder generator).

• **Period-doubling bifurcation (flip)**

A period-doubling bifurcation occurs only with periodic solutions. At the bifurcation point, a periodic trajectory with period T changes smoothly into one with period $2T$. An interesting fact is that the period-1 orbit daos not disappear – it just changes stability becoming unstable. The unstable periodic orbits created apart from stable orbits with doubled period in this kind of bifurcations will manifest themselves when analyzing chaotic attractors. As a matter of fact the infinite number of unstable periodic orbits embedded in a chaotic attractor were born in period doubling bifurcations leading to chaos.

• **Crisis**

The notion of crisis is associated with specific types of bifurcation phenomena in which orbits (attractors) co-existing in the state space collide with each-other or with an unstable orbit resulting in a qualitative change of behavior. This is the case fro example for so-called *blue sky catastrophe* in which the attractor disappears "in the blue sky" after collision with a saddle-type (unstable) periodic orbit. This is a global bifurcation.

1.6 Routes to chaos

Each of the local bifurcations may give rise to a distinct route to chaos if the bifurcations appear repeatedly when changing the bifurcation parameter. These routes are important because it is often difficult to conclude from experimental data alone whether irregular behavior is due to measurement noise or to underlying chaotic dynamics. Recognition of one of the typical routes to chaos in experiments is a good indication that the dynamics may be chaotic. In experiments we typically construct so-called *bifurcation trees* in which for changing parameter values we show corresponding behavior by plotting maxima of chosen variables, coordinates of points on a Poincaré section etc.. In discrete systems, one simply plots successive values of a state variable.

• **Period-doubling route to chaos**

When a cascade of successive period-doubling bifurcations occurs when changing the value of the bifurcation parameter it is often the case that finally the system will reach chaos. For this an infinite sequence of period-doubling is necessary – each of the successive bifurcations occurring at a smaller step of parameter variation. An infinite cascade of such doublings results in a chaotic trajectory of infinite period and a broad frequency spectrum. over a finite range of the bifurcation parameter because of a geometric relationship between the intervals over which the control parameter must be moved to cause successive bifurcations. Period-doubling is the most common type of routes to chaos and often is governed by a universal scaling law which holds in the vicinity of the bifurcation point to chaos μ_∞.

The ratio δ_k of successive intervals μ,

$$\delta_k = \frac{\mu_{2^k} - \mu_{2^{k-1}}}{\mu_{2^{k+1}} - \mu_{2^k}},$$

where μ_{2^k} is the bifurcation point for the period from $2^k T$ to $2^{k+1} T$. In the limit as $k \to \infty$, a universal constant called the *Feigenbaum number* δ is obtained:

$$\lim_{k \to \infty} \delta_k = \sigma = 4.6692\ldots.$$

The period-doubling route to chaos can be identified from a state-space plot (qualitative changes of orbits observed eg. on an oscilloscope), time series, power spectrum (successive appearance of spikes half-way between the already existing ones), or a Poincarémap (doubling number points on the section) when changing the bifurcation parameter value.

• Intermittency route to chaos

The route to chaos caused by saddle-node bifurcations comes in different forms, the common feature of which is a direct transition from regular motion to chaos. The most common type is the intermittency route and results from a single saddle-node bifurcation. Just after the bifurcation, the trajectory is characterized by long intervals of almost regular motion (called *laminar phases*) and short bursts of irregular motion. The period of the oscillations is approximately equal to that of the system just before the bifurcation. At the parameter passes through the critical value μ_∞ at the bifurcation point into the chaotic region, the laminar phases become shorter and the bursts become more frequent, until the regular intervals disappear altoghter. The scaling law for the average interval of the laminar phases depends on $|\mu - \mu_\infty|$ so chaos is not fully developed until some distance from the bifurcation point

Intermittency is best characterized in the time domain since its scaling law governs on the length of laminar phases.

• Torus breakdown route to chaos

The quasiperiodic route to chaos results from a sequence of Hopf bifurcations. In the first one a periodic orbit is born. In the second one this orbit bifurcates into a two-torus. The three-torus generated after the third Hopf bifurcation is not stable in the sense that it is destroyed by an arbitrarily small perturbation of the system (in terms of parameters) for which it disappears giving way to chaos.

1.7 Asymptotic behavior, attractors, limit sets—what can an electronic engineer see in practice ?

In this section we will try to clarify the introduced notions from an electronic engineer's point of view.

We should first note that engineers easily distinguish between linear and nonlinear circuits and systems, and in the everyday practice it is assumed in most cases that everything is linear (ie. in simple words a response of a system to a sum of signals is a sum of the responses to each of the signals separately and a response to a scaled copy of the input (multiplied by a constant) will be a scaled copy of the output (with the same scaling constant)). Linear systems are also "nicely behaved" – there is not much freedom in the classification of the possible responses[69,70,180,182,360,361]:

- all are almost all[i]solutions converge to a fixed point (unique!)

---

[i] Except the initial conditions in a set of measure 0

- all solutions are periodic (with an amplitude depending on the initial conditions,

- for systems with external periodic or quasi-periodic driving all (or almost all) solutions will converge to a periodic or quasi-periodic solution.

- all or almost all solutions are divergent (in reality this situation is not possible due to the finite energy supplied to the circuits and existing dissipation)

Typically such circuits as filters or amplifiers should belong to this class – they must have a unique operating point (stable fixed point in terms of dynamics). In all the design procedures one tries to maintain as close as possible the linearity of the circuitry with respect to the signals.

However there are many applications which require by definition of their functionality the existence of different types of behavior. From the above given classification it is clear that a linear circuit can not serve as a periodic signal generator (supplying the same waveform independently on the initial condition). Neither a linear circuit could serve as a bistable cell which clearly requires existence of two stable operating points (fixed points).

In fact there is a large variety of different circuit behaviors in the case of circuits with nonlinearities. Even considering autonomous nonlinear circuits we can encounter in practice for example the following types of behavior:

- as in the case of linear circuits all solutions converge to a unique operating point (fixed point) (this is the mode of operation of real analog RLC filters, amplifiers etc.)

- all solutions converge to one out of many equilibrium points (this is the mode of operation of bistable circuits, memory cells, threshold detectors, Schmitt triggers, sample-and-hold circuits etc.)

- All solutions converge to a unique periodic or quasi-periodic solution (this is the mode of operation of the oscillators, periodic signal generators etc.)

These are "normal" modes of operation every electronic engineer is accustomed with.

There are some more rarely met situations that are also known in practice: eg. depending on circuit parameters we can observe sub-harmonic solutions in some power circuits (eg. ferro-resonant circuits) (ie. various kinds of stable periodic solutions), one can observe so-called false synchronizations in the PLL circuits or (which means again existence of various stable periodic solutions depending on some parameter, initial condition or input signal choices).

1.7.1   How to recognize the behavior

In the simplest case in the time domain, an equilibrium point of an electronic circuit is simply a DC solution or operating point. This is a typical case when using an oscilloscope we can see the waveforms converging towards a constant level or in the XY mode towards a point on the screen.

Periodic solutions are more difficult to confirm. The time waveform may look periodic but we need better tools to confirm this fact. First we could look at phase plots (XY mode) - periodic solutions form closed curves in space and projections of such curves are easily visible on the scope screen. Further we can consider that every periodic signal $x(t)$ may be decomposed into a Fourier series—a weighted sum of sinusoids at integer multiples of a fundamental frequency Thus, a periodic signal appears in the frequency domain as a set of spikes at integer multiples *harmonics* of the *fundamental* frequency. The amplitudes of these spikes correspond to the coefficients in the Fourier series expansion of $X(t)$. The Fourier transform is an extension of these ideas to aperiodic signals; one considers the distribution of the signal's power over a continuum of frequencies rather than on a discrete set of harmonics[352,358]. The distribution of power in a signal $x(t)$ is most commonly quantified by means of the *power density spectrum*, often simply called the *power spectrum*. Using a spectrum analyzer we can readily make a read-out of at least a part of the signal spectrum. Processing the signals digitally we have to note the following. The simplest estimator of the power spectrum is the periodogram which, given N uniformly spaced samples $X(k/f_s), k = 0, 1, \ldots, N - 1$ of $x(t)$, yields $N/2 + 1$ numbers $P(nf_s/N), n = 0, 1, \ldots, N/2$, where f_s is the sampling frequency. $P(nf_s/N)$ is an estimate of the power in the component at frequency nf_s/N. By Parseval's theorem, the sum of the power in each of these components equals the mean squared amplitude of the N samples of $x(t)$. If $x(t)$ is periodic with period T, then its power will be concentrated in a DC component, a fundamental frequency component $1/T$, and harmonics. In practice, the discrete nature of the sampling process causes power to "leak" between adjacent frequency components; this leakage may be reduced by "windowing" the measured data before calculating the periodogram

Even more difficult situation occurs when observing quasi-periodic trajectories in experiments. Quasiperiodic behavior occurs in systems where two (or more) incommensurate frequencies are present. A periodically-forced or discrete-time dynamical system has a frequency associated with the period T of the forcing or sampling interval T of the system; if a second frequency is introduced which is not rationally related to T, then quasi-periodicity may occur.

A quasiperiodic function may typically be expressed as a countable sum of periodic functions with incommensurate frequencies (for example, $x(t) = \sin(t) + \sin(2\pi t)$ is a quasiperiodic signal). Time waveforms are very difficult to recognize from measurements. In the time domain, a quasiperiodic signal may look like an amplitude- or phase-modulated waveform. XY projection in some cases might help if we can identify a "donut-like" shape in the state space projections. This however is very difficult to visualize in higher dimensions. While the Fourier spectrum of a periodic signal consists of a discrete set of spikes at integer multiples of a fundamental frequency, that of a quasiperiodic solution comprises a discrete set of spikes at *incommensurate* frequencies. In principle, a quasiperiodic signal may be distinguished from a periodic one by determining whether or not the frequency spikes in the Fourier spectrum are harmonically related. In practice, it is impossible to determine whether a measured number is rational or irrational; therefore, any spectrum which appears to be quasiperiodic may simply be periodic with an extremely long period.

DC equilibrium, periodic, and quasiperiodic steady-state behaviors have been correctly identified and classified since the pioneering days of electronics in the 1920s. By contrast, the existence of more exotic steady-state behaviors in electronic circuits has only been acknowledged in the past twenty years. While the notion of chaotic behavior in dynamical systems has existed in the mathematics literature since the turn of the century, unusual behaviors in the physical sciences as recently as the 1960s were described as "strange". Probably most of the practicing engineers have also encountered "wild" types of behaviors – suddenly observing on the oscilloscope a "cloud" of waveforms or being not able to synchronize the scope at all. These bizzare behaviors in most practical cases are immediately judged as useless and usually the circuitry has to be redesigned without looking in much detail into what really happened in our experiment. Often the phenomena are accounted to be a result of noise or "strange couplings". I would dare to say that in most of these bizzare cases of "wrong design" the engineers observe in reality these strange waveforms that are neither constant nor periodic or quasi-periodic but are chaotic! I would even risk a statement that almost all circuits under specific circumstances (choices of parameters, initial conditions, input signals etc) can become chaotic! Today, we classify as *chaos* low-dimensional recurrent[j] motion in deterministic dynamical systems which exhibits both "randomness" and "order"[432].

---

[j] Because a chaotic steady-state does not settle down onto a single well-defined trajectory, the definition of *recurrence* must be used to identify post-transient behavior.

From a experimentalist's point of view, chaos may be defined as *bounded steady-state behavior which is not an equilibrium point, not periodic, and not quasiperiodic.* In the time domain waveforms look "random". Looking at an XY plot on the screen of an oscilloscope the trajectory winds around in a strange way filling the fragments of the space.

Repeated experiments show usually different waveforms as two trajectories started from almost identical initial conditions diverge exponentially and soon become uncorrelated (sensitive dependence on initial conditions).

"Randomness" manifests itself in the frequency domain as a broad "noise-like" Fourier spectrum.

Whether or not such a kind of behavior is interesting in practice we will try to analyze and give some answers in the next chapters.

In fact there are no general methods enabling the forecasting of asymptotic behavior of nonlinear circuits. It is very difficult to find design procedures which take into account system nonlinearities, In many cases the design is linear and the influence of real nonlinearities is tested either by bread-boarding or massive simulations.

At this moment the fundamental questions in analysis of possibly chaotic circuits are: understanding of the underlying phenomena, elaboration of new design methods to avoid unwanted chaotic/complex behavior, elaboration of new techniques in which chaos might prove useful.

To be able to analyze chaotic behavior three types of approaches are being combined:

- modeling of the circuitry and phenomena. Building an abstract mathematical model (dynamical system!) is a first step to understanding the behavior. To a certain extent analytical analysis of the model is also possible (mathematical proofs).

- simulation experiments using the developed model(s) (into this category we could put calculation of system responses for various sets of parameters, calculation of Lyapunov exponents, dimensions, entropy, construction of bifurcation diagrams etc.)

- laboratory experiments and analysis of experimental data.

Only a combination of these three approaches gives convincing results concerning dynamic behavior of the given nonlinear circuit.

CHAPTER 2

QUANTIFYING DYNAMIC BEHAVIOR

The mathematical notions introduced in the previous chapter enable us to distinguish between different kinds of dynamic behaviors. We are able to consider any observed phenomenon as an output of some system and classify it in general terms. Being confronted however with the measurements or observations taken in a real system we have great difficulty to tell what kind of stability has a given solution, whether it is a constant, periodic, quasi-periodic or chaotic one. We are not able to tell either is we have eg. the property of sensitivity to changes of initial conditions.

There are several tools developed which can help the researchers and engineers to quantify the observed behavior. A series of "number" characterisations can be introduced which are related to notions of stability/instability, asymptotic behavior, sensitivity to initial conditions etc. In this chapter we give an overview of these concepts. We will not present here however any numerical algorithms for their computations. Interested readers should consult the excellent books of Parker and Chua[358] or Nusse and Yorke[318].

The overview of notions for quantifying dynamic behavior will start with the concept of Lyapunov exponents.

2.1 Lyapunov exponents

We have already stressed commenting the definition of Lyapunov's stability concepts that they give only a local characterisation of solutions. In particular the notion of stability of Lyapunov tells us what happens to the solutions starting from nearby initial conditions. In other words this notion tells us what is the sensitivity to small changes of initial conditions – the solution is stable if for small changes of initial conditions the solutions will not deviate far from it. Moreover in the case of asymptotic stability one could say that despite the small perturbation of initial conditions the solutions will converge to the fiducial trajectory as $t \to \infty$.

As it could be seen from the comments below the definition of stability in the Lyapunov sense certain conclusions concerning stability could be drawn for systems which are linear or could be linearised in some sub-set of their state space. Theorem 2.4 gives us a characterisation of the stability of the systems in terms of eigenvalues of the Jacobian matrix or characteristic multiplies in the case of periodic matrices.

In a general case (ie. in the whole state space, different kind of system operators etc.) such a characterisation is not possible – eigenvalues can not be computed for nonlinear systems. Unlike linear systems in which the convergence/divergence directions are well defined by the eigenvectors corresponding to eigenvalues, these can not be defined in nonlinear systems.

To characterise the behavior of the neighboring trajectories in a general case we will introduce the notion of **Lyapunov exponents**. To introduce this notion we need to recall several concepts from the measure theory and some concepts of probabilistic systems. AS before we will proceed from the simplest concepts to the more involved ones.

Let us assume that the state space X has a defined probabilistic measure in it denoted by μ^a

Definition 2.1 (Invariant measure) *Let X, T, π – a dynamical system. The measure μ is called an* **invariant measure** *of the dynamical system π if for every $t \in T$, $t > 0$ and any measurable subset E of X the set $\pi(E, -t)$ is measurable and $\mu(\pi(E, -t)) = \mu(E)$.*

This means that the measure of a set (here E) is equal to the measure of any pre-image (backwards in time — at time $-t$) of this set via the system operation π.

Definition 2.2 (Measure preserving operation) *The operation $f : X \to X$ is called* **measure preserving** *if for every measurable set $A \subset X$ the set $f^{-1}(A)$ is also measurable and we have $\mu(A) = \mu(f^{-1}(A))$.*

An invariant measure in general could be decomposed (split) into several partial measures each being invariant again. If this is not the case we say that the measure is ergodic.

Definition 2.3 (Ergodic measure) *Let (X, T, π) – a dynamical system. A dynamical system is called* **ergodic** *with respect to the measure μ if the measure $\mu(A)$ of any invariant set A is equal 0 or 1. We say that the measure μ is* **ergodic** *with respect to the system operator π if the system defined by π is ergodic with respect to μ.*

For ergodic measures we have the following theorem:

---

[a]The measure μ of the space X is called probabilistic if $\mu(X) = 1$

Theorem 2.1 (Ergodic theorem) *If the measure μ is ergodic with respect to the dynamical system defined by π then for every continuous function f the equality:*

$$\lim_{T \to \infty} \int_0^T f(\pi(x_0, t)) dt = \int \mu(dx) f(x) \tag{2.1}$$

is satistied for almost all initial conditions x_0, with respect to the measure μ

This theorem provides an extremely important information – it tells in other words that for almost every initial conditions (except a set of measure 0) the time average of the function f is equal to its space average.

One has however to be very cautious as in general many ergodic measures and theorems may not be informative about typical behaviors (eg. for unstable periodic orbit measure concentrated on it is ergodic).

It can be also concluded[122] that the trajectory of almost every point passes through every set of positive measure, and stays in this set for a time which is asymptotically proportional to the measure of this set. This property is called **asymptotically equal trajectory distribution**.

Theorem 2.2 (Oseledec[108]) *Let $f : X \mapsto X$ – a measure preserving transformation with a measure μ which is ergodic with respect to f. Further let $T : X \mapsto M_{n \times n}$[b] be a measurable transformation such that:*

$$\int \mu(dx) log^+ \|T(x)\| < \infty \tag{2.2}$$

where $log^+ \xi = max(0, log\xi)$. Let $T_x^n \triangleq T(f^{n-1}(x)) \cdots T(f(x)) T(x)$. Then for almost all x with respect to μ there exists the limit:

$$\Lambda_x \triangleq \lim_{n \to \infty} (T_x^{n*} T_x^n)^{\frac{1}{2n}} \tag{2.3}$$

where T_x^{n*} denotes the conjugate matrix ($(T_x^{n*} T_x^n)$ is a positive definite matrix).

Having defined the matrix Λ_x we can now define the Lyapunov exponents:

Definition 2.4 (Lyapunov exponents) *The logarithms of the eigenvalues of the matrix Λ_x are called **Lyapunov exponents**. We will denote them by $\lambda_1 \geq \lambda_2 \geq \cdots \lambda_n$. They are almost everywhere constant with respect to measure μ.*

In simple words the Lyapunov exponents represent averaged in time properties of stretching and squeezing corresponding to the observation of deformations of a small ball centered at some initial condition when it evolves along

---

[b]Here $M_{n \times n}$ denotes the set of all $n \times n$ matrices.

the trajectory passing through this initial point. It should be pointed out here that the stretching/squeezing directions change when we move in the state space (which is not the case of linear systems). Oseledec theorem tells us that these averages do not depend on the choice of the initial point.

Let $\lambda_1' > \lambda_2' > \cdots$ be the non-repeating (different) Lyapunov exponents. Let E_s^i be the subspace of $\mathbb{R}^n$ corresponding to the eigenvalues of Λ_x which are less then or equal to $exp(\lambda_i')$. Then $\mathbb{R}^n = E_x^1 \supset E_x^2 \supset \cdots$ and the following theorem is true:

Theorem 2.3 *For almost all x with respect to measure μ we have:*

$$\lim_{n \to \infty} \frac{1}{n} log ||T_x^n u|| = \lambda_i' \tag{2.4}$$

if $u \in E_x^i \setminus E_x^{i+1}$. In particular for all vectors not belonging to the subspace E_x^2 this limit is equal to the largest Lyapunov exponent.

This theorem is extremely important as it provides a basis for all numerical algorithms used to estimating the largest Lyapunov exponent.

Let us consider next some properties of Lyapunov exponents for discrete- and continuous-time systems. We will consider the discrete-time dynamical systems first as some part of the analysis will be used for continuous-time systems.

2.1.1　Lyapunov exponents for a discrete-time dynamical system

Let us consider now a discrete-time dynamical system $(\mathbb{R}^n, \mathbf{Z}, f)$, where $f : \mathbb{R}^n \times \mathbf{Z} \mapsto \mathbb{R}^n$ is a differentiable function. Let us denote by $T(x)$ the matrix $(\frac{\partial f_i}{\partial x_j})$ of the partial derivatives of f_i calculated at the point x. For the n-th iterate F^n of the operation f one can calculate the partial derivatives using the chain rule, namely:

$$\frac{\partial (f^n)_i}{\partial x_j} = T(f^{n-1}(x)) \cdots T(f(x))T(x) \tag{2.5}$$

If μ is an ergodic measure for the transform f then the assumptions of the Oseledec theorem are satisfied and the Lyapunov exponents can be defined properly as logarithms of the eigenvalues of the matrix Λ_x defined by equation (2.3).

Let us take an infinitely small perturbation of the initial condition $\delta x(0)$. After n iterates it will become:

$$\delta x(n) = T_x^n \delta x(0) \tag{2.6}$$

On the basis of the Oseledec theorem for almost every $\delta x(0)$ we have

$$\delta x(n) \approx \delta x(0) e^{n\lambda_1} \tag{2.7}$$

The "sensitivity" to changes of initial conditions in an obvious way corresponds to the largest Lyapunov exponent λ_1.

2.1.2 Lyapunov exponents for a continuous-time dynamical system

Let us consider a continuous-time dynamical system $(\mathbb{R}^n, \mathbb{R}, \pi)$. Let us apply the Oseledec theorem (Theorem 2.2) to the operator $\pi_t : \mathbb{R}^n \mapsto \mathbb{R}^n$ defined by $\pi_t(x) \triangleq \pi(x, t)$. Let us denote by T_x^t the matrix of partial derivatives of the the operator π_t at the point x. If μ is an ergodic measure then the following limit exists:

$$\Lambda_x \triangleq \lim_{t \to \infty} (T_x^{t*} T_x^t)^{\frac{1}{2n}} \tag{2.8}$$

The Lyapunov exponents of the continuous-time dynamical system are defined in a similar way as before as the logarithms of the eigenvalues of Λ_x.

2.1.3 Lyapunov exponents and the type of attractor

There exists a direct link between the Lyapunov exponents and the type of attractor observed in the system. A stable fixed point in a continuous-time dynamical system will have all the Lyapunov exponents negative. If the trajectory on an attractor does not contain any fixed point then at least one Lyapunov exponent must be equal 0 m[176]. An attractor which is a closed orbit will have one Lyapunov exponent equal 0 and all the remaining ones negative. Quasi-periodic attractors (depending on the dimensionality) will have more than one Lyapunov exponent equal 0 and all the remaining ones negative. As already mentioned before there are also different attractors with some strange properties which in fact have positive Lyapunov exponents.

Apart from Lyapunov exponents there exists a number of different quantitative characterizations which can help us to identify the different types of motions – asymptotic behaviors.

2.2 Attractor dimension

Many authors[44,114,162,164] consider as a criterion for classification of asymptotic behavior of the system the dimension of the associated attractor. Existence of a so-called **strange attractor** ie. an attractor which has non-integer dimension, is considered as a criterion of existence of complicated possibly chaotic behavior.

There exist several notions of dimension which are being applied to calculation of dimension of attractors. We will recall here the most frequently utilized ones.

Definition 2.5 ("Capacity" – box-counting dimension) *The "capacity" defined by Kolmogorov is defined by:*

$$d_C = \lim_{\varepsilon \to 0} \frac{ln N(\varepsilon)}{ln(1//\varepsilon)} \tag{2.9}$$

where: the attractor in an n-dimensional state space has been covered by n-dimensional hypercubes of the side length ε . $N(\varepsilon)$ is the minimal number of cubes needed to cover the attractor completely.

Capacity has also another interesting interpretation. If we are interested in localization of an attractor in the state space with the accuracy ε, then it is sufficient to specify the position of N hyper-cubes covering it. From the equation (2.9) it follows that for small ε $ln N(\varepsilon) \approx d_C ln(1//\varepsilon)$. This means that the dimension gives us the amount of information necessary to specify the location of the attractor with a prescribed accuracy.

Definition 2.6 (Hausdorff dimension) *To define the Hausdorff dimension of an attractor we need to introduce a covering with n-dimensional hyper-boxes of a variable size ε_i. Let us introduce next*

$$l_d = \lim_{\varepsilon \to 0} l_d(\varepsilon) \tag{2.10}$$

where: $l_d(\varepsilon) = inf \sum_i \varepsilon_i^d$, and inf goes over all the coverings satisfying $\varepsilon_i \leq \varepsilon$. Hausdorff has proved that there exists a critical value $d = d_H$, above which $l_d = 0$, and below which $l_d = \infty$. This critical value d_H is called the Hausdorff dimension[114].

Definition 2.7 (Information dimension) *Information dimension is a generalization of the notion of capacity. It takes into account additionally the relative probability of passage of the trajectory through a given box in the covering (various regions of the attractor are visited by the trajectories with various frequencies – some are visited very often, some very rarely):*

$$d_I = \lim_{\varepsilon \to 0} \frac{I(\varepsilon)}{ln(1//\varepsilon)} \tag{2.11}$$

where:

$$I(\varepsilon) = \sum_{i=1}^{N(\varepsilon)} P_i ln \frac{1}{P_i} \qquad (2.12)$$

and P_i is the probability of a trajectory passing through the i-th box. $I(\varepsilon)$ specifies the amount of information gained from the system by a measurement with the accuracy ε.

2.2.1 Topological entropy

In the probability theory the entropy[164,466] is a measure of the amount of information gained (or uncertainty removed) during an experiment. The entropy of a dynamical system has to define the uncertainty of predicting the future behavior of the trajectories on the basis of the observation of its behavior in a given time interval. In the systems having attractors like fixed points or closed orbits (periodic solution) we can observe the phenomenon of "memory loss" in terms of initial conditions because many trajectories starting from different initial conditions converge to the attractor (ie. in the system having one attracting fixed point all trajectories starting from any initial condition will converge to this point) - the entropy of such a system is equal 0. In a chaotic system the situation is different: close (nearly identical) initial conditions lead to very different asymptotic behaviors (although trajectories converge towards the same chaotic attractor!). In such a case the longer we observe the trajectory the more information about its behavior we get. In such a situation the criterion for existence of chaos would be positive entropy.

To be able to calculate the value of entropy for dynamical systems let us introduce a proper mathematical setting and definitions for so-called topological entropy[376,387,466].

Let (X, ρ) be a compact metric space and α an open covering of $X : \alpha = \{A_i\}_{i=1}^n, X \subset \bigcup_{i=1}^n A_i, A_i$ - open sets.

Definition 2.8 *If $\alpha = \{A_i\}_{i=1}^n$ and $\beta = \{B_i\}_{j=1}^m$ are open coverings of the space X then the product of these two coverings is defined by:*

$$\alpha \bigvee \beta \triangleq \{A_i \bigcap B_j, i = 1, \ldots, n, j = 1, \ldots, m\} \qquad (2.13)$$

A sub-covering β of a covering α consists of a chosen subset of sets A_i belonging to α ie. $\beta = \{A_{i_k}\}_{k=1}^m$ $(a \leq i_k \leq n)$

Definition 2.9 (Entropy of a covering of a set) *For every open covering α of the space X let $N(\alpha)$ denote the number of sets in the minimal sub-covering; sub-covering β of the covering α of X is called minimal if any other* *sub-covering has more elements then β. The number:*

$$H_0 \triangleq \log N(\alpha) \tag{2.14}$$

Is called the entropy of the covering α

Let now $\varphi : X \mapsto X$ – a continuous operation $((X, \varphi)$ - defines a discrete-time dynamical system). Let $\alpha = \{A_i\}$ be a covering of X. Let $\varphi^{-1}(\alpha) = \{\varphi^{-1}(A_i)\}$, where $\varphi^{-1}(A_i)$ is the inverse image of the set A_i via the operation φ.

Definition 2.10 (Topological entropy for maps) *The limit:*

$$h(\varphi, \alpha) \triangleq \lim_{n \to \infty} \frac{1}{n} H_0(\alpha \bigvee \varphi^{-1}(\alpha) \bigvee \ldots \bigvee \varphi^{-n+1}(\alpha)) \tag{2.15}$$

is called **topological entropy of the operation φ with respect to the covering α**. *The number:*

$$h(\varphi) \triangleq \sup_{\alpha} h(\varphi, \alpha) \tag{2.16}$$

is called the **topological entropy of the operation φ**. *The supremum goes over all possible open coverings α.*

Topological entropy $h(\varphi)$ characterizes the relative velocity of mixing of points in the space X by the operation φ.

The above definition can be easily generalized for flows. Let us consider a continuous-time dynamical system $(X, \mathbb{R}, \pi)$.

Definition 2.11 (Topological entropy for flows) *The topological entropy for the flow defined by $(X, \mathbb{R}, \pi)$ is the number:*

$$h(\pi) = h(\pi_1) \tag{2.17}$$

where the mapping $\pi_t : X \mapsto X$ is defined by $\pi_t(x) \triangleq \pi(x, t)$.

Theorem 2.4 *Let $(X, \mathbb{R}, \pi)$ – a continuous time dynamical system. Then $h(\pi_t) = |t| h(\pi_1)$.*

Topological entropy of systems having regular behavior (having attractors such as a stable fixed point, a periodic orbit or a quasi-periodic orbit) is equal zero. In opposition to this the topological entropy of chaotic systems is positive.

2.2.2   *Temporal auto-correlation function*

Apart from Lyapunov exponents the temporal auto-correlation function provides one more measure of sensitivity to changes of initial conditions. Temporal auto-correlation functions is a measure of similarity of a trajectory at a time t to its value at some future time $(t + \tau)$ and is defined by:

$$AC(\tau) = \frac{1}{t_2 - t_1} \int_{t_1}^{t_2} x(t) \cdot x(t + \tau) dt \qquad (2.18)$$

Changing τ we can find the measure of self-similarity of a trajectory during its evolution in time. The Wiener – Khinchin theorem links the states of $AC(\tau)$ with the Fourier transform of the spectral power density function of the signal $x(t)$. Because the spectral power density functions of the signals constant in time, periodic or quasi-periodic (with incommensurate frequencies) have discrete spectra, then: $AC(\tau) \neq 0$ for $\tau \to \infty$, For a chaotic signal (in the sense of having continuous power spectrum in some interval) we have

$$AC(\tau) = 0 \text{ for } \tau \to \infty \qquad (2.19)$$

The temporal self-similarity of a trajectory is diminishing with time and vanishes for very long time intervals between the observation moments. The existence of a continuous power density spectrum over some frequency band and which follows the vanishing autocorrelation function for trajectories is sometimes considered as yet another criterion of existence of chaotic behavior[200,201]. The link between the temporal auto-correlation function and the power density spectrum is the basis of a criterion of existence of chaos developed by Ünal[448].

Analytical calculation of Lyapunov exponents, entropy, dimension or auto-correlation functions if possible at all can be carried out for vary rare examples of systems. In most cases one has to rely on numerical calculations and use of very sophisticated algorithms eg.[358]. In all cases the algorithms use only a finite-time sample of the observed trajectories and are performed with limited precision often using approximations, numerical integration and differentiation, matrix operations etc. One has to take the results of such numerical calculations with extreme care and thoroughly analyze the results before accepting their interpretation.

The classification of attractors and their various characterizations are given in Table 1.

It should be noted also that various characterizations of system behavior might lead to controversies in interpretation. For example Grebogi *et al.*[167] show examples of systems which possess a strange attractor (in terms of the fractal dimension) which is not chaotic as it has no positive Lyapunov exponents. Several examples of this kind are also shown by Brown and Chua [46].

CLASSIFICATION OF ATTRACTORS

Asymptotic behavior of trajectories	Attractor type		Lyapunov exponents	Dimension
	Continuous-time system	Poincaré map		
fixed point	point	-	$0 > \lambda_1 \geq ... \geq \lambda_n$	0
periodic trajectory	closed curve	point	$\lambda_1 = 0$	1
sub-harmonic (k-periodic)	curve	set of k points	$0 > \lambda_2 \geq ... \geq \lambda_n$	1
quasi-periodic (2-frequencies)	torus	closed curve	$\lambda_1 = \lambda_2 = 0$, $0 > \lambda_3 \geq ... \geq \lambda_n$	2
quasi-periodic (k-frequencies)	k-torus	k-1 torus	$\lambda_1 = ... = \lambda_k = 0$, $0 > \lambda_{k+1} \geq ... \geq \lambda_n$	k
chaotic	Cantor set like	Cantor set like	$\lambda_1 > 0, \sum_i \lambda_i < 0$	non-integer
hyper-chaotic	Cantor set like	Cantor set like	$\lambda_1 > ... \lambda_k > 0,$ $\sum_i \lambda_i < 0$	non-integer

Table 2.1: Classification of attractors following their basic properties and characterizations.

2.3   Chaos - definition problems

The concept of chaos is as old as our human history – people used to associate with this notion the properties of "disorder". Eg. the ancient Greeks believed that our planet appeared out of "chaos" – the initial disorder before divine introduction of some orderly behavior. Similar notions of chaos can be seen also in the Chinese and Japanese written texts dating back a thousand years ago.

On the other hand, without looking for an explanation of the origins of the Universe people were always observing very bizzare phenomena in their everyday life. They observed the flow of water in rivers, movement of clouds in the sky, the smoke and vapors during cooking, lightnings, hurricanes, vibrations of the ground or various useful objects and tools etc. Quite often these phenomena were later described as being strange, chaotic or bringing chaos.

At some point of human history we started making more precise observations and looking into mechanisms and origins of various phenomena. Thus Newton looking at the movement of the celestial bodies found out that sometimes the planets do not behave as expected – they do not move on closed orbits! – and formulated the so-called three-body problem. Navier and Stokes tried to understand the origins of the turbulent behavior in fluids and formulated nice mathematical models which give some insight into the phenomena. Already Poincaré in his works concerning qualitative theory of differential equations spotted that even simple physical systems described by low-order differential equations can exhibit very complicated oscillations which are neither periodic nor quasiperiodic and are not approaching any of these asymptotically. These solutions have all the properties of random trajectories despite the fact they were solutions of fully deterministic systems fully specified by underlying ordinary differential equations with fixed initial condition.

Still, for a very long time this bizzare types of behavior often referred to as chaos or sometimes described as complex behavior were not at all understood and even not properly defined!

Both terms – chaos and complexity were used to describe situations escaping our understanding, problems we could neither describe in common terms or that were to complicated to cope with.

Observing the behavior one can easily establish some basic properties: it is bounded in space (trajectories do not escape), the behavior looks highly irregular, there is not a repetition pattern neither in time nor in space – at least during the time of observation.

How can we characterize chaos and complexity? How can we say that some process is chaotic? How do we distinguish signals produced by a chaotic system from any other types of signals?

Many authors adopt a definition by negation – which is not well behaved either approaching a fixed point, a periodic or quasi-periodic solution is considered as chaos. But this does not help us much in investigations. What are however the specific properties of chaos which distinguish this type of motion from all the others? We know what chaos is not ! but we do not know what it is (apart from intuition...).

Proper problem statement and mathematical notions help us to overcome this problem to some extent. The apparatus of dynamical system theory and signal analysis tools provide a good insight into the problem of understanding the nature of chaos. But still researchers cannot agree on one commonly accepted definition. Brown and Chua[46] referring to the literature of the subject give a non-exhaustive list! of different definitions. Namely, various authors consider that the system is chaotic when eg.:

- It has a Smale horseshoe

- It has positive topological entropy

- It has positive Kolmogorov entropy

- It has a positive Lyapunov exponent

- Its sequences have positive algorithmic complexity

- It has a dense set of periodic orbits, is topologically transitive, and has sensitive dependence on initial conditions

- It is topologically transitive, and has sensitive dependence on initial conditions

- The power spectral density of related time-series has a component which is absolutely continuous with respect to Lebesgue measure

- The dimension of the attractor is non-integer (attractor is a fractal)

Nearly all of these definitions have some links with two fundamental properties[46]: (1) rapid loss of correlation between the future and the past, (2) rapid loss of information over time. Definitely positive Lyapunov exponents, sensitive dependence on initial conditions, positive entropy – all these notions involve in some way loss of information or propagation of even negligible errors.

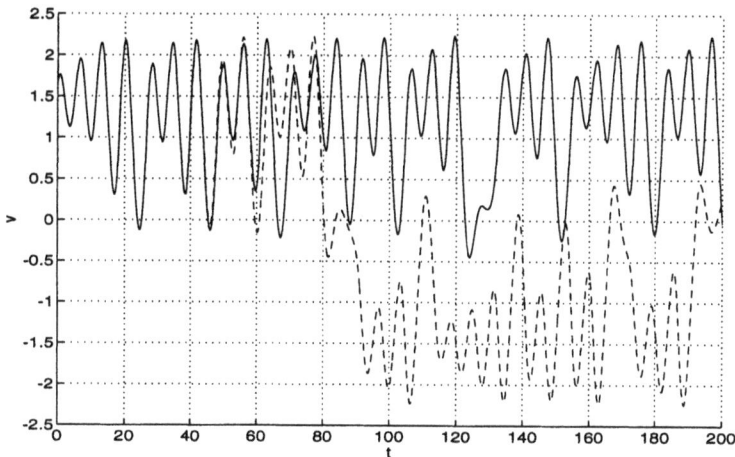

Figure 2.1: Illustration of the sensitive dependence on initial conditions. Two trajectories of Chua's oscillator starting from nearly identical initial conditions eventually separate resulting in different behavior. Here the difference is 0.001 in the first component of the initial condition only.

This means that trajectories of a chaotic system starting from nearly identical initial conditions will eventually separate and become uncorrelated (but they will always remain bounded in space). Mathematically speaking the trajectories are unstable in the Lyapunov sense i.e. it is not true that given an initial condition x_0:

$$\forall \delta > 0, \forall x(0) \text{ such that } |x(0) - x_0| \quad < \quad \delta \Rightarrow \exists \varepsilon(x(0)) \text{ such that}$$
$$|x(t, x(0)) - x(t, x_0)| < \varepsilon \quad (2.20)$$

Fig.2.1 shows an example of such trajectories - very small change of initial conditions causes (after some period when they remain close to each-other) eventual separation.

Sensitive dependence on initial conditions is very important from the practical point of view. In typical situations we can specify the initial conditions for a dynamical system only with some finite accuracy ε. If two initial conditions are closer to each-other then ε then they are not distinguishable in measurements. The trajectories starting from these initial conditions will diverge after a finite time and become uncorrelated (if the system is chaotic obviously).

Whatever high precision we use in measurements (experiments) the behavior of trajectories is not predictable – the solutions look virtually random despite the fact they are produced by a deterministic system.

CHAPTER 3

CHAOTIC SIGNAL ANALYSIS AND PROCESSING

In most experimental situations we have at our disposition only one measured signal from the considered system. In many cases it is an electric signal even when the considered systems is not an electronic one. Electric sensors are widely used to measure temperature, velocity, acceleration, light intensity, activity of the muscles, heart, brain etc. Modern techniques of data acquisition measure the signals in a specific way – they are being sampled both in time and space (A/D conversion, finite word-length effects, quantization, roundoff, overflow) which makes it possible to store them and process using computers.

What can be done in a situation when we have taken some measurements and want to interpret the results. How do we know that the signals represent chaotic state of the system? Is it possible to distinguish between chaos and noise? What conclusions can be drawn about the nature of the system and its dynamics? – these are the simples questions to ask when we encounter measured signals.

In this chapter we will give some answers to the above mentioned questions and problems. In particular we will consider reconstruction of system dynamics and spectral analysis of possibly chaotic signals.

3.1 Reconstruction of system dynamics from measured time series

A paradigm for state reconstruction is provided by the embedding theory. It says that typically points on an attractor in the state space correspond one-to-one with measurements of a limited number of variables and which follows the state space can be identified by measurements. Two types of embeddings are relevant for system identification on the basis of measurements: the topological embeddings and differentiable embeddings.

3.1.1 Topological embeddings

Let us consider an n-dimensional Euclidean space R^n, $x \in R^n$. Let F_O be a continuous function from R^n to R^m,

$$y = F_O(x) \tag{3.1}$$

Let $A \subset R^n$ be an attractor of a dynamical system. $F_O(A)$ is an image of the attractor A in R^m via the observation (measurement) function F_O. F_O is bijective (one-to-one) if for any $x_1, x_2 \in A$, $F_O(x_1) = F_O(x_2)$ implies $x_1 = x_2$. For a bijective function F_O there exists an inverse function F_O^{-1}.

A bijective map on A which is continuous and has a continuous inverse is called topological embedding. For topological embeddings we have the following theorem:

Theorem 3.1 [356] *Assume that A is a compact subset of R^n of box-counting dimension D_0. If $m > 2D_0$, then almost every C^1 function $F = (f_1, ..., f_m)$ from R^n to R^m is a topological embedding of A into R^m.*

In particular the delay coordinates can serve for constructing the topological embedding. In this case we have the results developed by Takens and Sauer *et al.*.

Theorem 3.2 (Takens embedding theorem) *with extensions by Sauer et al.*

Assume that a continuous time dynamical system has a compact invariant set A (eg. a chaotic attractor) of a box-counting dimension D_0, and let $m > 2D_0$. Let τ be the time delay. Assume that A contains only a finite number of equilibria, a finite number of periodic orbits of period $p\tau$ for $3 \leq p \leq m$, and that there are no periodic orbits of period τ and 2τ. Then, with probability one, a choice of the measurement function h yields a delay-coordinate function H which is bijective from A to $H(A)$.

$$
\begin{aligned}
y &= [y(t - \tau), \ldots y(t - m\tau)] \\
&= [f(x(t - \tau), \ldots, f(x(m\tau)] \\
&= H(x) \\
&= [h_1(x), \ldots, [h_m(x)]
\end{aligned}
\tag{3.2}
$$

3.1.2   Differentiable embedding

Theorem 3.3 (Sauer et al.[398]) *Assume that a continuous time dynamical system has a compact invariant smooth manifold A of a dimension d, and let $m > 2d$. Let τ be the time delay. Assume that A contains only a finite number of equilibria, a finite number of periodic orbits of period $p\tau$ for $3 \leq p \leq m$, and that there are no periodic orbits of period τ and 2τ. Assume that the Jacobians of the return maps of those periodic orbits have distinct eigenvalues. Then, with probability one, a choice of the measurement function h yields a delay-coordinate function H which is bijective from A to $H(A)$.*

$$
y = [y(t - \tau), \ldots y(t - m\tau)]
$$

$$\begin{aligned} &= [f(x(t - \tau)), \ldots, f(x(m\tau)] \\ &= H(x) \\ &= [h_1(x), \ldots, [h_m(x)] \end{aligned} \tag{3.3}$$

Differentiable embeddings offer two advantages compared to topological ones[356]. First there is a uniform upper bound on the stretching done by H and H^{-1}. Such H functions are referred to as bi-Lipschitz. The dimension of sets are preserved under bi-Lipschitz maps. Second, all Lyapunov exponents on an attractor are reproduced in the reconstruction. For the purpose of the present study we are interested in the consequences of the embedding theorems for the synchronization/transmission problem. These consequences can be summarized in the following result:

Theorem 3.4 (Main existence theorem) *If the assumptions of the embedding theorems are satisfied it is always possible to reconstruct the state of the system and synchronize (eg. by forcing the states) an exact copy of it on the basis of measurement of a single (scalar) output variable.*

Comments:

1. Embedding theorems offer only the existence result. Construction of an inverse of an embedding function is an open problem - no general solution or algorithm is available.

2. The above stated result are in close relation with the observability problem known from control theory. Observability issues are well developed for linear systems - only a limited number of results for nonlinear cases exist. Observers provide "the missing tool" for reconstruction.

3.1.3 Examples of reconstruction

Let us analyse now an example of reconstruction of an attractor from a measured time series of 1 million points obtained from numerical integration of Chua's circuit equations. For comparison we have drawn the original attractor in Fig.3.1 To be able to reconstruct the dynamics we further used only a recording of the v_{C1} voltage. To be able to apply the delay coordinate method we had to find a suitable time delay. This can be done using different approaches[384] eg. calculating autocorrelation function of the considered trajectory, finding higher-order correlations, estimating so-called fill factors or wavering products etc. For our example we used the approach proposed in[129] calculating so-called mutual information curve for varying time-delay and choosing the time-delay corresponding to its first minimum. Figure 3.2 shows results of calculation of the mutual information using a moving windows of various lengths - 512,

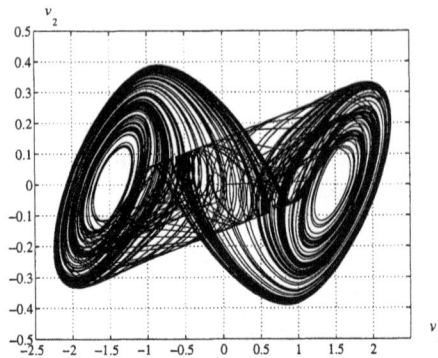

Figure 3.1: Two-dimensional projection of the Double scroll attractor observed in Chua's circuit.

1024 and 2048 samples of the measured time-series. Inspection of the mutual information curves permits to estimate the delay of 17 to 20 samples.

Let us now examine the reconstructed state-space plots. Figure 3.3 shows examples of several choices of time delays for the reconstruction. When the delay is too small the reconstructed attractor becomes squeezed and the dynamics is not well visualized — Fig. 3.3a gives an example for the delay of only 3 samples. Fig.3.3b shows the reconstructions based on 20 samples delay as indicated by the most accurate mutual information curve. In this case the attractor structure is easily seen. When the delay used for reconstruction becomes too large the delayed copies of signal become more and more uncorrelated — this situation is depicted in Fig. 3.3c and d. The trajectories shown in these two figures do not resemble at all the original double scroll structure.

3.2 Observers for chaotic systems

Several approaches to state reconstruction are known from control theory - they are referred to as the problem of constructing state observers which using given system output enable finding the state. Among many proposed solutions of this problem probably the best known is Kalman filter approach and Luenberger (asymptotic) observers developed for linear systems and extended also to nonlinear cases.

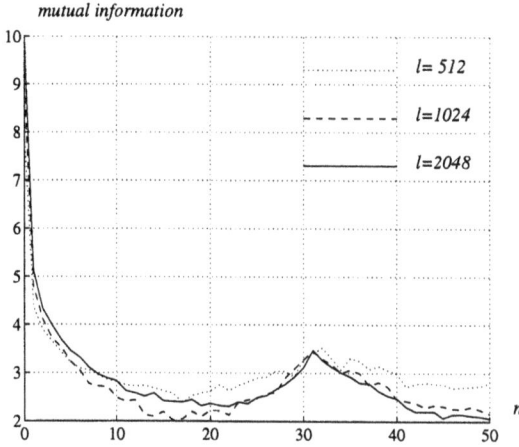

Figure 3.2: Mutual information as a function of the time delay. The dotted line corresponds to calculations using the moving window of 512 samples, the dashed line - to 1024 samples and the continuous curve to 2048 points respectively. The minimum for the continuous curve is around delay time of 20 samples, for the other two curves we find the minimum near 17 samples.

3.2.1 Linear observers

A. Exact reconstruction

Let us consider as an example an n-dimensional system

$$\frac{dx(t)}{dt} = Ax(t) + Bu(t) \tag{3.4}$$

with an output defined by

$$y(t) = Cx(t) \tag{3.5}$$

where: $x(t) \in R^n$, $u(t) \in R^r$, $y(t) \in R^m$, $x(0) = x_0$.

We say that system 3.4 is observable if its state at some time instant $x(t)$ can be uniquely determined on the basis of measurement of u and y on an interval $[t - q, t]$. This task can be carried out successfully if the so-called observability matrix defined by

$$S = \begin{bmatrix} C \\ CA \\ \vdots \\ CA^{n-1} \end{bmatrix} \tag{3.6}$$

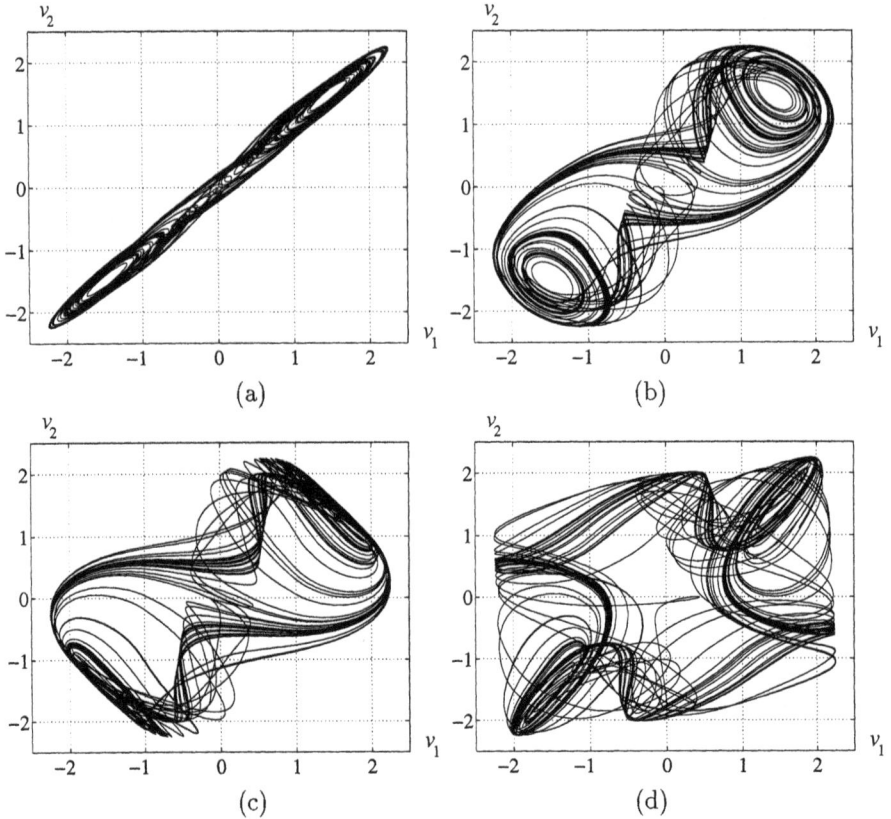

Figure 3.3: Reconstruction of double scroll attractor using measured v_{C1} time series. Successive figures show attractors reconstructed using time delays of 3, 20, 30 and 60 samples respectively. When time delay is too small the reconstructed attractor is squeezed, when it is too large the reconstruction is bad — the geometric structure of the attractor is lost. For the choice of delay time as in (b) the double scroll structure is well reproduced.

has full order. Then the state can be calculated using:

$$x(t) = M_q^{-1} \int_0^q e^{-sA^T} C^T (y(t - s)$$

$$+ \ C \int_0^s e^{-lA} Bu(t - s + l) dl) ds \tag{3.7}$$

where the matrix M_t is defined by:

$$M_t = \int_0^t e^{-sA^T} C^T C e^{-sA} ds \tag{3.8}$$

B. Asymptotic reconstruction

We will consider the system 3.4. Let us introduce now an auxiliary system called the Luenberger observer:

$$\frac{dw(t)}{dt} = Fw(t) + Hu(t) + Gy(t) \tag{3.9}$$

$$\frac{d\hat{x}(t)}{dt} = My(t) + Nw(t) \tag{3.10}$$

where F, G, H, M, N are constant real matrices, $\hat{x}(t) \in R^n$, $w(t) \in R^s$. Let us introduce a new variable

$$e(t) = w(t) - Px(t) \tag{3.11}$$

where $e(t) \in R^s$, P - a real matrix.

If

$$[\ G \quad F \ ] \begin{bmatrix} C \\ P \end{bmatrix} = PA \tag{3.12}$$

and

$$[\ M \quad N \ ] \begin{bmatrix} C \\ P \end{bmatrix} = I \tag{3.13}$$

and $H = PB$ then

$$\frac{de}{dt} = Fe(t) \tag{3.14}$$

$$\hat{x}(t) - x(t) = Ne(t) \tag{3.15}$$

Assuming that F is exponentially stable we have

$$\lim_{t \to \infty} |\hat{x}(t) - x(t)| = 0 \tag{3.16}$$

3.2.2   Chaos observers

So *et al.*[422,423] proposed two special types of observers allowing the reconstruction of the state variables of a chaotic system on the basis of observation of a scalar output. As above the proposed methods assume exact knowledge of an accurate mathematical description of system dynamics. For the sake of simplicity we will present a discrete version of the procedures. Let us assume that the dynamic is described by

$$\begin{aligned} x_{k+1} &= F(x_k) \\ y_k &= g(x_k) \end{aligned} \tag{3.17}$$

where F, g are nonlinear functions $x_k \in R^n$.

Full order observer The observer is taken as

$$\hat{x}_{k+1} = F(\hat{x}_k) + C_k[y_k - \hat{y}_k] \tag{3.18}$$

where $\hat{y}_k = g(F(\hat{x}_k))$ and C_k is a time-dependent n-dimensional control column vector which has to be adjusted at every iterate. then

$$x_{k+1} - \hat{x}_{k+1} = F(x_k) - F(\hat{x}_k) - C_k[g(F(x_k)) - g(F(\hat{x}_k))] \tag{3.19}$$

Linearizing about $\hat{x}_k$ gives

$$\delta x_{k+1} = [DF(\hat{x}_k) - C_k Dg(F(\hat{x}_k))DF(\hat{x}_k)]\delta x_k \tag{3.20}$$

where D denotes the derivative. If the magnitudes of the eigenvalues of the matrix defined by the RHS of the above equation are less then one at every iterate then the observer will asymptotically reconstruct the state vector.

Important notes:

- Observers can be considered also from a different viewpoint as inverse systems as considered by Hasler and his collaborators or response systems in Pecora-Carroll configuration!! (comp. Chapter 10)

- The concepts and methodologies described above provide tools for synchronizing ANY two identical systems - no matter the dimensionality or complexity, number of positive Lyapunov exponents etc. (comp. Chapter 10)

3.3   Characterization of chaos by unstable periodic orbits

One of the fundamental properties of chaotic attractors is existence of a countable (infinite) set of unstable periodic orbits within the attractor. These orbits are invisible in experiments as they are unstable but at the same time they constitute an invisible "skeleton" — the actual chaotic trajectory eventually passes arbitrarily close to any of these orbits.

Let us assume that we were able to detect some (if not all) of such unstable periodic orbits using experimental data only.

If this were feasible we could further find an approximation to the curvatures of any nonlinear multidimensional Poincaré map using a continuous polygonal surface made of hyper-planes in such a way that these hyper-planes are tangent to the graph of the map at the unstable periodic points and their slopes are determined by eigenvalues of the Jacobian matrices calculated at these points. One can obtain any needed accuracy of approximation as there exists a countable infinite number of unstable periodic orbits with growing periods and these orbits are dense on the asymptotic strange set — recovering more and more unstable cycles we obtain better approximations.

The main features of the characterization in terms of unstable periodic orbits are:

- Periodic orbits and their eigenvalues are topologically invariant — different representations of the same system (up to a smooth transformation of coordinates) must preserve their topological properties (a fixed point must remain a fixed point in any representation and the same applies to periodic orbits),

- Periodic orbits constitute a "skeleton" for the attractor — they determine its spatial layout,

- The eigenvalues of closed orbits are metric invariants — they describe the scaling between different pieces of the attractor.

- There exists a hierarchical ordering of unstable periodic orbits — short cycles give good approximations of the strange set.

- Periodic orbits are robust — they vary slowly with smooth parameter variations. The same applies to their eigenvalues.

- Unstable periodic orbits can be successfully extracted from experimental data - specific computational methods have been developed for this purpose and implemented in computer programs.

How to find unstable periodic orbits using experimental data?

The numerical procedure for processing of the chaotic signal assumes that we have a series of successive points $\{x_i\}$, $i = 0, 1, ...N$ on the system trajectory and taking any of these points x_m we search forward for the smallest positive integer k, such that $||x_{m+k} - x_m|| < \varepsilon$, where ε is the specified accuracy. It is further claimed that the orbit detected in this manner lies close to the unstable periodic orbit whose period is approximated by that of the detected sequence.

This approach has several drawbacks. Firstly, the results strongly depend on the choice of ε and the length of the measured time series. Further, they depend on the choice of norm and number of state variables analyzed; we used all three state variables, normalized the size of the attractor and used the Euclidean norm. Secondly, the stopping criterion ($||x_{m+k} - x_m|| < \varepsilon$) in the case of discretely sampled continuous-time systems is not precise enough. This means that one can never be sure of how many orbits have been found or whether all orbits of a given period have been recovered.

In many applications , however, it is sufficient to find only some of the unstable periodic orbits embedded in the attractor — this is the case for example in the approach to controlling chaos described by Ott, Grebogi and Yorke[11,13] (comp. Chapter 11).

For the purpose of this study we have developed a set of computer programs for detecting unstable periodic orbits. The Lathrop-Kostelich procedure has been refined by means of a stopping condition based on the distance of the initial point x_m from the evolving trajectory — not from distinct points belonging to it which are sampled discretely in time (D/A conversion of measured signals) or computed via numerical integration. (Thus we avoid the problem of of not detecting an orbit when x_m falls between two successive points along the trajectory). This slightly slows down the computations but the results are more reliable - the problem of not finding some of the orbits due to distance mismatch can be avoided.

Among unstable periodic orbits calculated using the described procedure there are several groups of nearly identical ones. It is very important to introduce a criterion for distinguishing different periodic orbits. This can be done on the basis of calculating the distance between the orbits. The distance between orbits Γ_1 and Γ_2 is defined as:

$$d_{orb} = \max_{x_i \in \Gamma_1} [\min_{x_k \in \Gamma_2} ||x_k - x_i||] \qquad (3.21)$$

Two orbits whose distance is smaller then the prescribed threshold are considered equal. During the experiments we found that when using long time series (typically 100000 basic periods of the circuit) some of the orbits were found repeatedly from 50 to 100 times. In our experiments we varied the

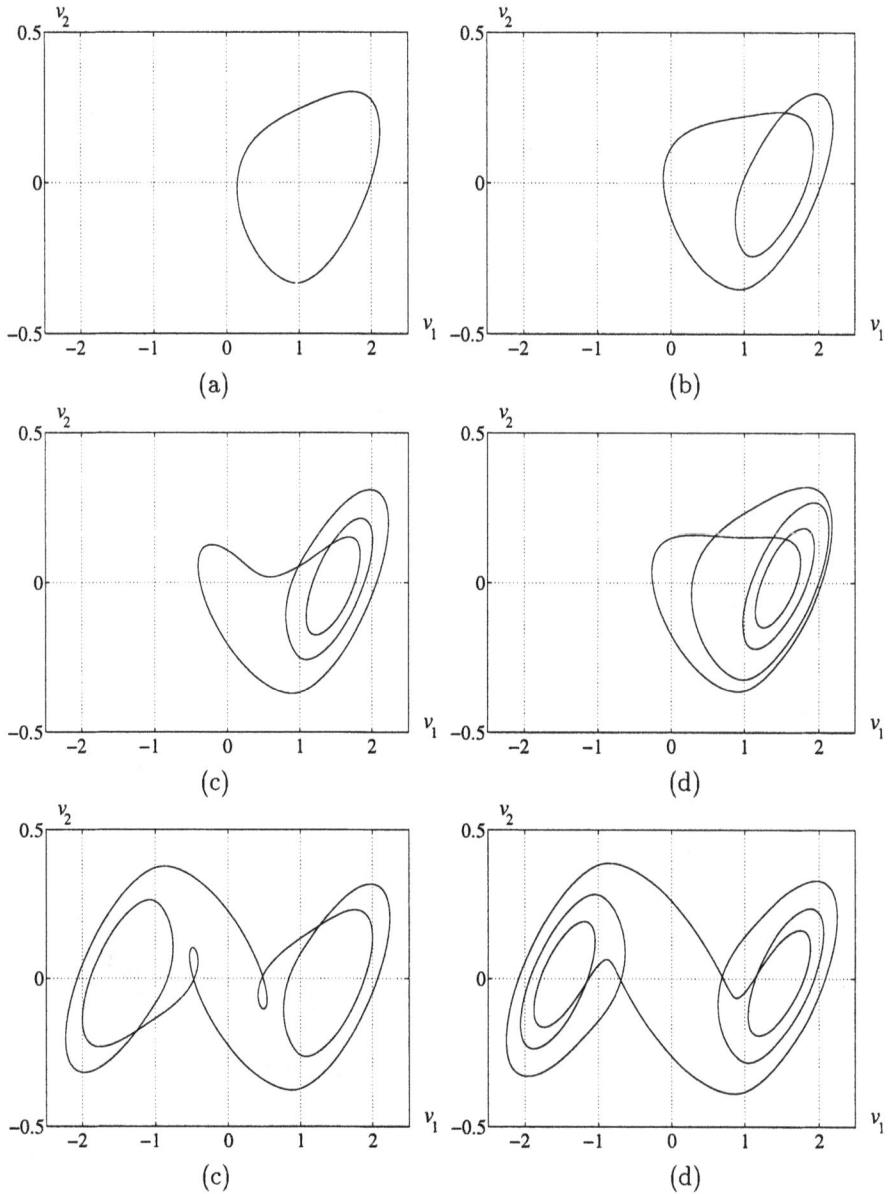

Figure 3.4: Unstable periodic orbits uncovered from the double scroll attractor using the same data series as in the example of reconstruction of the attractor described above.

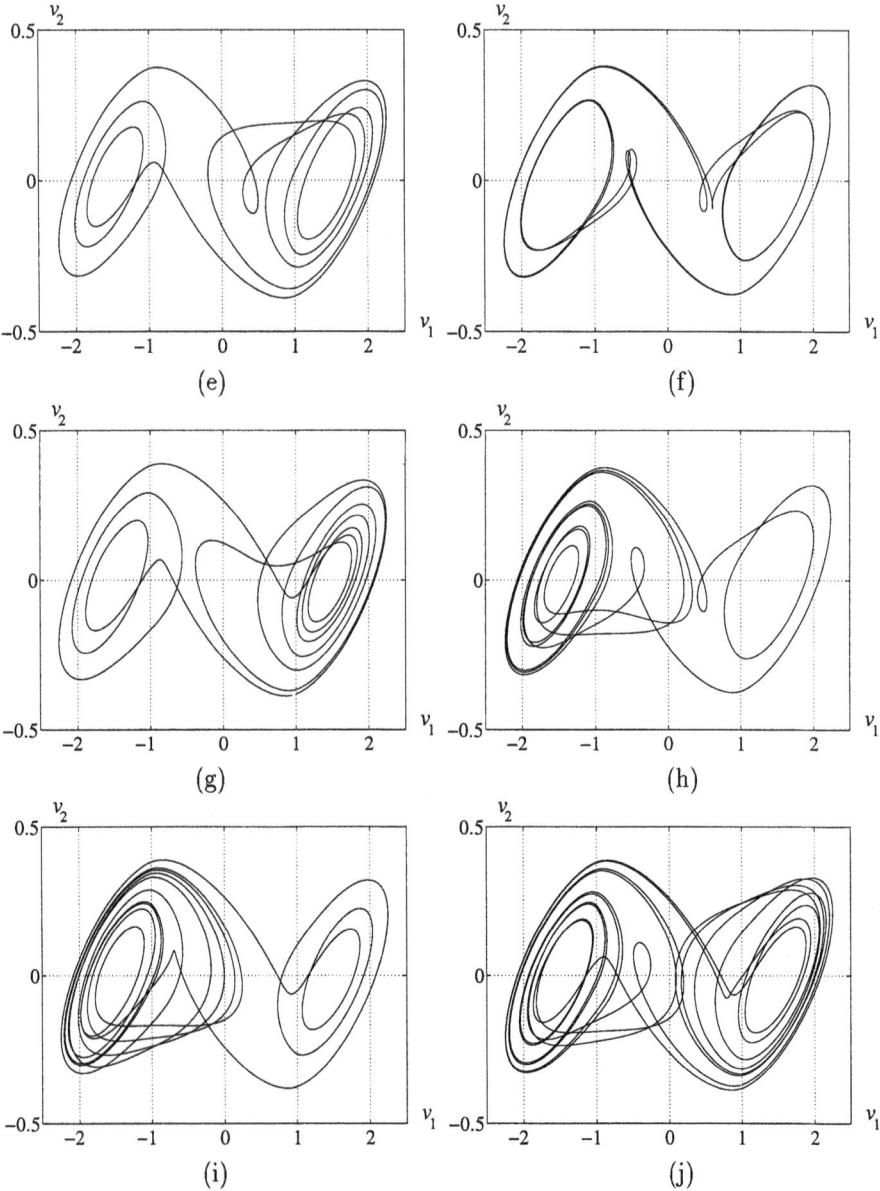

Figure 3.5: Continued

ε between 0.000001 and 0.001 and fixed the threshold for distinguishing between the orbits on 0.001. Although with greater ε more orbits with given period were detected most of them were later recognized as identical - there was no significant difference in the number and shape of different unstable periodic orbits found. Figure 3.4 shows some typical unstable orbits uncovered from a time series obtained by integration of equations describing dynamics of Chua's circuit. Fig.3.4a-d show simplest orbits - period-1, period-2, period-3 and period-4 winding around one of the system equilibria. In experiments it was possible also to find their symmetric counterparts around the other equilibrium as well as many longer period orbits. Fig.3.4c and d show symmetric orbits (winding around all equilibria). More complex unstable periodic orbits are shown in Fig.3.4e-j.

Examples of analysis available using our computer programs are show in Plates 11 and 12. Plate 11 shows projections of the attractor (here Chua's double scroll attractor). Plate 12 shows in color some of the detected unstable periodic orbits embedded within the double scroll attractor.

CHAPTER 4

ELECTRONIC CIRCUITS GENERATING CHAOS – A BRIEF OVERVIEW

During the last decade there was an avalanche of papers published in conference proceedings and journals describing new discoveries of chaotic behavior in a variety of electrical and electronic circuits and systems. There are several review papers presenting the state of the art in this domain[28,180,280,361] and also several special issues of journals devoted uniquely to this subject[472,473,474,475,476,477].

The investigations are carried out in three main directions:

- Experimental studies (simulation and laboratory);

- Mathematical modeling and analytic approaches;

- Applications.

Let us describe briefly the most interesting results.

In experimental investigations it was confirmed that almost all classes of electric and electronic circuits and systems can exhibit chaotic or complex behavior under some parameter choices. These circuits can be classified following several criteria. For example taking into account the circuit functionality and construction chaos has been confirmed in the following classes of circuits:

- simple RLC circuits[74,207,208,285,286,327,392],

- oscillators of various kinds[63,445,452], in particular described by Duffing equation[190,191,228,231,235,236], or Van der Pol equation[349,390,391,418],

- switched-capacitor circuits[109,383],

- phase-locked loops (PLL)[110,111,112,169,258,259],

- digital filters[77,133,75,76,101],

- flip-flops[118,429]

- adaptive filters[272,273,388],

- power converters and power supplies[43,178],

- power circuits[307].

As necessarily for chaos generation the circuit has to be nonlinear it is interesting to look where the nonlinearities come from in the circuits exhibiting chaotic behavior. The most commonly encountered elements introducing nonlinearities are:

- various kinds of diodes[18,20,82,208,210,270,285,315,316,370,371,464],

- thyristors[287,307],

- transistors[370],

- operational amplifiers (either the saturation characteristic is used directly or Op-Amps are used in realizations of special characteristics of resistors or transfer functions) [25,45,73,132,252,279,282,284,311,312,327,463,467,468,469],

- nonlinear inductors[47,72,359],

- Josephson junctions[195,233,234,395,452].

- in digital systems the nonlinear characteristics come from rounding, truncation and overflow operations.

From the physical or construction point of views most of these systems are very different but interestingly enough from the mathematical point of view they share a similar model and many common properties from the qualitative dynamics point of view. In the most common cases the mathematical models encounters are difference equations (in most cases first- or second-order), ordinary differential equations of autonomous or non-autonomous types.

Mathematical developments can thus be applied not only to a single circuit (its model) but in many cases to whole families of circuits (eg. Chua's circuit family, Saito's hysteretic circuit family etc.). A good examples of analysis of families of chaotic circuits can be found in[253,74,393,41]).

Mathematical investigations have led to a development of a "toolkit" useful in analysis of complicated behavior[1,173]. This toolkit contains now the methods originating from the qualitative theory of ordinary differential equations, tools from stochastic processes theory, bifurcation theory, perturbation theory The description of standard analysis methods can be found in a number of monographs eg. Guckenheimer and Holmes[171], Thompson and Stewart[432], Devaney[102], Bergé *et al.* [27], or Fisher and Smith[121]. Here we will repeat in the next chapter only the essential notions which are needed for understanding of the rest of the chapters of this book.

Commonly used mathematical methods for proving the existence of complicated behavior relay on the developments and theorems concerning system behavior in the neighborhood of homoclinic orbits. These methods developed in the sixties by Shil'nikov for autonomous systems and Melnikov for non-autonomous ones have been first rediscovered in the eighties and generalized to many situations interesting from the engineering applications point of view. Shil'nikov method often used for autonomous systems[3,4,49,143,144,405,406,407,415] was thoroughly developed by the mathematicians from the University of Nice (C.Tresser[11,13,436,437]), also by Glendinning[156,157] and Gaspard [142,143,144,145]. It remains certainly the most widely used precise method of proofs.

The Melnikov method finds wide-spread applications in analysis of systems with external forcing (non-autonomous). Wiggins in his book[454] has thoroughly analyzed various applications of this method. Useful developments and extensions are also due to Gruendler[170] and Salam[394].

Among other methods of analysis the method of return maps (point-wise transformations) is also of paramount importance. Since the introduction of this method by Poincaré the theory of such transformations has been sufficiently developed to find its way to engineering applications. The book by Nejmark[308] contains an excellent description of the method and various applications. Several interesting developments and applications of mapping methods can be found in the works of Mira and his collaborators [123,126,220,293,294,295].

Among vividly developing areas of analysis we should mention:

- **Symbolic dynamics**[303,309,194,195]

  The idea behind the symbolic dynamics method is to characterize the behavior of trajectories in some invariant set by specifying so-called series of symbols associated with trajectory behavior. To achieve this one chooses a set of local cross-sections of trajectories in the invariant set and then the symbols are assigned following the order of passage of the trajectory through the sections labeled by symbols (eg. letters or numbers). In many cases such a choice of sections is possible which guarantees that the relation between the trajectory behavior and symbol series is one-to-one - well defined and there is a unique correspondence - trajectory–symbol series. The description/characterization of the invariant set in terms of symbolic series can be further reformulated in terms of Markov chains. Markov chain apparatus can be used for proving existence of aperiodic (chaotic) trajectories on the basis of finite time observations[37]. Symbolic dynamics has been widely used to analyze the maps of the Smale horseshoe type.

- **Knot theory**[194,195,196]

  Knot theory allows classification of various kinds of trajectories by assigning to them so-called "templates" or "knot holders"[196]. There exists a well-developed theory allowing analysis of existence of Smale horseshoes on the basis of the template knowledge.

- **Cell-to-cell mappings**[199]

  Cell-to-cell mappings is method of global analysis. It enables development of effective algorithms for finding periodic orbits, domains of attraction, basin boundaries etc.

The third direction in studies of complicated dynamical systems is applications. During the recent five years a number of applications have been proposed and possibly will develop into real engineering applications. The most interesting and encouraging results have been obtained in the areas of:

- noise and random signal generation[109],

- forecasting using chaotic models[115,116],

- image analysis and compression using fractals[24],

- computer graphics and animation[24],

- transmission of signals on a chaotic carrier,

- engineering and bio-medical applications of chaos control

In the sequel we will describe only chosen aspects from the long list of problems mentioned above using also a limited number of tools which we think are available for engineers in their everyday practice. We will also try to list as many as possible open problems hoping that the readers might help in finding good solutions and new applications.

CHAPTER 5

EMPIRICAL METHODS FOR STUDYING CHAOTIC
SYSTEMS

Computer simulations remain still the most widely used method for studying chaotic and complex behavior. There exist some extremely sophisticated
algorithms [44,158,162,163,164,168,358,459] that are now incorporated in generally
available software eg. PHASER[246], DYNAMICS[465,318], INSITE[362] or DY-
NAMICAL SOFTWARE[399]. These software packages offer the potential user
an extreme flexibility in the choice of integration algorithms, finding periodic
orbits, calculating stable/unstable manifolds, regions of attraction, calculation
of Lyapunov exponents, construction of bifurcation diagrams, graphical pre-
sentation of results etc. On the other hand several developments of application-
specific laboratory instruments for measurements in chaotic systems are being
developed. Let us mention here only a few: Oscilloscope adds-on for producing three-dimensional projections of plots[79,80,81], Poincaré maps, bifurcation
diagrams[149,150,151] or sophisticated spectrum/signal analyzers.

In this chapter using our RC-ladder chaos generator[324,327,329] as an example we will show results of typical empirical analyses.

5.1 Laboratory test equipment

A standard electronic engineer's tool for observing phenomena varying in time
(chaos obviously belongs to this class) is the oscilloscope. Most of oscilloscopes
have two built-in functions: using time base for observation of signal variation
in time and XY-mode in which one variable can be displayed as a function of
another variable. In this last mode we can display parametric curves with the
time as parameter such as Lissajous figures or projections of trajectories in the
state space. Some typical experimental results observed using an oscilloscope
will be described on an example in one of the subsequent sections. Visualization
of time-waveforms and state trajectories (projections) are standard operations.
Using simple additional circuitry it is possible to make more advanced tests
eg. observing Poincaré maps.

5.1.1   Simplest Laboratory Experiments

Comparing with other domains there is a great simplicity in carrying out laboratory experiments in electrical and electronic circuits and systems. The general purpose measuring equipment is easily available, sensing of voltages and currents as typical variables poses no problems, interface with computers is relatively simple to develop. Synthesis of electrical circuits described by given differential equations is also possible in most cases. Realization of nonlinear functions used in many experiments can be based on standard methods of synthesis[71,377]. Several types of nonlinear characteristic synthesizers are also available. Talking about standard test equipment available we have nowadays excellent oscilloscopes with many additional features (storage, digitization, mathematical options etc.) and spectrum analyzers which can be used in experiments with chaotic circuits. Below we will describe some of the results available using a standard oscilloscope with simple add-on circuitry enabling measurements of Poincaré maps.

5.1.2   Advanced trajectory observations using an oscilloscope

Several authors proposed various simple add-on equipment which enable observation of Poincaré cross-sections of attractors, 1-dimensional maps, bifurcation diagrams using a standard oscilloscope.

To explain how Poincaré sections are taken and how approximate one-dimensional maps are built using Poincar'e section signals let us look at Fig.5.1. First a suitable section plane (transversal) for system trajectories is chosen. Position of points of intersection on this plane defines the Poincaré map. Taking just one coordinate of these points (projecting them on one of the coordinate axes) and plotting the graph of a function "current coordinate(previous coordinate)" we obtain an approximate 1-D map as shown in Fig.5.1 b. To generate electronically a Poincare section of an attractor specific circuitry for detecting the passage (intersection) of the trajectory by the chosen plane in the state space is needed. This kind of Poincaré section can be implemented eg. as shown in Fig.5.2 which is a variation of circuitry proposed by Kennedy[243]. This circuitry allows adjustments not only of fixed levels but also the slopes of the section plane in the space. In this implementation it is possible to select a Poincaré plane of the form:

$$av_i + bv_j = U_0 \tag{5.1}$$

The slope is controlled by adjusting the potentiometer $R20$. The position of the Poincaré plane is fixed by setting the reference voltage U_0 (adjusting the

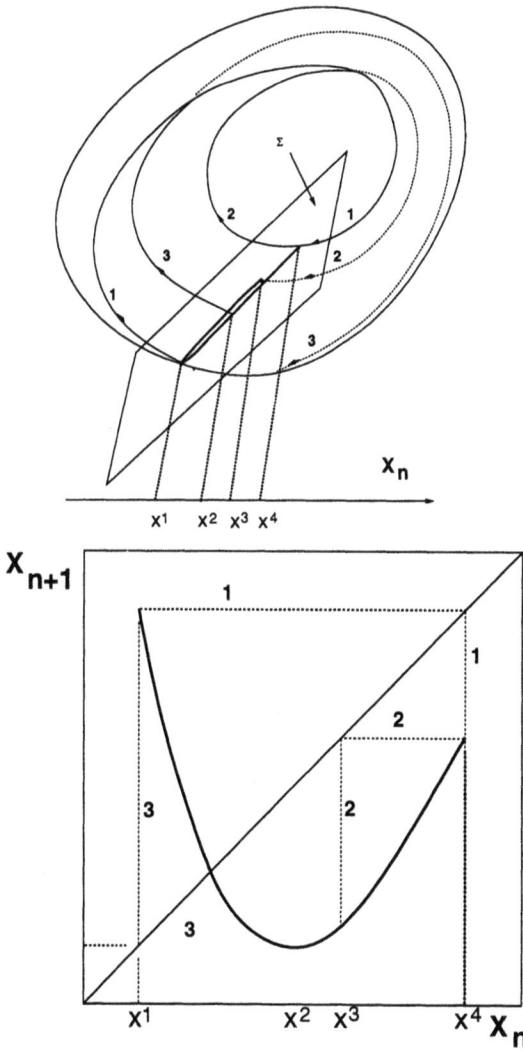

Figure 5.1: Construction of the Poincaré map - section plane is chosen and successive points of intersection of trajectories with this plane are memorized (a). Approximate 1-D map is constructed by projection of the points on the Poincaré plane onto a chosen axis (line) (b).

potentiometer R_{10}). The comparator detects the crossings of this level by a linear combination of variables chosen by the user. Suitable position of the switch S_2 allows the user to select positive or negative going crossings of the Poincaré plane. The logic output from the EX-OR gate triggers the cascade of two monostables which respectively produce two non-overlapping 'high' pulses of width approximately $\delta t \approx 4\mu s$. The second pulse is used to produce the d_τ output for Z-modulation of the oscilloscope to highlight the points belonging to the plane Σ. Outputs P_1 and P_2 from monostables are used also to trigger at suitable moments the two sample-and-hold circuits which enable us to produce signals suitable for observation of an approximate one-dimensional map. These signals are denoted $V_j(t_{k+1})$ and $V_j(t_k)$ in the circuit diagram. The values stored at successive moments correspond to a point and its image via the 1-D map when the delay between pulses is chosen accordingly.

Complete list of elements for this set-up in the following: AMP - amplifier LF347, CMP - comparator LM311, XOR - exclusive OR gate 1/4 74HCT86, MS1 and MS2 - two monostables - 74HCT123, SHX and SHY - sample-and-hold - LF398, C_4, C_5 - 10nF, C_6, C_7, C_8 - 1nF, R_{11}, R_{13}, R_{14}, R_{17} - $1k\Omega$, R_{12}, R_{16}, R_{21} - $10k\Omega$, R_{15} - 51Ω, R_{18} - 300Ω, R_{19} - $82k\Omega$, R_{20} - $200k\Omega$, D_1 - 1N4148, Q_1 - NPN transistor 2N2222.

Using this circuit it is possible to build a laboratory set-up for observing trajectories, Poincaré sections and approximate 1D maps as shown in Fig.5.3.

Using the proposed add-on circuits we have observed in laboratory experiments several attractors and their Poincaré sections. The results are shown in color Plates 6 - 10. Plates 6 through 8 show the trajectories observed in Chua's circuit undergoing a period-doubling sequence. Plates 9 and 10 show Poincaré sections taken at two different positions (shadowed in the top photographs). It is clear that the graphs of the Poincaré map changes when changing the position of the section plane.

5.1.3 *Oscilloscope observations in the RC-ladder chaos generator*

Let us describe in some detail our exemplary generator (Fig.5.4). Its dynamics are governed by a third-order Lur'e type equation of the form:

$$\frac{d\mathbf{x}(t)}{dt} = \mathbf{A}\mathbf{x}(t) + \mathbf{B}F[\mathbf{C}^T\mathbf{x}(t)] \qquad (5.2)$$

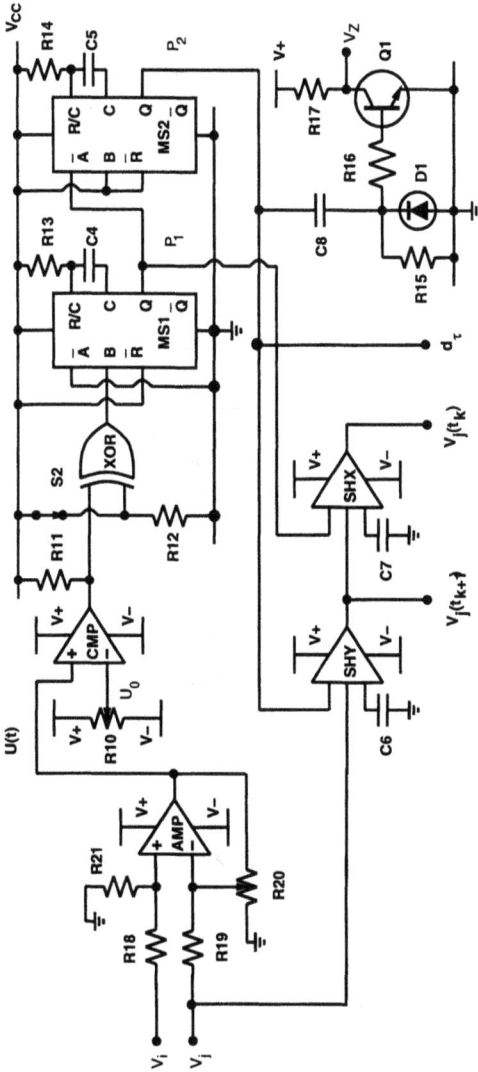

Figure 5.2: Simple realization of a circuitry for displaying one-dimensional maps using a standard oscilloscope.

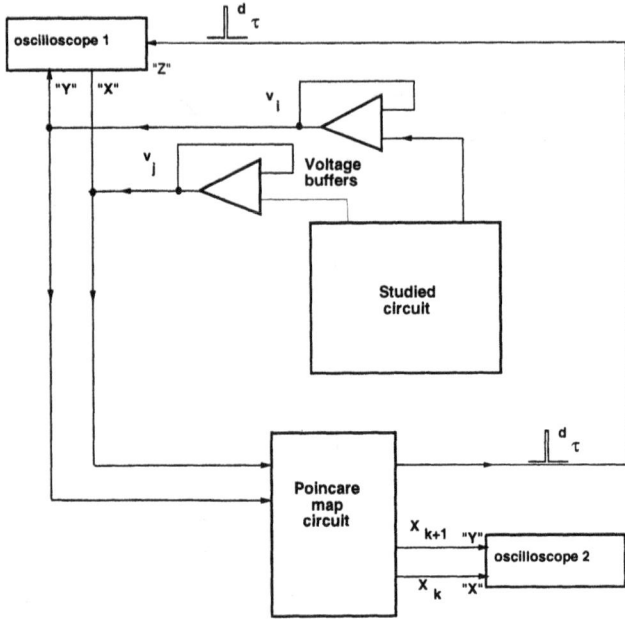

Figure 5.3: Laboratory set-up for observation of attractors, highlighting the Poincaré sections and observing an approximate one-dimensional map.

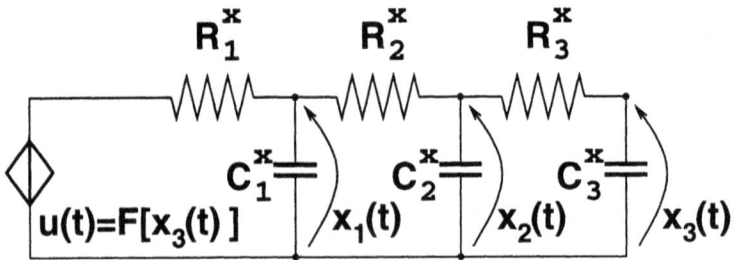

Figure 5.4: Circuit diagram of the RC-ladder chaos generator.

where:

$$A = \begin{bmatrix} \frac{-1}{R_1 C_1} + \frac{-1}{R_2 C_1} & \frac{1}{R_2 C_1} & 0 \\ \frac{1}{R_2 C_2} & \frac{-1}{R_2 C_2} + \frac{-1}{R_3 C_2} & \frac{1}{R_3 C_2} \\ 0 & \frac{1}{R_3 C_3} & \frac{-1}{R_3 C_3} \end{bmatrix} \qquad (5.3)$$

$$B = \begin{bmatrix} \frac{1}{R_1 C_1} \\ 0 \\ 0 \end{bmatrix} \qquad (5.4)$$

$$C = \begin{bmatrix} 0 \\ 0 \\ 1 \end{bmatrix} \qquad (5.5)$$

$$F(\sigma) = m_1 \sigma + \frac{1}{2}(m_0 - m_1)(|\sigma + \sigma_{bp}| - (|\sigma - \sigma_{bp}|) \qquad (5.6)$$

σ_{bp} - the coordinate of the breakpoint of the piecewise-linear characteristic (comp. Fig.6.1).

In the input-output convention this system can be considered as a linear system (RC-quadrupole) with the transfer function:

$$G(s) = \frac{G_1 G_2 G_3}{\begin{array}{c} s^3 C_1 C_2 C_3 + s^2 (G_1 C_2 C_3 + G_2 C_2 C_3 + G_2 C_1 C_3 + G_3 C_1 C_3 + \\ G_3 C_1 C_2) + s(G_1 G_2 C_3 + G_1 G_3 C_3 + G_1 G_3 C_2 + \\ G_2 G_3 C_3 + G_2 G_3 C_1 + G_2 G_3 C_2) + G_1 G_2 G_3 \end{array}} \qquad (5.7)$$

with a nonlinear feedback $F(\sigma)$.

The piecewise-linear characteristic was chosen because of the simplicity of hardware (op-amp) realization and also because in this case the analysis of qualitative behavior of solutions of the equation (5.2) is simplified to a great extent.

Qualitatively similar phenomena can be observed for other types of characteristics for example smooth ones of the form: $F(\sigma) = \sigma(a\sigma^2 + b\sigma + c)$.

In order to investigate the dynamic behaviors of the RC-ladder network with nonlinear feedback (Voltage-Controlled-Voltage-Source) we bread-boarded a laboratory circuit following the diagram shown in Fig.5.5. A standard implementation of the VCVS piecewise linear characteristic using diodes and operational amplifiers has been used.

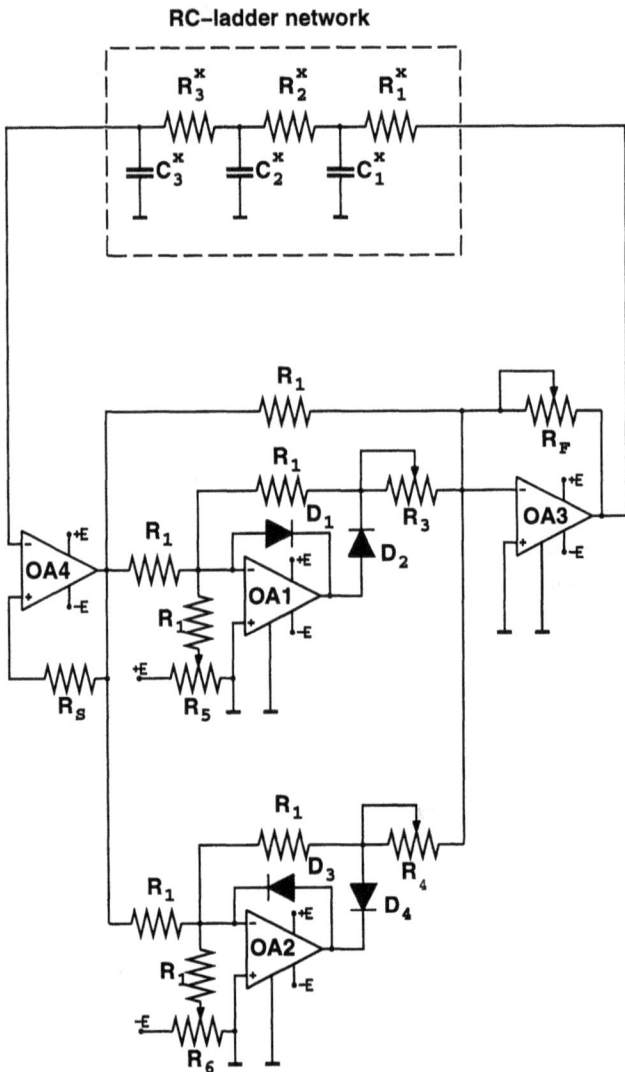

Figure 5.5: Circuit diagram of the laboratory test circuit for RC-ladder generator.

The CA3140 (RCA) op amps, 1N1141 diodes and the resistors $R_1 = R_3 = 10k\Omega$, variable resistors $100k\Omega$ have been used. Elements of RC ladder: $R_1^* = R_2^* = R_1^* = 1k\Omega$, $C_1^* = C_2^* = C_3^* = 0, 1\mu F$. Typical trajectories observed using a standard type oscilloscope are shown in the color Plates1-4. The photographs show typical period doubling sequence (period-one, period-two and period-four orbits, chaotic trajectory developed from the period doublings). Various chaotic trajectories and periodic windows are also shown. Plates 5 and 6 show experimental results obtained during measurements taken in the same circuit but with "dual" nonlinear characteristic ie. when the graph is reflected symmetrically against the x-axis. It should be noted that the attractors observed in this case are qualitatively different.

5.2   Simulation experiments for observing trajectories

To investigate the trajectory behavior in the state space and in particular the asymptotic properties of solutions we used standard numerical integration routines. All the numerical results obtained are in excellent agreement with the results of laboratory tests and confirm the existence of the earlier discovered phenomena.

Let us analyze the results of simulation experiments for the RC-ladder generator for various choices of circuit parameters. The Fig.5.6 shows the evolution of trajectories when varying the parameter m_1 of the nonlinear characteristic (with all remaining parameters of the system being fixed $R_1 = R_2 = R_3 = 1\Omega$, $C_1 = C_2 = C_3 = 1F$, $m_0 = -33.03$, $\sigma_{bp} = 0, 2$). The Fig.5.7 shows the evolution of trajectories of the system with changes of C_1 (with fixed $R_1 = R_2 = R_3 = 1\Omega$, $C_2 = C_3 = 1F$, $m_0 = -33, 03, m_1 = 330, 0, \sigma_{bp} = 0, 2$). In both cases we observe a typical route to chaos via period doublings (Feigenbaum sequence) – the asymptotic behavior changes from a period-one orbit, via period-two, period-four etc. We observe an infinite sequence of sub-harmonics with periods ω, 2ω, 4ω, 8ω etc. to apparently aperiodic orbits with very complicated structure. The chaotic trajectories observed in the case of changing m_1 are of two different kinds: with a single asymmetric "scroll" near one of the equilibria P^+ or P^- – Fig.5.6e,f) or and with two symmetrically positioned "scrolls" (near P^+ or P^- - Fig.5.6g,h). We will explain this phenomenon when analyzing the bifurcation diagram for m_1 – at some parameter value one observes so-called attractor merging[47]) of two co-existing attractors. Similar phenomena can be observed when varying C_1. In Fig.5.7 it is worthwhile to notice the existence of reversed Feigenbaum sequences (period halvings) leading from chaos to periodic trajectories of diminishing complexity. The abundance

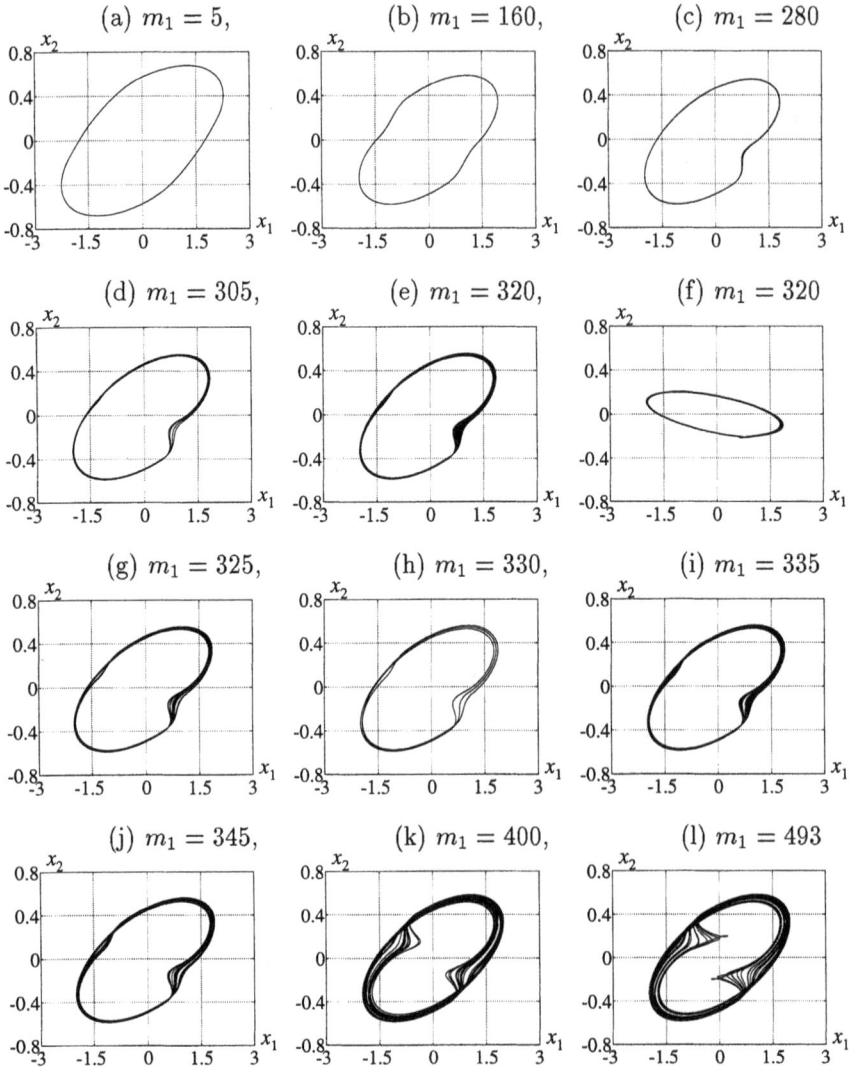

Figure 5.6: Trajectories of the RC-ladder generator obtained in numerical simulations when changing the values of the m_1 parameter with fixed $m_0 = -33,03$, $\sigma_{bp} = 0,2$ and all resistors and capacitors normalized to 1.

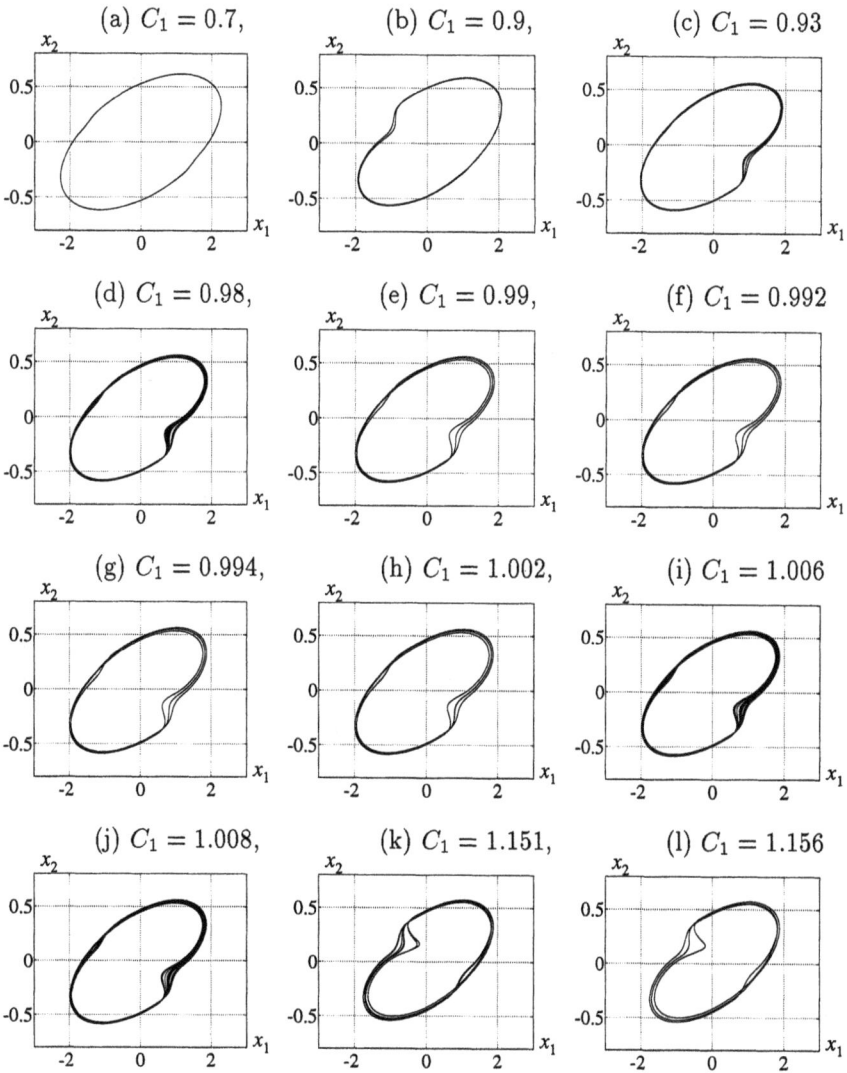

Figure 5.7: Trajectories of the RC-ladder generator obtained in numerical simulations when changing the values of C_1 with fixed $m_1 = 330$; $m_0 = -33,03$; $\sigma_{bp} = 0,2$ and all resistors and capacitors normalized to 1.

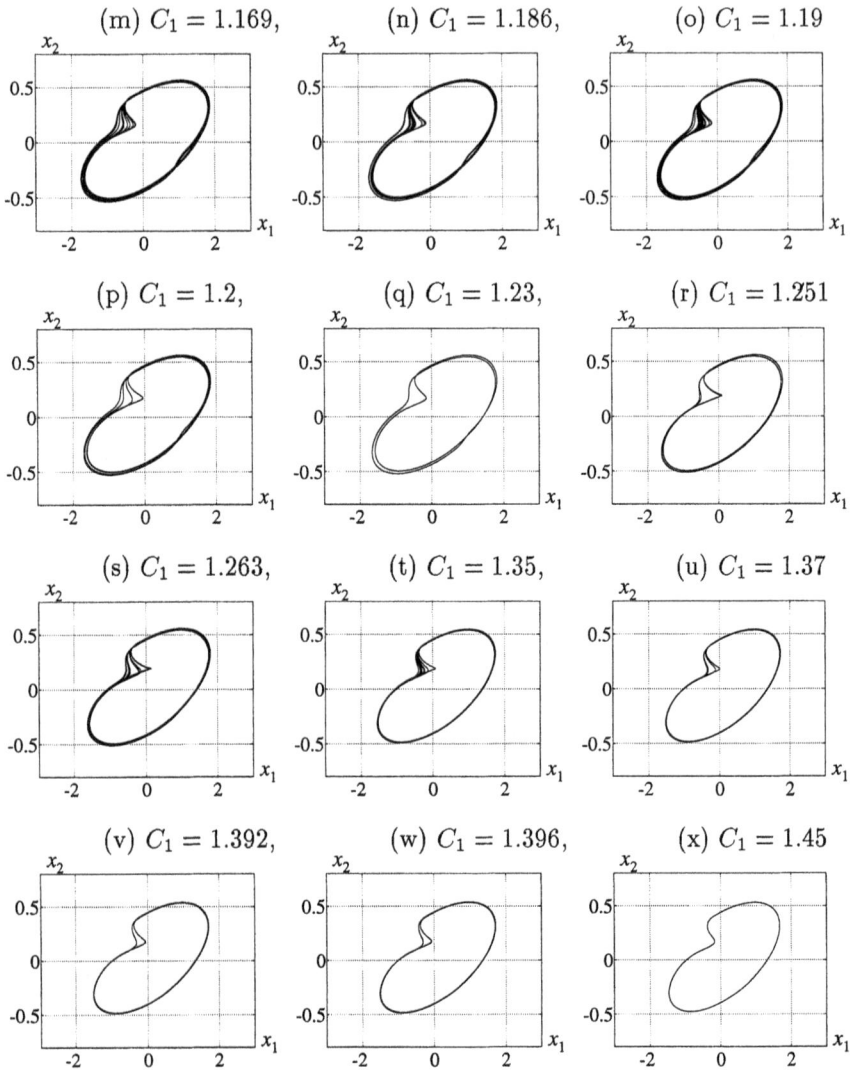

Figure 5.8: Trajectories of the RC-ladder generator obtained in numerical simulations when changing the values of C_1 with fixed $m_1 = 330$; $m_0 = -33,03$; $\sigma_{bp} = 0,2$ and all resistors and capacitors normalized to 1. — Continuation from Fig.5.7

of various trajectories shown in Fig.5.6 and 5.7 is astonishing.

Using typical pictures of systems behavior we have made an "art" picture to show typical behaviors on a background of stable/unstable manifolds and switching planes (Fig.5.9). It is interesting to notice the existence of divergent trajectories escaping to infinity. This kind of trajectories was not observed in laboratory experiments as all the op-amps for growing input signals will eventually saturate. Divergence would require a power supply of infinite power.

Observation of trajectories is the very simple but crude way of judging the system behavior. We would be interested in observing how the behavior depends on parameter changes but to be able to do it in fine detail. Introducing a Poincaré section plane and observing the position of the intersection points when varying the parameter with a very small step-size we can build so-called bifurcation diagrams depicting the dependence of the number of intersections and their position on the Poincaré plane on the actual value of a chosen parameter.

5.3 Investigating bifurcation phenomena

During the simulation experiments described in the previous section we could observe that typical trajectories intersect the S^+ plane (with $\dot{x}_3 > 0$) along the line FA – comp. Fig.5.9). This observation suggests that it is possible to choose this line on the Poincaré plane and construct a one-dimensional model of the dynamics. The positions of successive intersection points on this line determine the type of trajectory. To construct the bifurcation diagrams we have recorded just one coordinate of the successive point of intersection of system trajectories with the switching plane S^+. The state equations were integrated using a 4-th order Runge-Kutta method.

To eliminate the transient behavior the solutions during the interval equal 20 time constants of the RC section were discarded. Further the x_1 coordinates of successive trajectory piercings with the plane $x_3 = 0$ were memorized and plotted on the bifurcation diagram (the observation was carried out during ca. 100 time constants of the RC section).

At the beginning of each bifurcation analysis the initial conditions were imposed for the first simulation (parameter value) only. Later the successive experiments (integrations) were started from the last reached point of the previous analysis (as one says in jargon start from a point on the attractor). This approach enables close monitoring of the attractor evolution when the bifurcation parameter is changed.

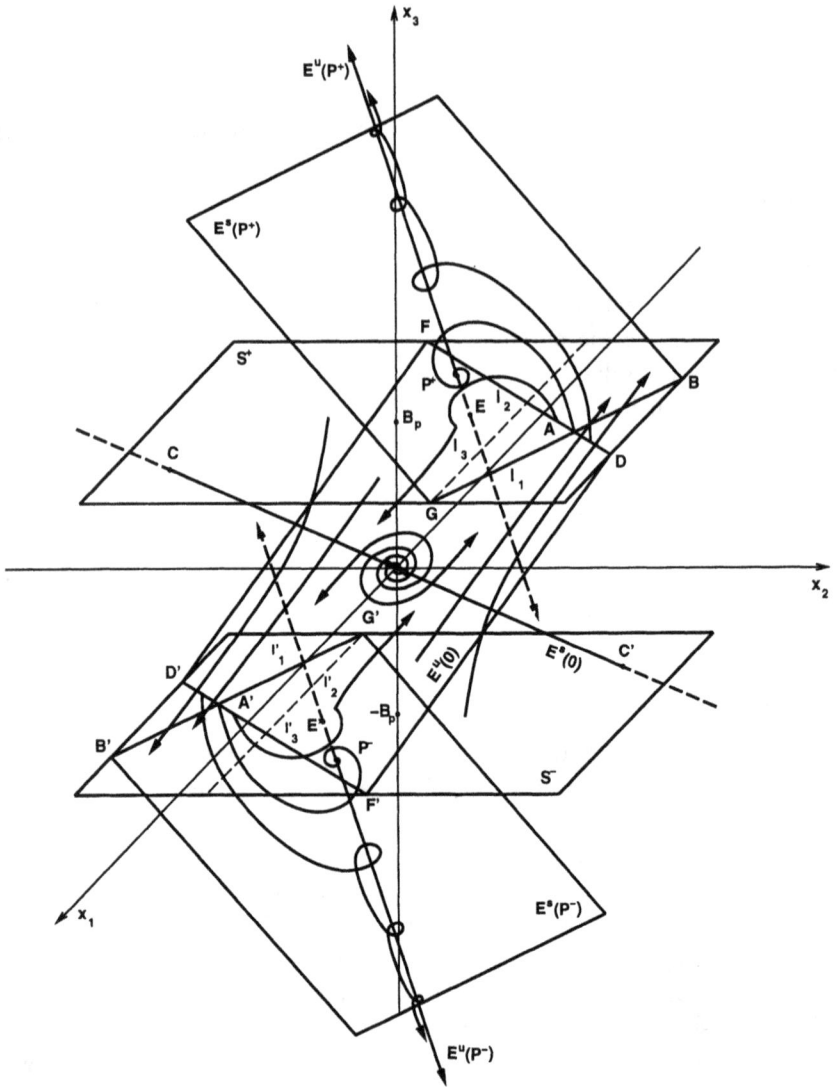

Figure 5.9: Typical behavior of the trajectories of the RC ladder chaos generator in 3-D on a background of the stable/unstable manifolds of equilibria and switching planes for the nonlinearity.

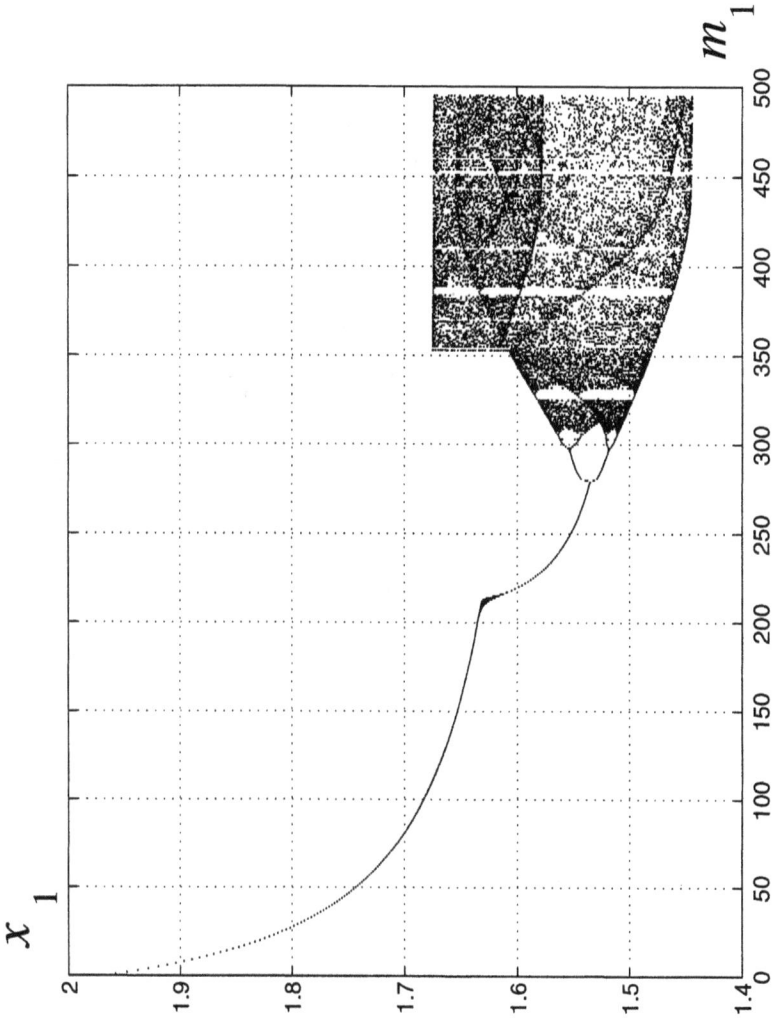

Figure 5.10: Bifurcation diagram for the choice of m_1 slope of the nonlinear characteristic. One can clearly see that there is a "missing branch".

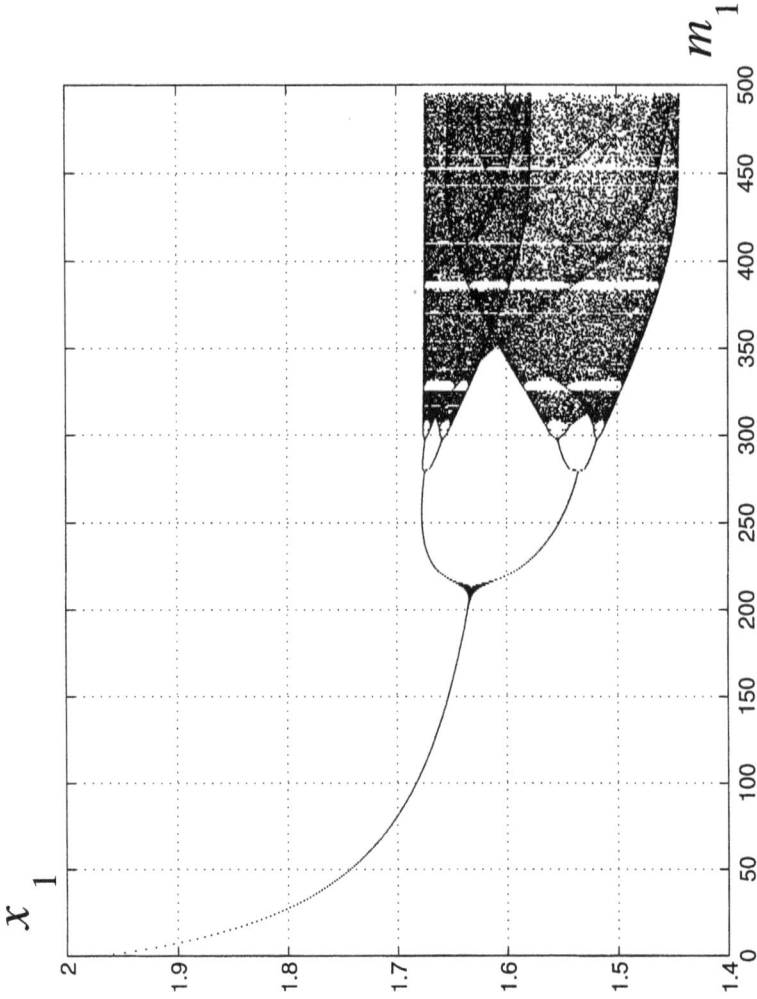

Figure 5.11: Full bifurcation diagram for the m_1 bifurcation parameter. The two branches corresponding to two co-existing attractors (symmetric against the origin) were constructed separately starting the experiments from different initial conditions. The sub-diagrams were later plotted together in this figure.

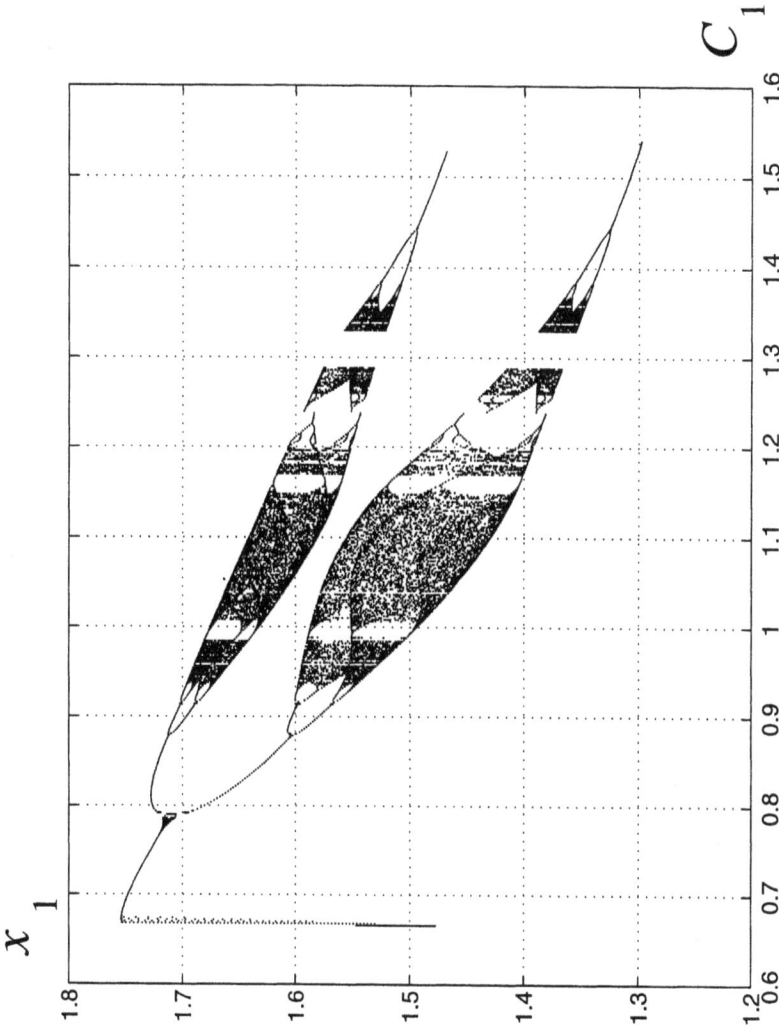

Figure 5.12: First part of the bifurcation diagram obtained when varying the parameter C_1 (with fixed $R_1 = R_2 = R_3 = 1\Omega$, $C_2 = C_3 = 1F$, $m_0 = -33,03$; $m_1 = 330,0$; $\sigma_{bp} = 0,2$)

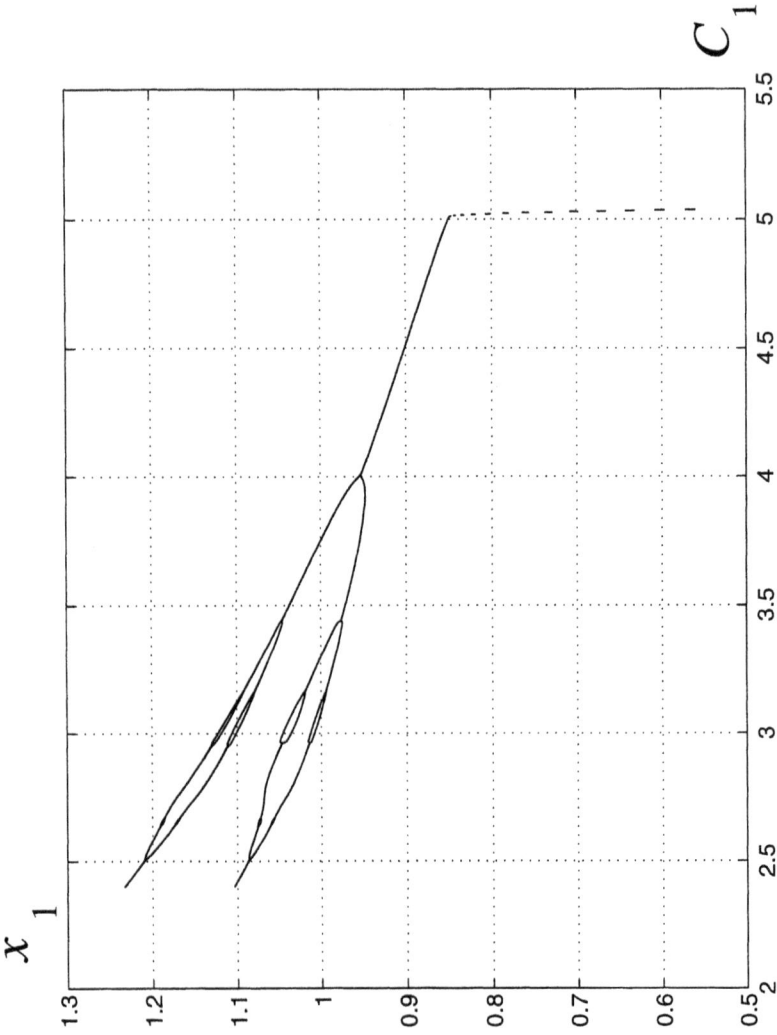

Figure 5.13: Second part of the bifurcation diagram obtained when varying the parameter C_1 (with fixed $R_1 = R_2 = R_3 = 1\Omega$, $C_2 = C_3 = 1F$, $m_0 = -33,03$; $m_1 = 330,0$; $\sigma_{bp} = 0,2$)

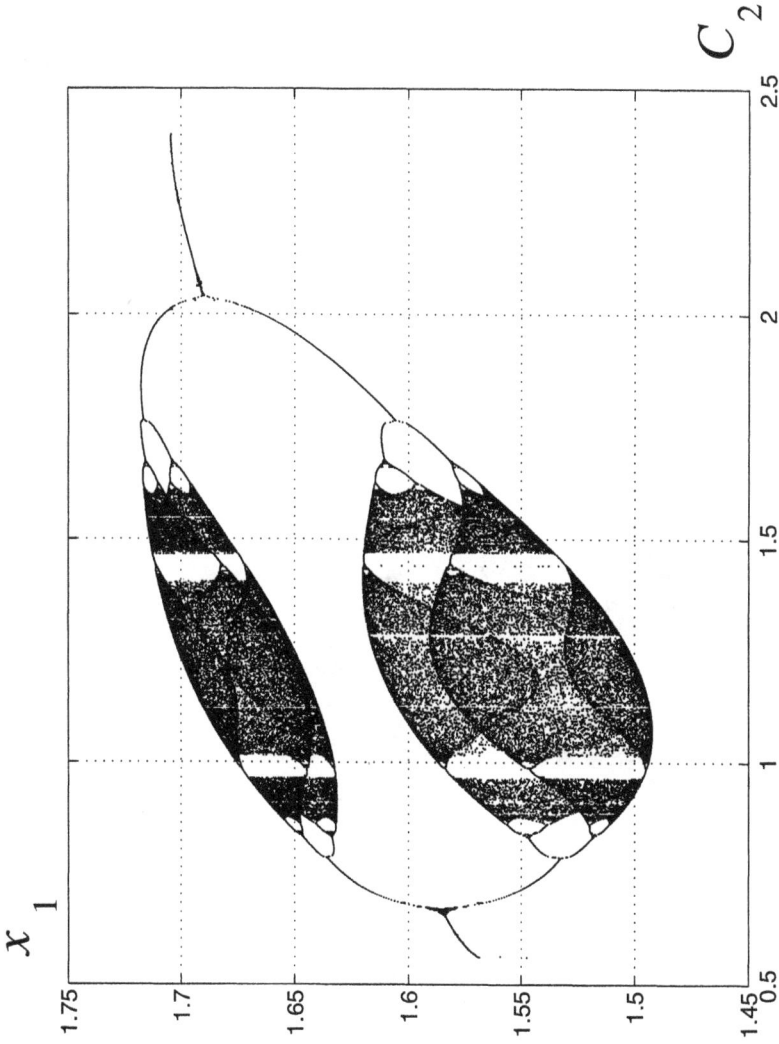

Figure 5.14: Bifurcation diagram obtained when varying the parameter C_2 (for fixed $R_1 = R_2 = R_3 = 1\Omega$, $C_1 = C_3 = 1F$, $m_0 = -33,03$; $m_1 = 330,0$; $\sigma_{bp} = 0,2$)

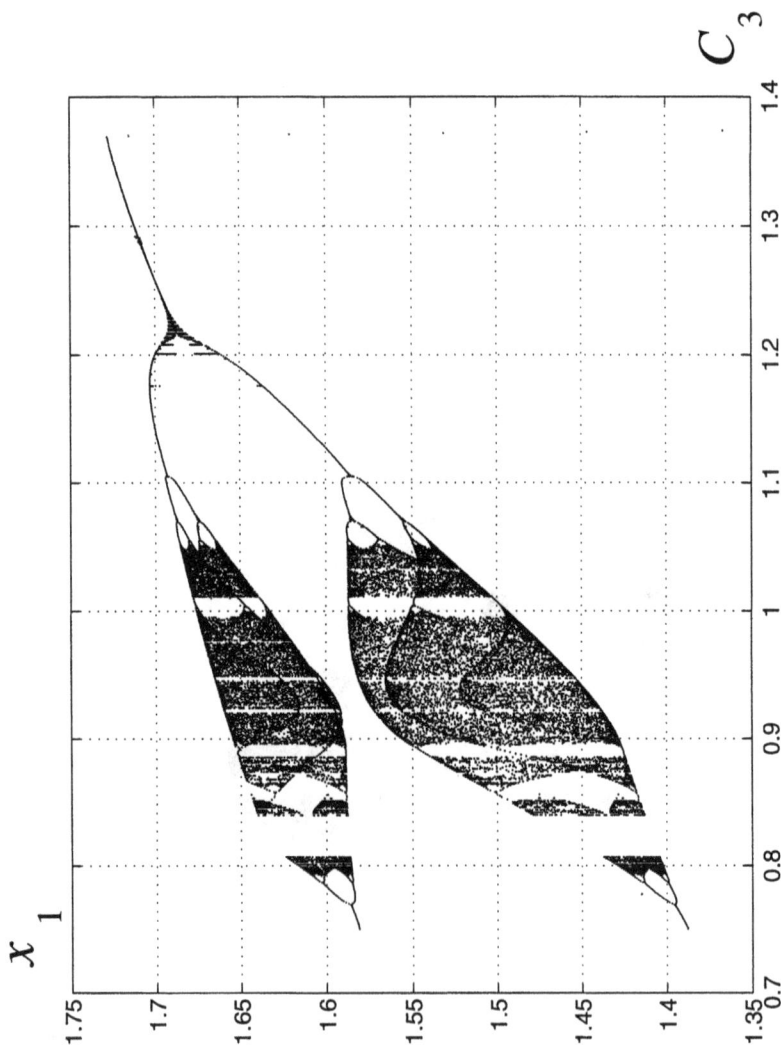

Figure 5.15: Bifurcation diagram obtained when varying the parameter C_3 (with fixed $R_1 = R_2 = R_3 = 1\Omega$, $C_1 = C_2 = 1F$, $m_0 = -33, 03$; $m_1 = 330, 0$; $\sigma_{bp} = 0, 2$)

Depending on the type of trajectory observed for a given parameter value the number of intersection points changes – from a single point for a period-one orbit to a densely filled interval for a chaotic trajectory.

The results of first bifurcation experiments were pretty unexpected. We found a threshold value for the m_1 parameter for which there was a jump of the size of the chaotic attractor (for $m_1 \approx 353$) which was born via a period doubling sequence. Analysis of the diagram suggested existence of a second branch of the diagram (see Fig.5.10). Starting the experiments with different (symmetric) initial conditions we were able to see visualize the missing part of the bifurcation diagram representing the attractor coexisting in the system in some parameter m_1 interval. The complete bifurcation diagram is shown in Fig.5.11. Two coexisting attractors are born in the symmetry breaking bifurcation from a single attractor (periodic orbit) of the same kind. These attractors evolve into chaos when changing the bifurcation parameter and eventually merge for $m_1 \approx 353$.

The bifurcation diagrams shown in figures 5.12 - 5.15 were constructed in the way described above for the m_1 bifurcation diagram. The branches representing coexisting attractors were constructed using simulation experiments starting from different initial conditions. In each diagram the two branches correspond to two co-existing attractors each having its own domain of attraction.

In all the bifurcation diagrams (Fig.5.11 - 5.15) one can observe several typical kinds of behavior - bifurcation sequences. The simples bifurcation – creation of a periodic orbit is linked with so-called Hopf bifurcation (or generalized Hopf bifurcation)[102,171]. Another common type of observed bifurcation is the symmetry breaking bifurcation (or pitchfork) in which one stable periodic orbits splits into two stable orbits of the same period and one unstable orbit – all of the same period as the orbit from which they were created. These are easily visible as the first bifurcations in the constructed trees.

The most interesting bifurcation phenomenon is definitely the infinite sequence of period doublings (Feigenbaum sequence[47,119,283]) leading to generation of chaotic behavior. The period doublings were described by Feigenbaum as universal behavior[119] – typical for a large number of physical systems. Certainly it remains the best studied route leading to chaos and is often considered as a partial proof of existence of chaos in the system. In-depth analysis of period doubling sequences can be found in the works of Alligood and Yorke[6], Crawford and Omohundro[90], Feigenbaum[119] or Mallet-Paret and Yorke[269].

It is important to notice here the existence of "crisis" phenomenon[166] – abrupt disappearance of the attractor. In the C_1 or C_3 diagrams it is easy to

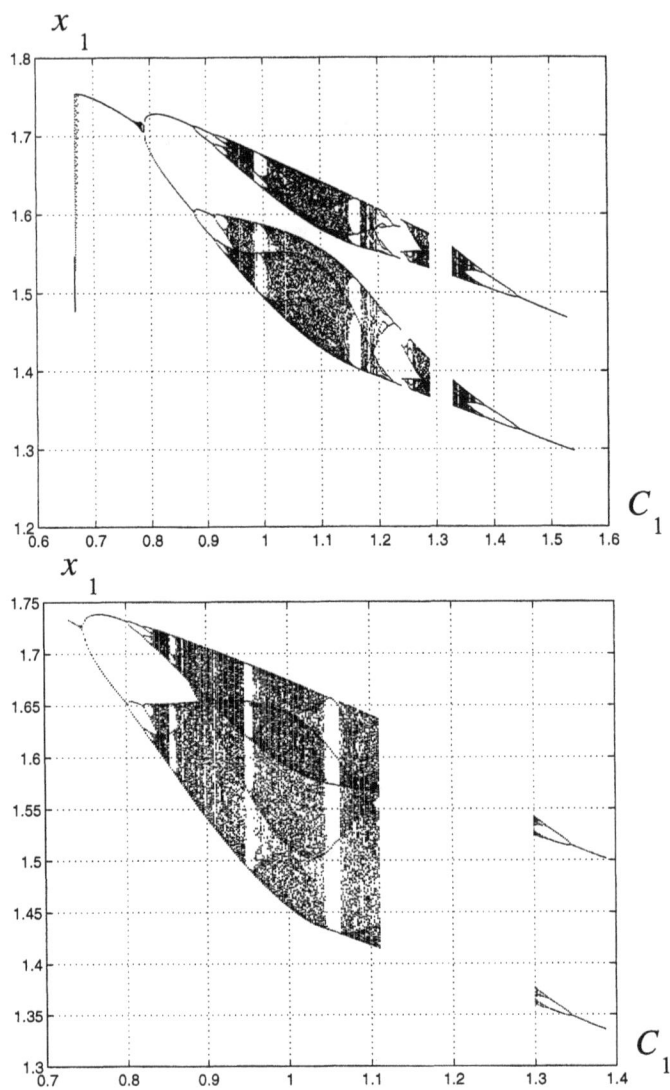

Figure 5.16: Comparison of the bifurcation diagrams obtained for variable C_1 for two different values of $m_1 = 400$ and $m_1 = 330$. The bifurcation sequences are different in each of the cases. (eg. attractor merging).

find the points at which the bifurcation diagram ends abruptly and reappears after an "empty" interval. We called this phenomenon "divergent windows". Yet another crisis-type phenomenon can be seen eg. in the m_1 bifurcation diagram. There exists a border line at which the attractor disappears and does not reappear again. Such a phenomenon is called in the literature a boundary crisis[47]. The mechanisms of crisis phenomena can be explained in part using approximate one-dimensional maps which will be discussed later.

In the parameter range where we observe chaotic oscillations one can observe also a number of periodic windows – regions of existence of periodic orbits lying within chaotic intervals[47,283].

In all presented bifurcation diagrams because of the symmetry of the system equations one can observe identical bifurcation sequences on both branches representing co-existing attractors. Both co-existing orbits always bifurcate for the same parameter value. Disturbing the symmetry of the nonlinear characteristic causes substantial changes in bifurcation sequences observed for each of the coexisting attractors. The symmetry breaking bifurcation is not symmetric any more – one of the branches shows "delayed" bifurcation sequence with respect to the other. The parameter values for which we observe similar phenomena eg. period doublings on each branch are not identical any more.

The Fig.5.16 shows how the bifurcation phenomena observed when varying C_1 change when changing m_1. For larger values of m_1 we observe merging of the two co-existing chaotic attractors[47]) – there is a qualitative change in the observed phenomena.

Most of the presented bifurcation diagrams show bi-directional sequences, re-merging bifurcation sequences [29] often referred to as "bubbles". The beginning of any bifurcation tree is linked with creation of a period-one orbit – Hopf or Hopf-like bifurcation. In the analyzed RC-ladder chaos generator such type of bifurcation can be observed for two distinct parameter values. By a continuous change of parameter values it is possible to pass from one Hopf bifurcation value to the other – this is the reason of creation of a re-merging sequence.

It is possible to calculate the parameter values for which one can observe a Hopf-type bifurcation at the equilibrium 0. To do this let us calculate the stability sectors for the linearized system obtained by replacing the nonlinear function $F(\sigma)$ in equation (5.2) by a linear one $F(\sigma) = m\sigma$, ie. let us find out for which slopes of the piecewise linear characteristic (m) the linear system is stable.

Choosing as varying one of the parameters of the RC quadrupole (with fixed remaining parameters equal 1) we have obtained the following equations for the borders of stability sectors of the linearized equations as function of parameters:

$$\text{for } R_1 = \frac{1}{G_1} \quad 1 \; > m > -\frac{12R_1^2 + 14R_1 + 3}{R_1} \tag{5.8}$$

$$\text{for } R_2 = \frac{1}{G_2} \quad 1 \; > m > -\frac{6R_2^2 + 15R_2 + 8}{R_2} \tag{5.9}$$

$$\text{for } R_3 = \frac{1}{G_3} \quad 1 \; > m > -\frac{3R_3^2 + 16R_3 + 10}{R_3} \tag{5.10}$$

$$\text{for } C_1 \quad 1 \; > m > -\frac{3C_1^2 + 16C_1 + 10}{C_1} \tag{5.11}$$

$$\text{for } C_2 \quad 1 \; > m > -\frac{6C_2^2 + 15C_2 + 8}{C_2} \tag{5.12}$$

$$\text{for } C_3 \quad 1 \; > m > -\frac{12C_3^2 + 14C_3 + 3}{C_1} \tag{5.13}$$

For fixed slope of the piecewise linear characteristic at 0 there exist two values of the bifurcation parameter for which the equilibria loose stability. These two parameter values determine the interval of instability. Passing of one of these values results in a Hopf bifurcation - the equilibrium looses stability and a periodic oscillation is born.

It is interesting to notice the identity of the equations (5.13) for $C_1 - R_3, C_2 - R_2$ and $C_3 - R_1$.

CHAPTER 6

ANALYTICAL ANALYSIS OF QUALITATIVE BEHAVIOR

Let us consider now a dynamical system described by the following ordinary differential equation:

$$\frac{d}{dt}\mathbf{x}(t) = \mathbf{A}\mathbf{x}(t) + \mathbf{B}F[\mathbf{C}^T\mathbf{x}(t)] \tag{6.1}$$

where :

$$\mathbf{A} = \begin{bmatrix} 0 & 1 & 0 \\ 0 & 0 & 1 \\ -1 & -Q & -P \end{bmatrix} \tag{6.2}$$

$$\mathbf{B} = \begin{bmatrix} 0 \\ 0 \\ 1 \end{bmatrix} \tag{6.3}$$

$$\mathbf{C} = \begin{bmatrix} 1 & 0 & 0 \end{bmatrix} \tag{6.4}$$

$P, Q > 0$, $F : \mathbb{R} \to \mathbb{R}$ - piecewise-linear function (eg. as shown in Fig.6.1)

$$F(\sigma) = \begin{cases} m_0\sigma & \text{for } |\sigma| < \sigma_{bp} \\ m_1\sigma - \sigma_{bp}(m_1 - m_0) & \text{for } \sigma \geq \sigma_{bp} \\ m_1\sigma + \sigma_{bp}(m_1 - m_0) & \text{for } \sigma \leq -\sigma_{bp} \end{cases} \tag{6.5}$$

The equations describing dynamical behavior of many physical systems can be transformed into the form (6.1). For example a number of systems generating chaotic oscillations like dual Chua's circuit[279,282,283], one of prototype Rössler systems[385], or the RC-ladder chaos generator[322,327], belong to this class of systems. Properties of solutions of systems belonging to this class were the subject of many studies. Burkin and Leonov[48,49] Leonov[263], Georgiev[154] have elaborated a number of criteria concerning existence of periodic solutions for this class of systems. In[327] using the results of Burkin and Leonov the class of nonlinear (piecewise-linear) characteristics guaranteeing existence of periodic solutions in the system (6.1) have been obtained. Leonov[264] has extended for the class (6.1) known Bendixons' criterion on nonexistence of limit cycles. Grabowski[160] studied the conditions for existence of homoclinic orbits and application of Shilnikov's theorem for proving existence of chaos.

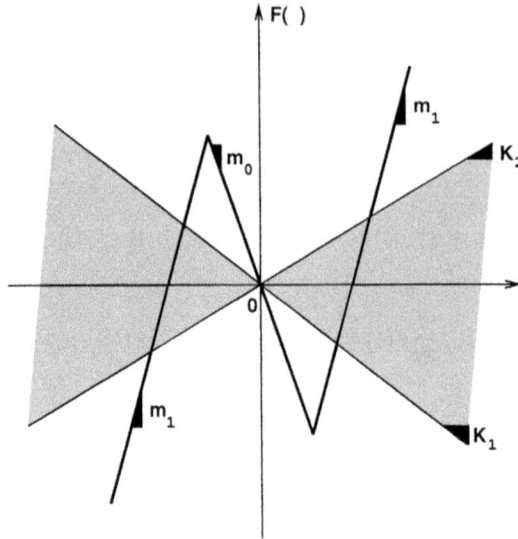

Figure 6.1: Piecewise-linear characteristic of the function $F(.)$

Using the results of Grabowski[159] it is possible that for the systems of the considered class the Ajzerman's conjecture is satisfied ie. the nonlinear system (6.1) is asymptotically stable if the nonlinear characteristic F satisfies the condition:

$$K_1 \leq \frac{F(\sigma)}{\sigma} \leq K_2 \qquad (6.6)$$

where : K_1, K_2 are the borders of the Hurwitz sector of the linearized system with $F(\sigma) = K\sigma$. The asymptotic stability region of the system (6.1) is darkened in the Fig.6.1. The considered piecewise-linear characteristic is not contained within this region and does not satisfy the condition (6.6).

6.1   Analysis of system geometry

Specific choice of the nonlinear characteristic in the form of a piecewise-linear function facilitates the analysis of the system behavior in the state space in a great extent. We will show below the basic geometric properties of solutions. (comp. Fig.5.9).

- **PROPERTY 1** The switching planes for the nonlinearity:

$$S^+ = \{(x, y, z) \in \mathbb{R}^3 : x = \sigma_{bp}\} \qquad (6.7)$$

$$S^- = \{(x, y, z) \in \mathbb{R}^3 : x = -\sigma_{bp}\} \tag{6.8}$$

divide the state space $\mathbb{R}^3$ into three regions:

$$D_0 = \{(x, y, z) \in \mathbb{R}^3 : x \leq |\sigma_{bp}|\} \tag{6.9}$$

$$D_1 = \{(x, y, z) \in \mathbb{R}^3 : x > \sigma_{bp}\} \tag{6.10}$$

$$D_{-1} = \{(x, y, z) \in \mathbb{R}^3 : x < -\sigma_{bp}\} \tag{6.11}$$

The trajectories of system (6.1) are composed from sections being solutions of the linear systems in each of the three domains D_0, D_1, D_{-1}. In the domain D_0:

$$\frac{d}{dt}\mathbf{x}(t) = (\mathbf{A} - m_0 \mathbf{B}\mathbf{C}^T)\mathbf{x}(t) \tag{6.12}$$

in the domain D_1 :

$$\frac{d}{dt}\mathbf{x}(t) = (\mathbf{A} - m_1 \mathbf{B}\mathbf{C}^T)(\mathbf{x}(t) - \mathbf{x}_p) \tag{6.13}$$

In the domain D_{-1} :

$$\frac{d}{dt}\mathbf{x}(t) = (\mathbf{A} - m_1 \mathbf{B}\mathbf{C}^T)(\mathbf{x}(t) + \mathbf{x}_p) \tag{6.14}$$

where :

$$x_p = [\sigma_p \ 0 \ 0]^T \tag{6.15}$$

$$\sigma_p = \frac{m_1 - m_0}{1 - m_1}\sigma_{bp} \tag{6.16}$$

$$\tag{6.17}$$

• PROPERTY 2

For the choice of the piecewise-linear characteristic as shown in Fig.6.1 there are three fixed points in the system: the origin O, $P^+(\sigma_p, 0, 0)$ and $P^-(-\sigma_p, 0, 0)$. In each of the domains D_0, D_1 and D_{-1} there is exactly one equilibrium point. The dynamical behavior and stability properties of solutions in each of the linear domains are prescribed by the eigenvalues γ, $\alpha \pm \beta i$, which are the solutions of the characteristic equation:

$$\lambda^3 + P\lambda^2 + Q\lambda + 1 - m_i = 0 \text{ where } \begin{cases} m_i = m_0 & \text{in the central region} \\ m_i = m_1 & \text{in the outer region} \end{cases} \tag{6.18}$$

Between the eigenvalues we have the following relations:

$$\alpha = -\frac{1}{2}(P + \gamma), \ \beta = \frac{1}{2}\sqrt{3\gamma^2 + 2P\gamma + 4Q - P^2} \qquad (6.19)$$

• PROPERTY 3

In the domain D_0 the trajectory behavior is defined by the set of eigenvalues:

$$\gamma_0, \alpha_0 \pm \beta_0 i, \gamma_0 < 0, \alpha_0 > 0 \qquad (6.20)$$

The stable manifold $E^s(0)$ of the equilibrium 0 is generated by the eigenvector:

$\begin{bmatrix} 1 & \gamma_0 & \gamma_0^2 \end{bmatrix}^T$.

$$E^s(0) = \{(x, y, z) \in \mathbb{R}^3 : x = \frac{y}{\gamma_0} = \frac{z}{\gamma_0^2}\} \qquad (6.21)$$

The unstable manifold $E^u(0)$ of the origin 0 is spanned by the eigenvectors:

$\begin{bmatrix} 1 & \alpha_0 & \alpha_0^2 - \beta_0^2 \end{bmatrix}^T , \begin{bmatrix} 0 & \beta_0 & 2\alpha_0\beta_0 \end{bmatrix}^T$.

$$E^u(0) = \{(x, y, z) \in \mathbb{R}^3 : (\gamma_0^2 + P\gamma_0 + Q\gamma_0)x + (\gamma_0 + P)y + z = O\} \quad (6.22)$$

At the equilibria P^+ and P^- the eigenvalues are as follows:
$\gamma_p, \alpha_p \pm \beta_p i, \gamma_p > 0, \alpha_p < 0$. The stable manifold $E^s(P^\pm)$ is spanned by the vectors :

$\begin{bmatrix} 1 & \alpha_p & \alpha_p^2 - \beta_p^2 \end{bmatrix}^T , \begin{bmatrix} 0 & \beta_p & 2\alpha_p\beta_p \end{bmatrix}^T$.

$$E^s(P^\pm) = \{(x, y, z) \in \mathbb{R}^3 : (\gamma_p^2 + P\gamma_p + Q)(x \mp \sigma_p) + (\gamma_p + P)y + z = 0\} \qquad (6.23)$$

The one-dimensional unstable manifold is generated by the eigenvector:
$\begin{bmatrix} 1 & \gamma_p & \gamma_p^2 \end{bmatrix}^T$ and is defined by:

$$E^u(P^\pm) = \{(x, y, z) \in \mathbb{R}^3 : x \mp \sigma_p = \frac{y}{\gamma_p} = \frac{z}{\gamma_p^2}\} \qquad (6.24)$$

In this way the dynamic properties of the system are defined by the set of six eigenvalues: $\gamma_0, \alpha_0 \pm \beta_0 i, \gamma_0 < 0, \alpha_0 > 0, \gamma_0, \alpha_0 \pm \beta_0 i, \gamma_0 < 0, \alpha_0 > 0$; thus the considered system belongs to the "dual double scroll" family proposed by Parker and Chua[363] (this means that there exists a linear similarity transform bringing (6.1) to the form proposed in the work of Parker and Chua; (comp. also the work of Komuro[253] about the normal forms of piecewise-linear systems). The relative position of the

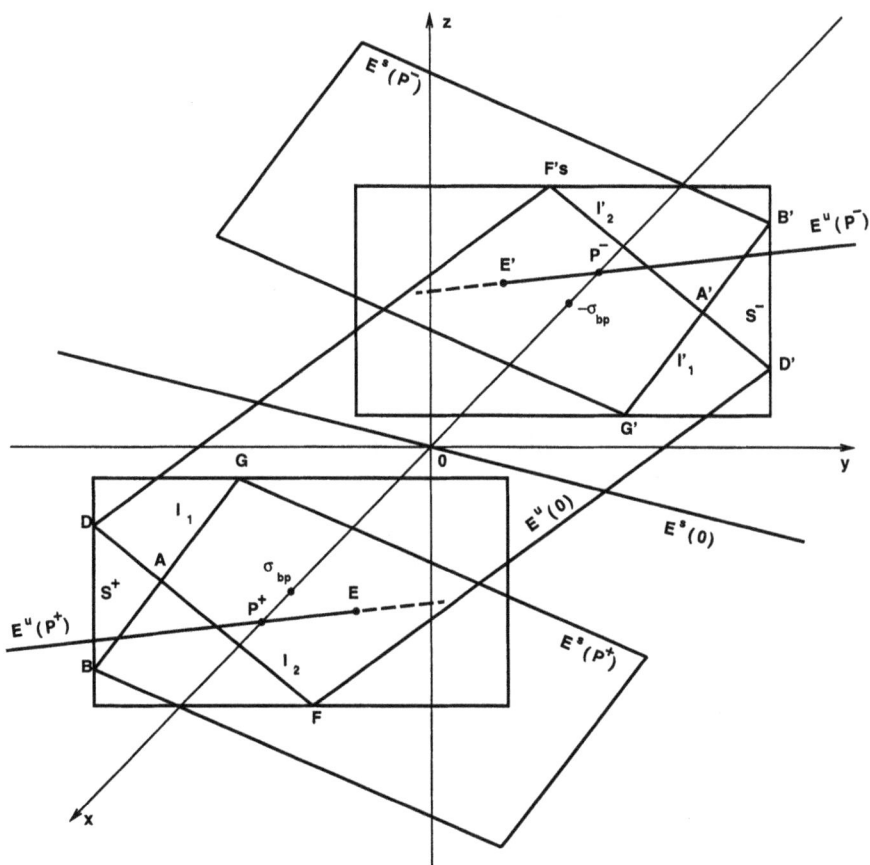

Figure 6.2: Relative position of the stable and unstable equilibria and the switching planes in the system described by the equation (5.2).

stable and unstable manifolds of the system equilibria and the positions of the switching planes of the nonlinearity is shown in Fig.6.2 (comp. also Fig.5.9).

- **PROPERTY 4**

The plane S^+ (respectively S^-) is divided into two half-planes by the line:

$$L_{tan} = \{(x, y, z) \in \mathbb{R}^3 : x = \sigma_{bp}, y = 0\} \qquad (6.25)$$

along which the trajectories in the domain D_0 are tangent to S^+; Respectively

$$L_{tan}^- = \{(x, y, z) \in \mathbb{R}^3 : x = -\sigma_{bp}, y = 0\} \qquad (6.26)$$

along which the trajectories in the domain D_0 are tangent to S^-.
In this way the half-plane:

$$S_{in}^+ = \{(x, y, z) \in \mathbb{R}^3 : x = \sigma_{bp}, y > 0\} \qquad (6.27)$$

is the set of initial conditions for system trajectories entering from D_0 into D_1, and the set

$$S_{out}^+ = \{(x, y, z) \in \mathbb{R}^3 : x = \sigma_{bp}, y < 0\} \qquad (6.28)$$

is the set of initial conditions for system trajectories passing from the domain D_1 into the domain D_0. (In a similar way we can define the sets S_{in}^- i S_{out}^-). Further division of the plane S^+ is done by the stable manifold $E^s(P^+)$ along the line:

$$L_1 = \{(x, y, z) \in \mathbb{R}^3 : x = \sigma_{bp}, (\gamma_p + P)y + z = 0\} \qquad (6.29)$$

The half-plane

$$S_{ret}^+ = \{(x, y, z) \in \mathbb{R}^3 : x = \sigma_{bp}, (\gamma_p + P)y + z < 0\} \qquad (6.30)$$

is the set of initial conditions for trajectories returning into the central domain and the complementary half-plane is the set of escape points ie. all initial points of trajectories divergent to infinity. In an analogous way we can introduce the division of the plane S^- .

• **PROPERTY 5**

The state equations in each of the domains of linearity can be solved analytically. The general form of solutions is as follows:

$$\begin{bmatrix} x(t) \\ y(t) \\ z(t) \end{bmatrix} = \begin{bmatrix} \psi(t; x(0); Px(0) + y(0); Qx(0) + Py(0) + z(0)) \\ \psi(t; y(0); Py(0) + z(0); (m-1)x(0)) \\ \psi(t; z(0); -Qy(0) + (m-1)x(0); (m-1)y(0)) \end{bmatrix} \quad (6.31)$$

where:

$$\psi(t; \xi_1; \xi_2; \xi_3) = \kappa_1 e^{\gamma t} + \kappa_2 e^{\alpha t} \cos(\beta t) + \kappa_3 e^{\alpha t} \sin(\beta t) \quad (6.32)$$

$$\kappa_1 = \frac{\xi_1 \gamma^2 + \xi_2 \gamma + \xi_3}{(\alpha + \gamma)^2 + \beta^2} \quad (6.33)$$

$$\kappa_2 = \xi_1 - \kappa_1 \quad (6.34)$$

$$\kappa_3 = \frac{1}{\beta}[\xi_2 + (\alpha + \gamma)\xi_1 + (\alpha - \gamma)\kappa_1] \quad (6.35)$$

In the domain D_0 in the above formulas we have to substitute: α_0, β_0 ,γ_0, m_0, and in the domain D_1 (or resp. D_{-1}) we have to put: α_p, β_p ,γ_p, m_1 and shift the solution by $-\sigma_p$ (σ_p) in the x coordinate.

• **PROPERTY 6**

Because the system is piecewise-linear it is natural to consider half-return Poincaré maps between the nonlinearity switching planes S^+ i S^- (comp the works of Kahlert and Rössler[217,219], Parker and Chua[363], Chua, Komuro and Matsumoto[73]). these planes do not constitute global sections of trajectories – there exist trajectories which do not cross them at all. considering the map from S^+ into itself induced by the trajectories of the system (6.1) in the domain D_1 we will take into account only those initial conditions which belong to $S_{in}^+ \cap S_{ret}^+$, ie. those giving birth to trajectories will pass later into D_0 (in particular the segment of the line L_1 – see Fig.6.2). this reasoning is based on the properties of the map Π_0 – if the general assumption ($\bullet$) is satisfied, then all trajectories passing from D_0 do D_1 (D_{-1}) must pass very close to this line. To construct the return map (Poincaré map let us consider two half-maps:

Π_1 : S^+ into itself in the domain D_1

Π_0 : S^+ into S^- by the system trajectories in the domain D_0.

Because the system is symmetric, we will obtain analogous half-maps Π_0^- : $S^- \to S^+$ and $\Pi_1^- : S^- \to S^-$ (these are just reflected with respect to the origin).

- **PROPERTY 7**

  The considered system has the **volume contraction** property. This means that its divergence is always negative:

  $$\text{div}\vec{V} = \frac{\partial \dot{x}}{\partial x} + \frac{\partial \dot{y}}{\partial y} + \frac{\partial \dot{z}}{\partial z} = -P \qquad (6.36)$$

  Let us consider a ball of initial conditions at $t = 0$ having the volume $\Omega(0)$. After some time t its volume will become:

  $$\Omega(t) = \Omega(0)e^{-Pt} \qquad (6.37)$$

  This means that the volume is contracting in time. This property excludes the possibility of existence of an attractor of a T^2 torus-type which is equivalent to nonexistence of quasi-periodic solutions.

6.2   Multilevel oscillations

Yakubovich in his work[213] introduces a very interesting characterization of oscillatory (recurrent) solutions proposing so-called $(-\alpha, \beta)$- oscillations. Both periodic, quasi-periodic and aperiodic solutions belong to this class.

Let us consider the class of feedback systems described by a Lur'e equation

$$\frac{d\mathbf{x}(t)}{dt} = \mathbf{A}\mathbf{x}(t) + \mathbf{B}F[\mathbf{C}^T\mathbf{x}(t)] \qquad (6.38)$$

where : $\mathbf{A}$ - matrix $n \times n$, $\mathbf{B}, \mathbf{C}, \mathbf{x}(t) \in \mathbb{R}^n$ $F : R \to R$ - locally Lipschitz function, $F(0) = 0$,
Let $G(s) = \mathbf{C}^T(\mathbf{A} - s\mathbf{I})^{-1}\mathbf{B}$ – transfer function of the system.
Definition 6.1 ($(-\alpha, \beta)$- oscillations) *The solution* $\mathbf{x} = \mathbf{x}(t)$ *is called* $(-\alpha, \beta)$- *oscillatory* ($\alpha \geq 0$, $\beta \geq 0$, $\alpha + \beta > 0$) *if it is bounded and the output signal* $\mathbf{y}(t) = \mathbf{C}^T\mathbf{x}(t)$ *infinitely many times changes sign when* $t \to \infty$ *and there exists a series* $\{t_n\}$, $t_n \to \infty$, *such that* $y(t_n) \notin [-\alpha, \beta]$.

The following theorem specifies the conditions for existence of this kind of oscillations.
Theorem 6.1 (Existence of $(-\alpha, \beta)$- oscillations[213]) *Let* $det\mathbf{A} \neq 0$, *there exists the derivative* $F'(0) = \frac{F(\sigma)}{d\sigma}|_{\sigma=0}$ *and*

1. *the equation* $\Phi(y) = y + G(0)F(y) = 0$ *has the only solution* 0 *(ie. the system has a unique equilibrium* 0*),*

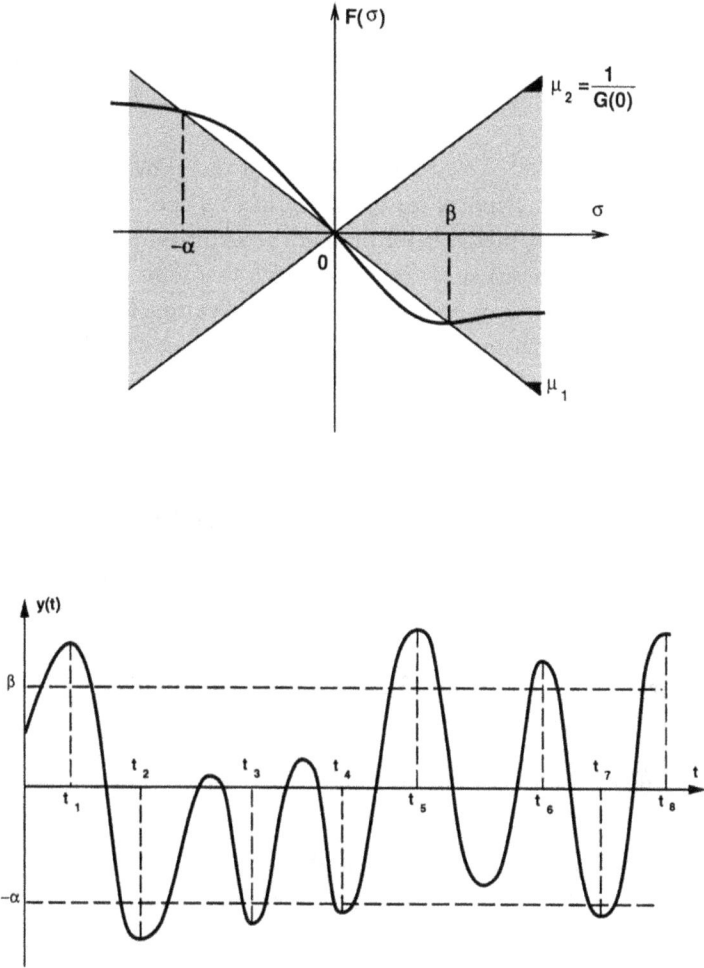

Figure 6.3: Nonlinear characteristic guaranteeing existence of $(-\alpha, \beta)$- oscillations (a), typical waveform of the $(-\alpha, \beta)$- oscillatory solution (b); μ_1 i μ_2 denote the borders of the Hurwitz sector for the linear system $\frac{dx(t)}{dt} = [A + \mu BC^T]x(t)$ (the simplest case of a single stability sector is depicted).

2. *the linear system $\frac{d\mathbf{x}(t)}{dt} = [\mathbf{A} + \mathbf{B}F'(0)\mathbf{C}^T]\mathbf{x}(t)$ is unstable and has no periodic solutions.*

3. *all solutions of the equation (6.38) are bounded*

Then there exist the numbers α and β such that almost all solutions (all except those starting from the set of initial conditions of measure 0) are $(-\alpha, \beta)$ oscillatory.

Yakubovich[213] gives several versions of the above theorem and also some frequency-domain criteria for existence of this specific kind of solutions. He gave also effective methods for finding the numbers α i β . The Fig.6.3 shows graphical interpretation of the Yakubovich theorem.

This theorem has however limited application to systems with a single equilibrium. It can be generalized in the following way. Let us introduce new definition:

Definition 6.2 (Multilevel oscillations[161]) *Let us assume that the Lur'e system considered above has only isolated equilibria, ie the equation:*

$$\Phi(y) = y + G(0)F(y) = 0 \tag{6.39}$$

has distinct solutions $y_i, i = 0, \pm1, \pm2 \ldots$.
The solution $\mathbf{x}(t)$ is called a multilevel oscillation if $y(t) = \mathbf{C}^T\mathbf{x}(t)$ infinitely many times passes the level y_0 or y_{-1} or y_{+1} or y_{-2}, etc. and $\lim_{t\to\infty} y(t) \neq y_i$

Theorem 6.2 (Criterion for existence of multilevel oscillations)
Let $\det\mathbf{A} \neq 0$ and

1. *the equation (6.39) has distinct solutions $y_i, i = 0, \pm1, \pm2 \ldots$, at which $\Phi(y)$ changes sign and $y_0 = 0$,*

2. *there exists $k \in \mathbb{R}$, such that $Re\lambda(\mathbf{A} + k\mathbf{BC}^T) < 0$ (the real parts of all eigenvalues of this matrix are all negative) and*

$$\left.\frac{F(\sigma) - k\sigma}{\sigma}\right|_{|\sigma|\to\infty} \to 0 \tag{6.40}$$

Then [a]either all solutions are multilevel oscillatory or

$$\Phi[y(t)]_{t\to\infty} \to 0 \tag{6.41}$$

---

[a]If the condition (6.40) is satisfied, the characteristic $F(\sigma)$ is "asymptotic to some straight line belonging to the Hurwitz sector" – which guarantees the boundedness of solutions

If we additionally assume the all equilibria are locally unstable then the condition (6.41) is satisfied only for initial conditions in a set of measure 0 belonging to the local stable manifolds of these equilibria. It is not possible to tell *a priori* around which levels the solutions will oscillate – they could oscillate around all or only some levels out of the large number of equilibria. The graphical interpretation of this criterion is shown in Fig.6.4.

It is a matter of a simple calculation and drawing a sketch of the characteristics to confirm that the RC-ladder chaos generator, Chua's oscillator and many other circuits satisfy the assumptions of the criterion for existence of multilevel oscillations. As in these systems there are three equilibria – they will be three-level oscillatory in general. It was confirmed by simulations that two-level oscillatory solutions also exist in these circuits. It should be noted however that it is not possible to distinguish between periodic and chaotic oscillations using the proposed criterion. Both definitions: of $(-\alpha, \beta)$- oscillations and multilevel oscillations comprise both types of solutions.

6.3 Chaos in the Shil'nikov sense

The existence of the specific type of orbits that are doubly asymptotic to system equilibria (in other words starting and ending at some equilibrium point) has very interesting implications on the overall system behavior. There are two types of such orbits:

- **homoclinic orbits** starting and ending at the same equilibrium point;

- **heteroclinic orbits** linking two different equilibrium points

Two basic methods of analytical proofs of existence of chaos, namely the method based on the Shil'nikov theorem for autonomous systems and the method based on Melnikov theorem for non-autonomous ones are closely linked with existence of homoclinic orbits in the system and in fact analyze the behavior of system trajectories in a close vicinity of such special orbits. Both methods have many variants. We will describe below one of many variants of the theorem of Shil'nikov. For the details of Melnikov method we refer the reader to the book of Wiggins[454] or papers generalizing Melnikov method to systems in $\mathbb{R}^n$ (originally the method was developed for systems with a small parameter in $\mathbb{R}^2$ only), or systems without the small parameter restriction[170,394].

Shil'nikov[2,3,143,144,405,406,407,408,409,410,411,412,413,414,415] and his collaborators have thoroughly elaborated the initial result valid only for neighborhoods of a saddle-node fixed points. One could say that Shilnikov's theorem and its

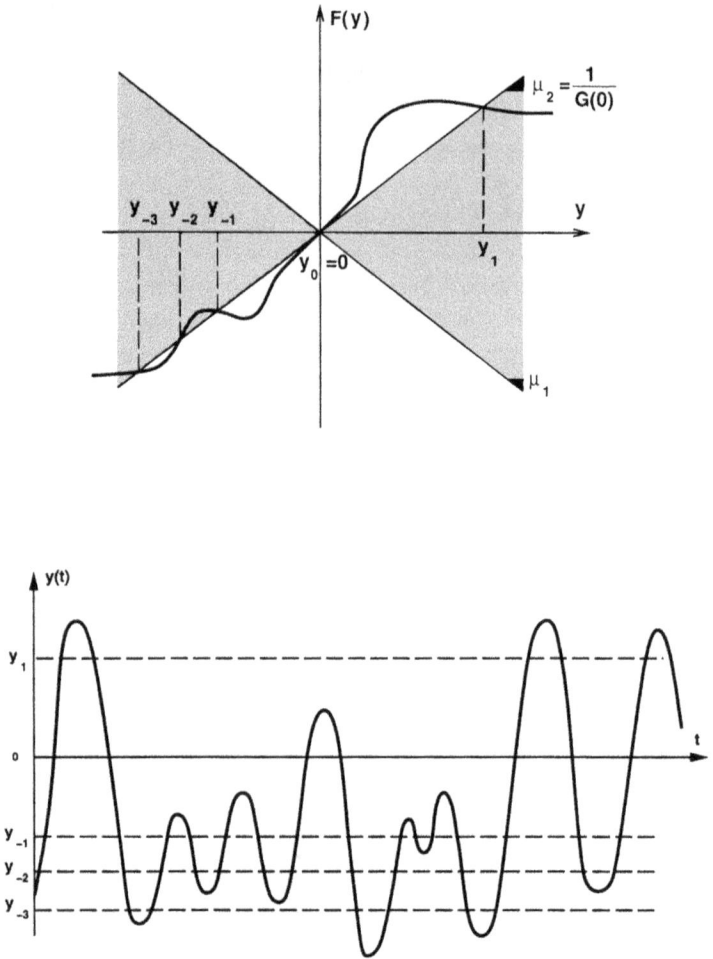

Figure 6.4: Nonlinear characteristic guaranteeing existence of multilevel oscillations (a); typical waveform of a multi-level oscillatory solution (b);

variants have become now a standard tool for proving complicated behavior in autonomous systems. Application of Shilnikov's theorem and interpretation of results is not an easy task[11,12,13,68,142,143,144,150,151,436,437]. Only the fact of existence of a countable but infinite number of homoclinic orbits shown in the cited works of Gaspard, Glendinninga and Tresser, could pose serious understanding problems to say nothing about the interpretation of existence of a Smale's horseshoe and infinite number of unstable periodic orbits.

Existence of chaos in the Shilnikov's sense has been confirmed for any known electronic systems like Chua's circuit and its family[73,74,153] or dual Chua's circuit[363]. The works of George[153] and Grabowski[160] the Shilnikov's theorem has been applied to a general class of piecewise linear systems with single nonlinearity.

Theorem 6.3 (Shil'nikov's theorem) [160,291,406,407]

Let us consider a differential equation of the form:

$$\dot{\mathbf{x}}(t) = f[\mathbf{x}(t), \mu] \qquad (6.42)$$

where: $f : \mathbb{R}^3 \times \mathbb{R} \rightarrow \mathbb{R}^3, f(0, \mu) = 0$ *for all* μ *in some neighborhood of 0. We will assume that* f *is sufficiently regular to satisfy the assumptions of the Hartman-Grobman*[171] *linearization theorem.*[b] *Let us assume the the Jacobian matrix* $\frac{\partial f(\mathbf{x}, \mu)}{\partial \mathbf{x}}$ *at* $\mathbf{x} = 0$ $\mu = 0$ *has the eigenvalues* γ, $\alpha \pm \beta i$, $\alpha, \beta, \gamma \in \mathbb{R}$, $\beta > 0$, α *and* γ *of opposite signs. Let us assume further that for* $\mu = 0$ *the system (6.42) has a homoclinic orbit with respect to the equilibrium 0. Then in the neighborhood of this homoclinic orbit for small perturbations of* μ *one can observe the following qualitative types of trajectory behavior:*

- *if* $|\frac{\gamma}{\alpha}| > 1$, *then for* $\mu > 0$ *the system will possess a stable periodic orbit the period of which tends to infinity for* $\mu \rightarrow \infty$, *and for* $\mu < 0$ *in the system there will be no recurrent trajectories.*

- *if* $|\frac{\gamma}{\alpha}| < 1$, *then for* $\mu = 0$ *in the system exist an infinite countable number of unstable periodic orbits and in the neighborhood of the of the homoclinic orbit the Poincaré map defined by trajectories of the system (6.42) on some chosen transversal section has a countable number of Smale's horseshoes which are structurally stable and persist for small* $|\mu|$.

---

[b]The Hartman-Grobman linearization theorem states that if the Jacobian matrix $Df(\mathbf{x}) = \frac{\partial f(\mathbf{x}, \mu)}{\partial \mathbf{x}}$ calculated at $\mathbf{x} = 0$ has no zero or purely imaginary eigenvalues then there exists a neighborhood U of the equilibrium point 0, in which the asymptotic behavior of solution of the nonlinear system and their stability type is the same as for the linear system $\dot{\mathbf{x}}(t) = Df(0)\mathbf{x}(t)$ (The detailed discussion of this theorem can be found eg. in[171]).

Let Ω_0 be a point belonging to the homoclinic orbit Γ_0, and S_T a plane perpendicular to the stable manifold of 0, containing Ω_0. Let us consider a map Π_S, generated by the solutions of the continuous-time system on S_T. Let us take the points $A_1 \in S_T$ and $A_1' = \Pi_S(A_1)$. Choosing A_1 in such a way that the distances from Ω_0 to A_1 and B_1' are equal we can state that $A_1 B_1' A_1' B_1$ is a rectangle. Let us choose next the points $A_2' = B_1'$ and $A_2 = \Pi_S^{-1}(A_2')$. Continuing this iterative process we will obtain a series of rectangles $A_i B_i' A_i' B_i$ on S_T. Let now R be a plane perpendicular to the local unstable manifold of 0, containing the point Ω_1 belonging to this manifold. The transformation mapping the points on the plane S_T to the points of first intersection of trajectories starting from them with the plane R R transforms the family of rectangles on S_T into a family of spiral regions. The mapping transforming the points of the plane R into the points of first intersections of trajectories starting from them with the plane S_T transforms the family of spiral regions in such a way that they intersect the initial rectangles.

The map generated by the solutions of the continuous system on S_T, $\Pi_S(A_i B_i' A_i' B_i)$ has the properties of the Smale horseshoe map.

- **Comment 1**:

  The fact that the assumptions of the first part of the Shilnikov's theorem are satisfied is often considered as a proof of existence of chaotic behavior. This reasoning is based on the properties of the Smale's horseshoe map[171]:

  - The map has an invariant set Λ $(\Lambda \subset \mathbb{R}^2)$, such that:
    * Λ is a Cantor set;
    * Λ contains a countable infinity of saddle-type (unstable!) periodic orbits with arbitrarily long periods;
    * Λ contains an uncountable number of non-periodic bounded orbits;
    * Λ contains a dense orbit;
  - Smale's horseshoe is structurally stable.

- **Comment 2**:

  The existence of Smale's horseshoes does not imply the existence of an attractor. In other words even if we prove that some system is chaotic in the sense of Shil'nikov it might be impossible to observe chaotic behavior in experiments in the predicted parameter range.

- **Comment 3**:

  For $|\frac{\gamma}{\alpha}| < 1$ i $\mu > 0$ in the system exists an infinite but countable number of complicated homoclinic orbits for the parameter values $\mu_1, \mu_2, \mu_3, ... \mu_n$, $... \to 0$ (compare the papers of Gaspard[143,144,145] and Glendinning[156,157]).

6.3.1 Homoclinic orbits in the RC-ladder chaos generator

Taking into account the geometric considerations carried out in the previous sections we will analyze the existence homoclinic and heteroclinic trajectories in the RC-ladder chaos generator in function of the bifurcation parameter m_1 for fixed values of other parameters of the circuit. The equations describing the dynamics of the circuit (5.2) for fixed: $R_1 = R_2 = R_3 = 1$, $C_1 = C_2 = C_3 = 1$, $m_0 = -33,03$, $\sigma_{bp} = 0,2$ and $m_1 = var.$ are transformed into the form (6.1) with the same type of nonlinearity, with $P = 5$, $Q = 6$.

- **Trajectories homoclinic to the origin** 0

  Let us consider the neighborhood of the point 0. If a trajectory homoclinic to this point exists it must be pieced together of three segments being solutions of the linear systems in respective domains: a spiraling divergent segment on the plane $E^u(0)$ (until the moment of its intersection with one of the switching planes S^- or S^+); a section of the stable manifold $E^s(0)$ between the points 0 and $D^0 = E^s(0) \cap S^+$ (or S^- resp.) and a segment of trajectory in the outer region which begins at the point H^0 on the line $L = E^u(0) \cap S^+$ (resp. S^-) and ends exactly at the point $D^0 = E^s(0) \cap S^+$ (resp. S^-), as shown in Fig.6.5. The problem of finding the homoclinic trajectory relies on finding the value of the bifurcation parameter (which fixes the position of the point H^0) and the return time T after which the trajectory starting from H^0 after the time T will arrive at the point D^0. The coordinates of the point D^0 are uniquely defined by: $D^0(x, y, z)$ where $\frac{x}{\gamma_0^2 + 3\gamma_0 + 1} = \frac{y}{\gamma_0 + 1} = z$, $z = \pm\sigma_{bp}$. Knowing the equation of the line L and the general formulas for solutions we can show[160], that this problem has no solutions for any value of m_1 with fixed $P = 5$, $Q = 6$, $m_0 = -33.03$ and $\sigma_{bp} = 0.2$. This means that for this set of circuit parameters in system (6.1) there are no homoclinic trajectories to the equilibrium 0.

- **Homoclinic trajectories at equilibria** P^+ (P^-)

  At the equilibria P^+ (P^-) the homoclinic trajectories if they exist they must start on the local unstable manifold and target, after evolution in

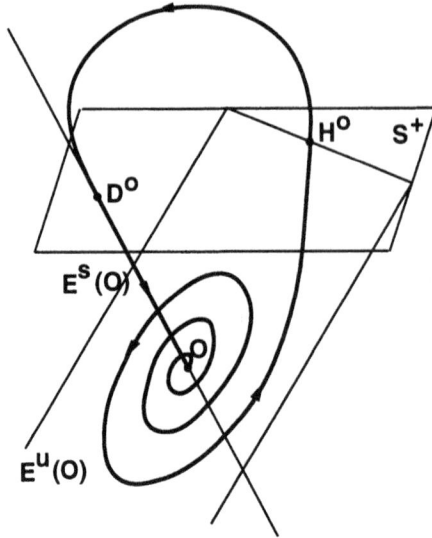

Figure 6.5: Simplest homoclinic trajectory of the origin 0

the central region, exactly on the stable manifold P^+ (P^-) (see Fig.6.6). Knowing the equation for system solutions and the coordinates of points D^P and H^P we can look for a value of the bifurcation parameter such that these equations are satisfied (for fixed $P = 5$, $Q = 6$, $m_0 = -33,03$ and $\sigma_{bp} = 0,2$).

Using a computer it is possible to find a few first values of the bifurcation parameter (there is an infinite number of them!) describing the position of the point H^P and the return time T for which there exist homoclinic orbits to the equilibria P^+ and P^-, eg.: $m_1 = 215,934$, $m_1 = 412,3$, $m_1 = 1215,0$, $m_1 = 1680,5$.

Typical position of simplest homoclinic orbits in the state space with respect to the stable and unstable manifolds of the equilibria is shown in Fig.6.7. Both homoclinic trajectories exist for exactly the same parameter value (by virtue of the symmetry of the system) and are symmetric against the origin. With growing bifurcation parameter value we have observed that the size of these homoclinic orbits grows and they approach each other near the origin. It is probable that the "collision" of these orbits is linked with the phenomenon of bringing together the two Rössler type spiral attractors into a single attractor encircling all three equilibria.. This phenomenon can be easily identified on

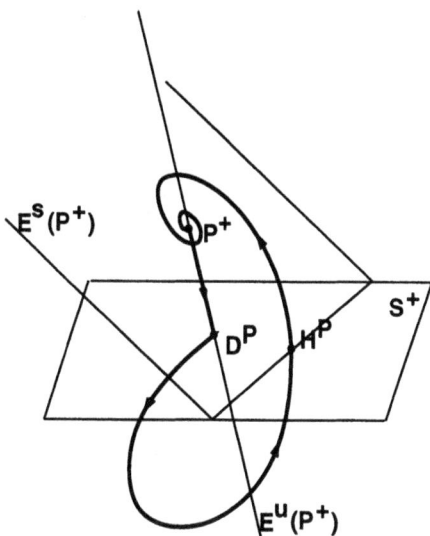

Figure 6.6: Homoclinic trajectory for equilibrium P^+

the m_1- bifurcation diagram in one of the previous chapters[435].

The condition $|\frac{\gamma}{\alpha}| > 1$ is satisfied for $m_1 > 281, 0$, so the values of m_1, as in 2, 3, 4 above, the Shilnikov's theorem implies the existence of chaotic motion. For the values of m_1, as under point 1 above, the second part of Shilnikov's theorem concerning existence of periodic orbits is satisfied.

Let us note that only the value of $m_1 = 412, 3$ belongs to the interval of parameters for which we observed in the system non-divergent orbits (comp. the bifurcation diagram shown in Fig.5.11).

The relation between existence of chaos in the Shil'nikov sense and the existence of an attracting limit set existing for the same set of system parameters remains still an open problem.

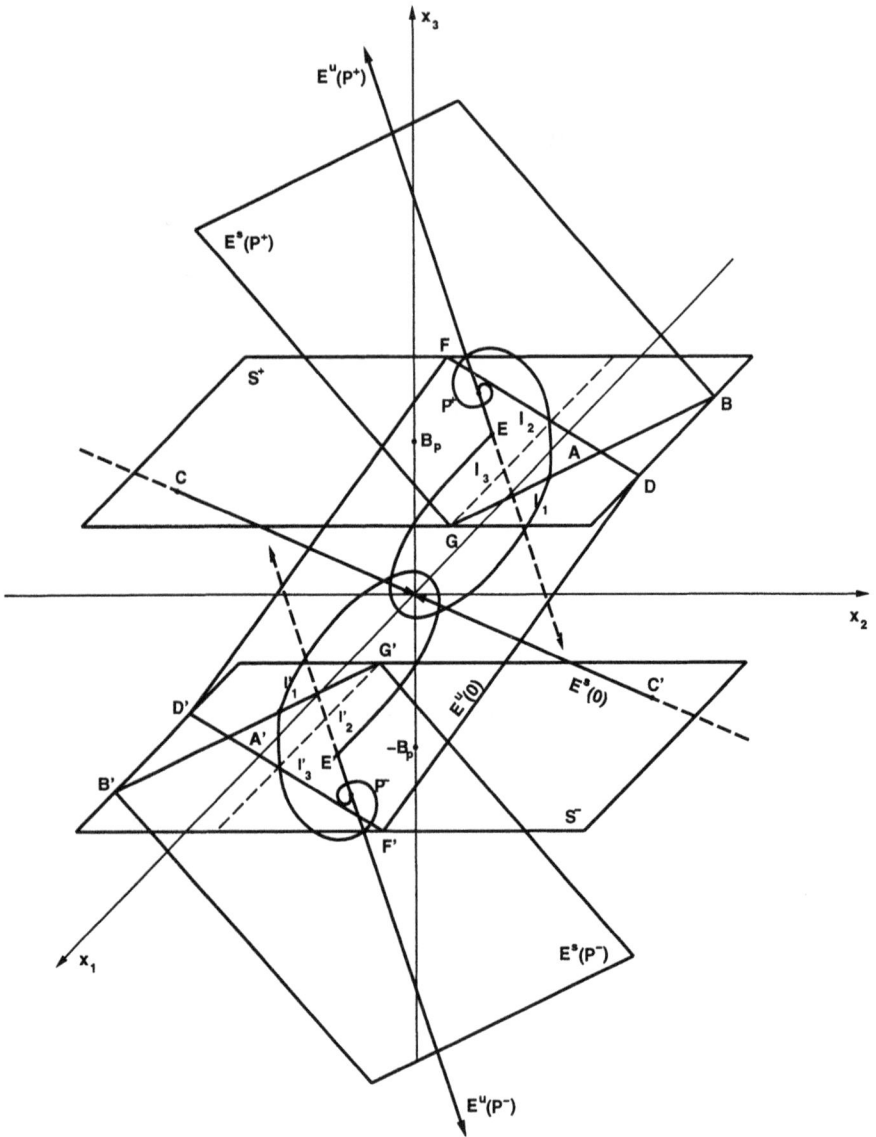

Figure 6.7: The location of the pair of simplest homoclinic trajectories in $\mathbb{R}^3$ for the case of the RC-ladder chaos generator.

CHAPTER 7

POINT MAPPINGS

Point mappings constitute one of most widely used methods of analysis of oscillatory systems. These methods allow reduction of the dimensionality of the considered problem thus simplifying the analysis considerably. The basis of this method comes from early works of Poincaré. The method of point mappings in its many variants (return map, stroboscopic map etc.) is now one of the best known mathematical methods for studying chaotic behavior[73,171,217,301,432]. The basic concept of the point mapping methods comes from simple observations of behavior of chosen points belonging to system trajectories in the phase-space. More precisely speaking we observe how points in the state space chosen to belong to specific manifolds (planes) or at specified time moments evolve (are transformed) via system trajectories. Choosing a point on a plane transversal to system trajectories we can observe when and where the trajectory will return to this plane thus defining so-called return maps (first return map when the trajectory returns to the section plane for the first time; higher order return maps defined by successive intersections). We define so-called stroboscopic map observing points on a trajectory equally spaced in time - measuring the position at discrete time moments (strobe). A special case of stroboscopic map can be found in systems with periodic external forcing when the observations are taken periodically with a period equal to that of the forcing.

In the case of piecewise-linear systems, with a natural choice of the transversal section plane as the switching plane of the nonlinear function, special cases of return maps can be formulated. They are called in this case half-maps. Rössler and his collaborators have obtained many interesting results concerning half-return maps[217,218,219,220,221,446,447]. Chua, Komuro and Matsumoto have applied this method in analysis of Chua's circuit[73] (double scroll). Kahlert[223,224,225] has carried out a detailed analysis of half-maps in Chua's circuit and has given an explanation of the chaos producing mechanism[222]. Poincaré half maps have been also constructed for the "dual double scroll" class of systems[363]. Point mappings have been also applied in analysis of the piecewise linear systems by Sparrow[425] and Brockett[40,41].

For point mappings which belong to a general class of discrete-time maps we have many interesting results concerning existence of chaos. Famous Li-Yorke "period three" theorem[265,266,267] is one of such results. Interesting criteria can be also found in the works of Baillieul, Brockett *et al.*[22], Kloeden[245] or Marotto[277].

Here we will consider construction of the point mappings for a family of systems governed by equations of the form 6.1.

7.1   Two-dimensional map and its properties

The planes S^+ and S^- in a natural way constitute local sections for trajectories of system (6.1). They are not global section planes because there exist sets of initial conditions producing divergent trajectories. (Fig. 5.9).

Several simulation studies have been carried out in order to find out if the trajectories intersect the section planes along one-dimensional manifolds. Successive magnifications shown in Fig.7.1 reveal the fine structure which is clearly two-dimensional.

It is possible to prove using a simple reasoning concerning the multipliers of successively created periodic orbits that the observed structure must be two-dimensional and the observations are not a result of calculation errors. The reasoning is as follows:

Let us assume that the system (6.1) has a stable periodic orbit and what follows the point mapping defined by the trajectories has a fixed point x^*. The eigenvalues of the Jacobian matrix calculated at x^* determine the stability type of the orbit. For a flip bifurcation (period-doubling) to take place one of the multipliers must go out of the unit circle through the point -1. The multipliers of the newly born 2-periodic orbit have to be within the unit circle to assure its stability but both of them must be positive (as the products of the previous, negative multipliers of the period-1 orbit). For the successive period-doubling bifurcation to take place again one of the multipliers must leave the unit circle through -1. This situation is depicted in Fig.7.2. In this figure we indicated a forbidden region for the multipliers. The lower part of the triangle is forbidden because in this case the mapping changes orientation which implies that the trajectories should intersect - this is however impossible in continuous-time systems. Thus the multipliers must go from the region RD to RU.

To achieve such a change of sign of multipliers when continuously changing a bifurcation parameter they must become complex in some range of parameter which excludes in turn existence of a one dimensional manifold along which

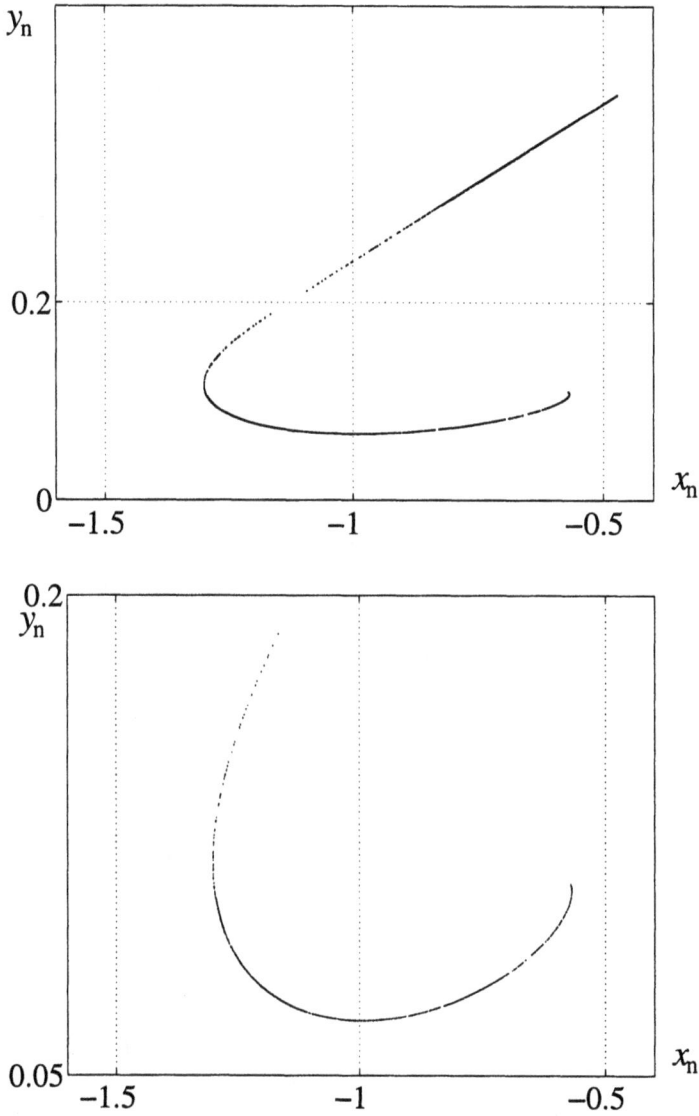

Figure 7.1: Typical trajectory of the point mapping defined by trajectories of the continuous-time system on the plane S^+. Successive figures show magnifications of a chosen detail revealing the fine "wafer" structure of the attractor.

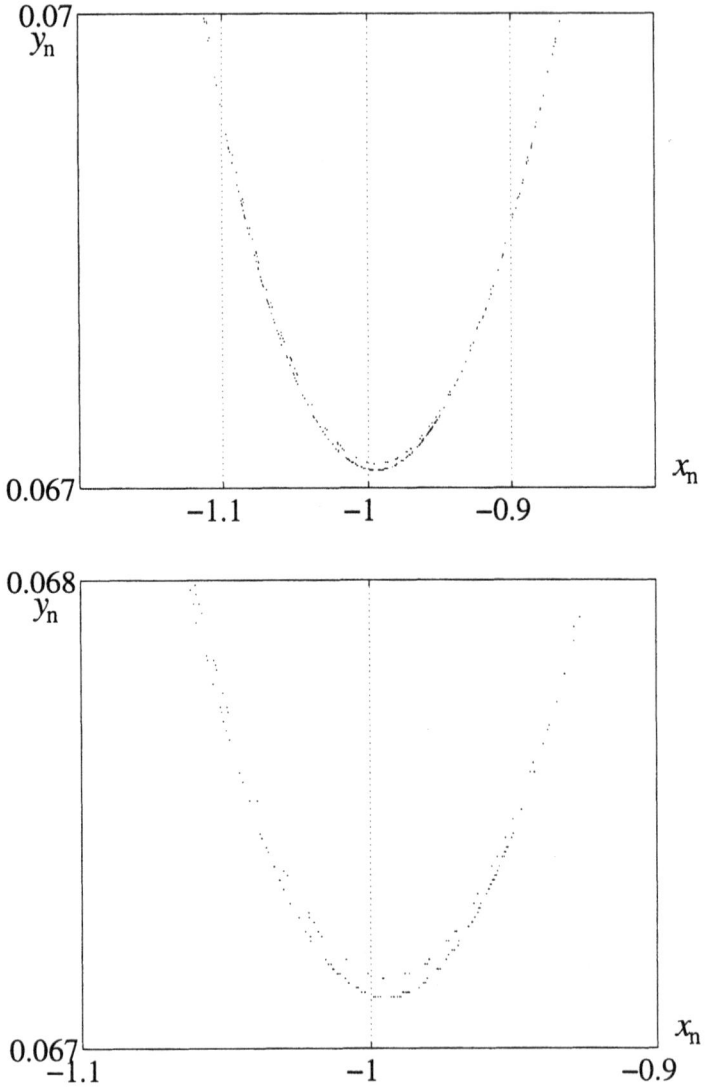

Figure 7.1: Continued

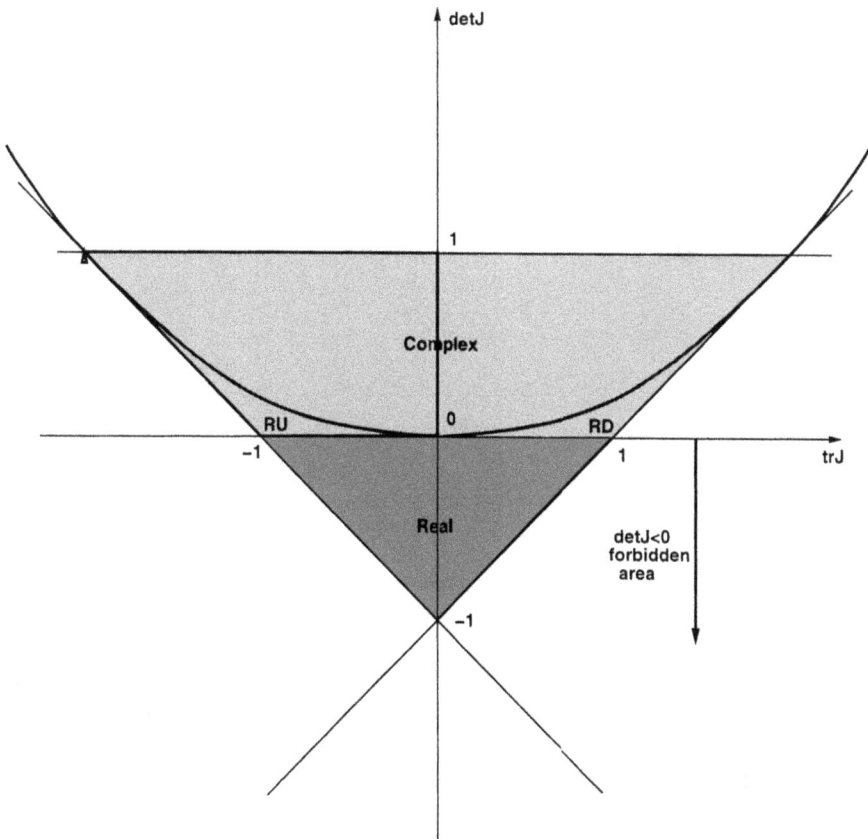

Figure 7.2: Stability regions of the periodic orbit as function of the trace and determinant of the Jacobian matrix.

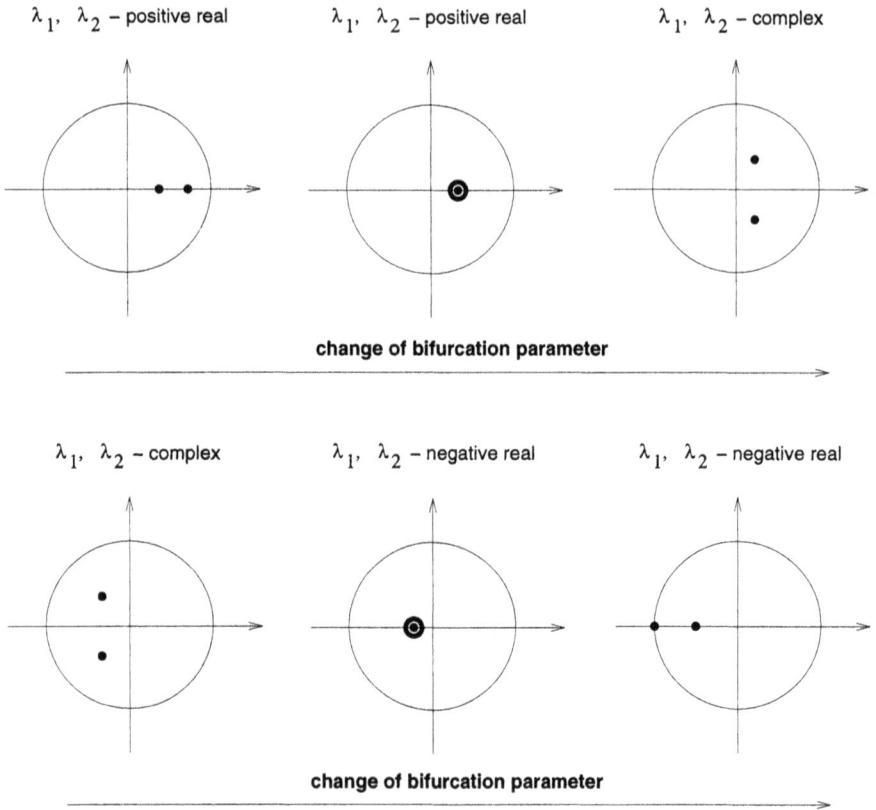

Figure 7.3: Evolution of the multipliers of a periodic orbit undergoing a period-doubling bifurcation.

successive iterates of the mapping would be arranged. The evolution of multipliers of a typical periodic orbit undergoing a period doubling is schematically shown in Fig.7.3.

7.2   Approximate one-dimensional map

In the previous section we have considered the fundamental properties of two-dimensional maps generated by trajectories of a third-order continuous-time system described by differential equations. The simulation experiments not only confirmed these properties but also suggest that in many cases the two-dimensional map is in fact nearly one-dimensional. Below we show the assumption under which construction of a one-dimensional map is valid and gives a good approximate description of system dynamics.

- **Fundamental assumption** In some neighborhood of the unstable fixed point 0 (eg. in the central region D_0) the stable eigenvalue of the linearized system (here γ_0) is dominant ie.

$$\left|\frac{\gamma_0}{\alpha_0}\right| \gg 1 \tag{7.1}$$

This means that trajectories in this region are very strongly attracted towards the unstable manifold (in our case in the domain D_0 they attracted towards $E^u(0)$).

This assumption is equivalent to existence of so-called strong stable foliation in the neighborhood of the equilibrium point 0 (comp. Guckenheimer and Williams[172]). The entire reasoning of Williams[455], concerning mathematical background of construction of an approximate one-dimensional map is also valid in the case of piecewise-linear systems.

The approximate one-dimensional map is composed of several simple mappings:

- $\Pi_1 : S^+ \cap E^u(0) \to S^+$ defined by system trajectories in the outer region. This mapping transforms line segments into deformed spiral segments.

- $\Pi_0 : S^+ \to S^-$ defined by the trajectories in the central region. Π_0 could be further decomposed into $\Pi_2 \circ \Pi_3$ (on the basis of the fundamental assumption (•) on existence of strong attraction towards $E^u(0)$).

  - $\Pi_2 : S^+ \to E^u(0)$ represents projection of S^+ onto $E^u(0)$ in the direction of the stable manifold $E^s(O)$,

  - $\Pi_3 : E^u(0) \to S^- \cap E^u(0)$ by the divergent trajectories on the plane $E^u(0)$ in the central region.

- **Properties of the half-map Π_1**

  Using general expressions (6.31)-(6.35) giving analytically the solutions of the system (6.1), let us investigate the geometric properties of the map $\Pi_1 : S^+ \to S^+$ in the outer region. To do this let us transform the system equations in the outer region into a normal form in such a way that the unstable manifold of $P^+ - E^u(P^+)$ be identical with the axis **z** of the new local coordinate system, and the stable manifold $E^s(P^+)$ be identical with the plane **x-y** in the new coordinate system and the equilibrium point P^+ becomes the origin of the new coordinate system.

The needed similarity transformation is defined by a matrix T and its inverse:

$$T_R = \begin{bmatrix} 1 & 0 & 1 \\ \frac{1}{2}(P+\gamma) & \frac{1}{2}\sqrt{3\gamma^2 + 2P\gamma + 4Q - P^2} & \gamma \\ \gamma^2 + P\gamma + Q & \frac{1}{2}(P+\gamma)\sqrt{3\gamma^2 + 2P\gamma + 4Q - P^2} & \gamma^2 \end{bmatrix} \quad (7.2)$$

$$T_R^{-1} = \frac{-1}{3\gamma^2 + 2P\gamma - 2Q + P^2} \times$$

$$\times \begin{bmatrix} 2\gamma(2\gamma + P) & 2(\gamma + P) & 2 \\ \frac{2\gamma(3\gamma^2 + 3P\gamma + 2Q)}{\sqrt{3\gamma^2 + 2P\gamma + 4Q - P^2}} & \frac{-4(P\gamma + Q)}{\sqrt{3\gamma^2 + 2P\gamma + 4Q - P^2}} & \frac{-2(P+3\gamma)}{\sqrt{3\gamma^2 + 2P\gamma + 4Q - P^2}} \\ \gamma^2 - P^2 + 2Q & -2(\gamma + P) & -2 \end{bmatrix} \quad (7.3)$$

The above expressions are valid both in the central and outer regions when substituting respective eigenvalue γ_0 or γ_p. The coordinates of the new and old systems are linked via the following formulas:

$$\begin{bmatrix} x(t) \\ y(t) \\ z(t) \end{bmatrix} = T_R^{-1} \begin{bmatrix} x(t) \\ y(t) \\ z(t) \end{bmatrix} - \begin{bmatrix} \sigma_p \\ 0 \\ 0 \end{bmatrix} \quad (7.4)$$

and the equations describing system behavior in the neighborhood of P^+ in the old and new coordinate systems are as follows:

$$\frac{d}{dt} \begin{bmatrix} x(t) \\ y(t) \\ z(t) \end{bmatrix} = \begin{bmatrix} \alpha_p & -\beta_p & 0 \\ \beta_p & \alpha_p & 0 \\ 0 & 0 & \gamma_p \end{bmatrix} \begin{bmatrix} x(t) \\ y(t) \\ z(t) \end{bmatrix} \quad (7.5)$$

To simplify our considerations further let us make a rotation transformation around the z in such a way that the plane S^+ becomes parallel to the axis x. Then the equations defining S^+ and $E^u(0)$ are simplified to:

$$S^+ = \{(x, y, z) \in R^3 : z = a_0 y + b_0\} \quad (7.6)$$

$$E^u(0) = \{(x, y, z) \in R^3 : x = a_1 y + b_1 z + c_1\} \quad (7.7)$$

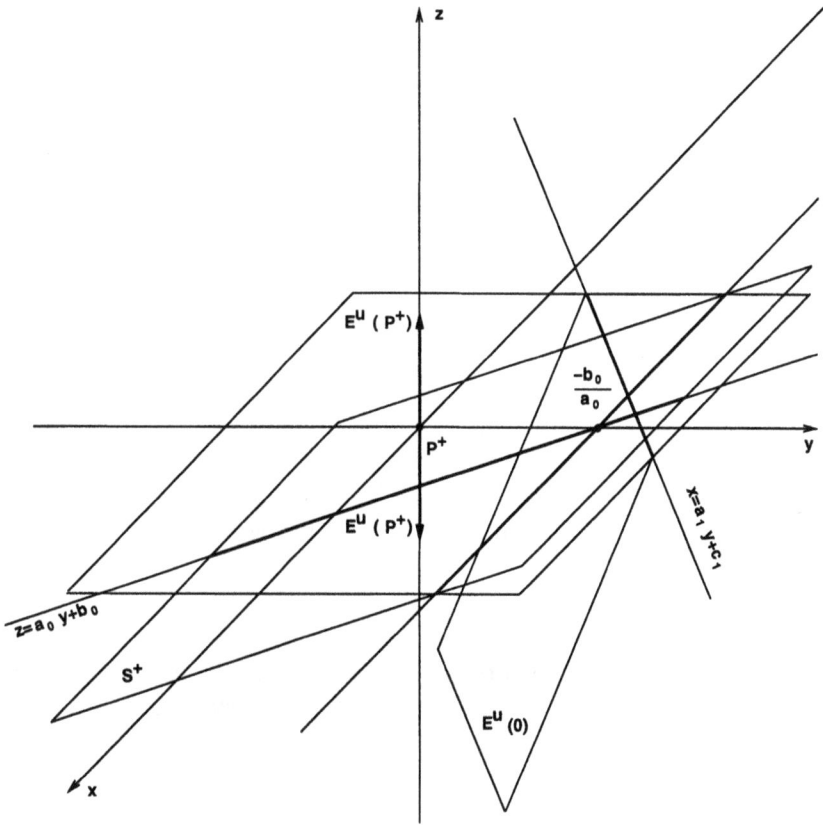

Figure 7.4: Position of the stable and unstable manifolds in the new coordinate system

Positions of these planes are shown in the new coordinate system in figure 7.4. Trajectories starting from the initial condition on $S^+ \cap E^u(0)$ (with parameterization of their position by $\mathbf{y}(0)$), ie.

$$
\begin{aligned}
\mathbf{x}(0) &= a_1\mathbf{y}(0) + b_1\mathbf{z}(0) + c_1 \\
&= \mathbf{y}(0)[a_1 + a_0 b_1] + c_1 + b_0 b_1 \quad\quad (7.8) \\
\mathbf{z}(0) &= a_0\mathbf{y}(0) + b_0 \quad\quad (7.9)
\end{aligned}
$$

After time $T = T(\mathbf{y}(0))$ the trajectory returns to the plane S^+ and pierces it at a point $X_T(\mathbf{x}(T), \mathbf{y}(T), \mathbf{z}(T))$ with:

$$
\begin{aligned}
\mathbf{z}(T) &= e^{\gamma_p T}\mathbf{z}(0) = e^{\gamma_p T}(a_0\mathbf{y}(0) + b_0) = a_0\mathbf{y}(T) + b_0 \quad\quad (7.10) \\
\mathbf{y}(T) &= \mathbf{y}(0)e^{\alpha_p T}cos(\beta_p T) \quad\quad (7.11)
\end{aligned}
$$

Combining the above equations we get:

$$
a_0 e^{\alpha_p T}\mathbf{y}(0)cos(\beta_p T) + b_0 = e^{\gamma_p T}(a_0\mathbf{y}(0) + b_0) \quad\quad (7.12)
$$

and

$$
\frac{e^{\gamma_p T} - 1}{e^{\alpha_p T}cos(\beta_p T) - e^{\gamma_p T}} = \frac{a_0}{b_0}\mathbf{y}(0) \quad\quad (7.13)
$$

Assuming further that $e^{\gamma_p T} \gg e^{\alpha_p T}cos(\beta_p T)$ (when $\alpha < 0$, $\gamma > 0$) the approximate formula for calculation of the return time becomes:

$$
T = T(\mathbf{y}(0)) = \frac{1}{\gamma_p}ln\left(\frac{1}{1 + \frac{a_0}{b_0\mathbf{y}(0)}}\right) \quad\quad (7.14)
$$

It should be stressed that this formula is valid only for the initial conditions from which trajectories return to the central region in particular these belong-. ing to the set:

$$
L_{2seg} = S^+ \cap E^u(0) \cap S_{in}^+ \cap S_{ret}^+ \quad\quad (7.15)
$$

For the initial points approaching the line $L_1 = S^+ \cap E^s(P^+)$ ($\mathbf{y}(0) \rightarrow -\frac{a_0}{b_0}$), the return time tends to infinity while for those points approaching the line L_{tan} ($\mathbf{y}(0) \rightarrow 0$), the return time goes to 0. These facts find full experimental confirmation. This is a special case of the return-time functions considered by Parker and Chua[363] and Chua *at al.*[73]. Substituting T we obtain a parametric formula for the curve on the plane S^+, into which is transformed the line $S^+ \cap E^u(0)$:

$$
\begin{aligned}
\mathbf{x}(T) &= (a_1\mathbf{y}(0) + b_1\mathbf{z}(0) + c_1)e^{\alpha_p T}sin(\beta_p T) \quad\quad (7.16) \\
\mathbf{y}(T) &= \mathbf{y}(0)e^{\alpha_p T}cos(\beta_p T) \quad\quad (7.17)
\end{aligned}
$$

In the polar coordinates we obtain:

$$r^2 = e^{2\alpha_p T}[(a_1 \mathbf{y}(0) + b_1 \mathbf{z}(0) + c_1)^2 sin^2(\beta_p T) + \mathbf{y}^2(0)cos^2(\beta_p T]) \quad (7.18)$$

This equation (parameterized by T) represents a deformed logarithmic spiral. (compare the "spiral image property"[73]).

We can state also that the pieces of the straight lines parallel to $S^+ \cap E^u(0)$ contained between the lines $L_1 = S^+ \cap E^s(P^+)$ and L_{tan} are transformed into segments of deformed spirals having a common point $\mathbf{x}(\infty) = \mathbf{y}(\infty) = 0$, and their ends belong to the straight line L_{tan} (which is invariant — transformed into itself with $T = 0$).

- **Conclusion 1** • The half-map $\Pi_1 : S^+ \cap E^u(0) \to S^+$ transforms the straight line sections (in particular L_{2seg}) into segments of deformed logarithmic spirals.

- **Properties of the half-map Π_0**

Behavior of system trajectories in the central region defines the properties of the half-map Π_0. Let us consider the general formulas defining the solutions with the initial conditions on the deformed spirals as described in the previous section. Using the fundamental assumption (•) we can simplify the considerations of Parker and Chua[363] and Chua *et al.*[73].

Let us observe first that the mapping Π_0 is defined by two kinds of movements: one defined by the stable manifold $E^s(0)$ and an unstable one spiraling outwards on the plane $E^u(0)$. Assuming that $|\frac{\gamma_0}{\alpha_0}| \gg 1$, we can treat the movement along the trajectories in the central region as a combination of a projection (approximately instantaneous in the limiting case) along $E^s(0)$ onto $E^u(0)$ and a slow spiral movements on this plane. Thus, assuming $|\frac{\gamma_0}{\alpha_0}| \gg 1$, the mapping $\Pi_0 : S^+ \to S^-$ (symmetrically $S^+ \to S^-$) can be decomposed into projection Π_2 of spirals from the plane S^+ onto $E^u(0)$ and Π_3 - mapping via the unstable spiral movement the points obtained from projection into the line $E^u(0) \cap S^-$ (respectively — S^+). Π_2 introduces further deformation of the spiral and changes the arrangement (succession) of points and Π_3 causes its squeezing. The direction Π_2 is defined by the eigenvector $e_1 = \begin{bmatrix} 1 & \gamma_0 & \gamma_0^2 \end{bmatrix}^T$. The projection of the point $X(\sigma_{bp}, y_0, z_0)$ in this direction onto the plane $E^u(0) = \{(x, y, z) \in R^3 : (\gamma_0^2 + P\gamma_0 + Q)x + (\gamma_0 + P)y + z = 0\}$ defined as $\Pi_2 : (\sigma_{bp}, y_0, z_0)) \to (x_P, y_P, z_P)$ where:

$$z_P = \frac{-\sigma_{bp}\gamma_0^2(\gamma_0^2 + P\gamma_0 + Q) - y_0\gamma_0^2(\gamma_0 + P) + z_0(2\gamma_0^2 + 2P\gamma_0 + Q)}{3\gamma_0^2 + 2P\gamma_0 + Q} \quad (7.19)$$

$$y_P = \frac{z_p - z_0}{\gamma_0} + y_0 \tag{7.20}$$

$$x_P = \frac{z_p - z_0}{\gamma_0^2} + \sigma_{bp} \tag{7.21}$$

The equations (7.19)-(7.21) define a linear transformation thus the image of a spiral in Π_2 is also a spiral. Second transformation — Π_3 is obtained combining $\Pi_1 \circ \Pi_2$ and is defined by trajectories starting from a spiral on $E^u(0)$ and ending on the line $E^u(0) \cap S^-$. It can be easily shown that in the equations defining the solutions (6.32)-(6.35) the constant $C_1 = 0$ in this case and the trajectories are simple spirals and there exists a stretching and squeezing directions as all trajectories must hit the switching plane S^-.

- **Conclusion 2** • The mapping $\Pi_0 = \Pi_2 \circ \Pi_3 : S^+ \to S^-$ transforms a segment of a deformed logarithmic spiral into a line section $S^- \cap E^u(0)$.

7.2.1   The deformed spiral map

Using the reasoning described in the previous section we were able to propose a one-dimensional approximate model of dynamics of the original continuous-time system. This model was called the "squeezed (deformed) spiral map"[331,332].

We will compare now the results obtained using the one-dimensional model with the ones obtained using the full system model (three dimensional). The model map can be constructed in the following way (Fig.7.5) :

- **Step 1.** Every point $X(x)$ from the unit interval $[0; 1]$ is mapped to a point $\underline{X}$ belonging to a spiral $r = Re^{-\alpha\theta}$ in such a way that the length of the arc $0\underline{X}$ is equal x (this is an idealization of the transformation of a line section by Π_1).

- **Step 2.** The points belonging to the spiral arc obtained in step 1 are now projected perpendicularly onto a chosen line l^* (with a slope $tg(\phi)$).

- **Step 3.** Finally the interval on the line l^* is normalized to the unit length.

Parameters R, α and ϕ determine the properties of the mapping. The construction of the map reveals the stretching and folding property characteristic for systems with complicated behavior and responsible for the formation of Smale's horseshoes[2,171]. The half-return map $\Pi_H : [0, 1] \to [0, 1]$ constructed as described above can be described mathematically by:

$$x_{n+1} = F(x_n) = \frac{R(1 - Ax_n)cos[-\frac{1}{\alpha}ln(1 - Ax_n) - \phi] + W_1}{W_2} \tag{7.22}$$

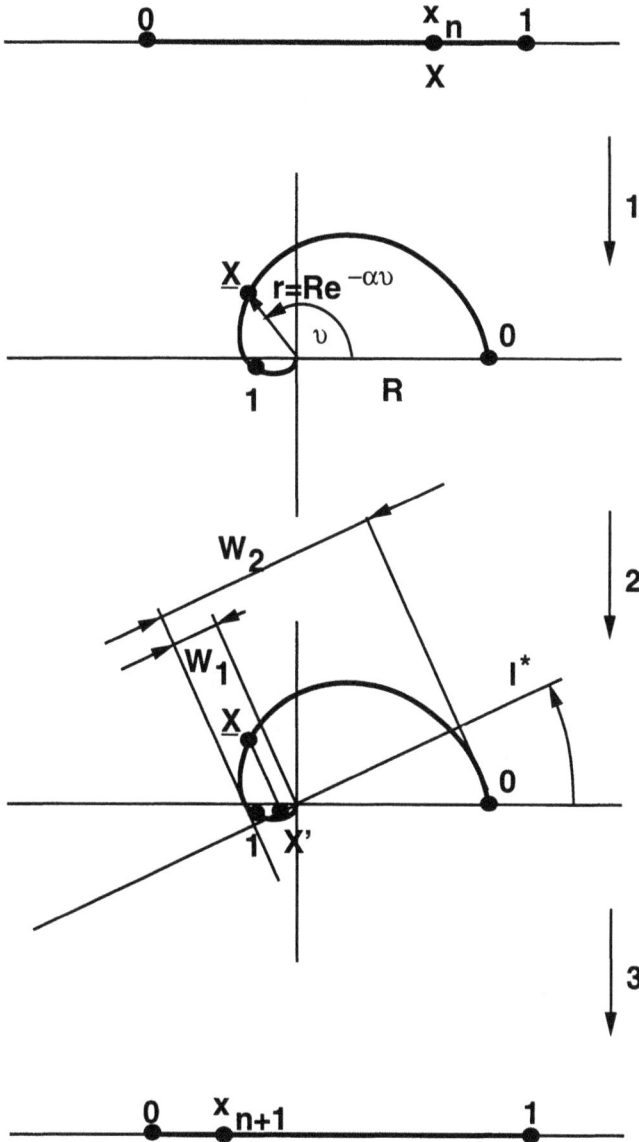

Figure 7.5: Construction of the squeezed spiral map showing formation of Smale's horseshoes.

where :

$$A = \frac{\alpha}{R\sqrt{1+\alpha^2}} = \frac{1}{L} \qquad (7.23)$$

L - the length of the logarithmic spiral. Normalizing constants:

$$W_1 = \frac{R}{\sqrt{1+\alpha^2}}e^{-\alpha[\pi+\phi-arctg(\alpha)]} \qquad (7.24)$$

$$W_2 = W_1 + W_3 \qquad (7.25)$$

$$W_3 = \begin{cases} Rcos(\phi) & \text{for } x_m < 0 \\ \frac{R}{\sqrt{1+\alpha^2}}e^{-\alpha[\phi-arctg(\alpha)]} & \text{for } x_m > 0 \end{cases} \qquad (7.26)$$

with:

$$x_m = \frac{1}{A}e^{-\alpha[\phi-arctg(\alpha)]} \qquad (7.27)$$

The return map Π_R is defined by:

$$\Pi_R = \Pi_H \circ \Pi_H \qquad (7.28)$$

The properties of this map depend on the choice of parameters α ($\alpha > 0$), R, ϕ . The equation (7.23) constitutes a constraint on R and α because $L \geq 1$.

7.2.2   Properties of the squeezed spiral map

The mapping Π_H (Π_R), defined by equations (7.22)-(7.27) belongs to the class of one-dimensional maps with multiple extrema (critical points). Examining the behavior of the first and second derivative of $F(x)$ (equation 7.22):

$$\frac{\partial F(x)}{\partial x} = -\frac{1}{W_2}cos[-\frac{1}{\alpha}ln(1-Ax) + \phi + arctg\frac{1}{\alpha}] \qquad (7.29)$$

$$\frac{\partial^2 F(x)}{\partial x^2} = \frac{A}{\alpha W_2(1-Ax)}sin[\frac{1}{-\alpha}ln(1-Ax) + \phi + arctg\frac{1}{\alpha}] \quad (7.30)$$

one can find the number and position of the extrema x_k :

$$x_k = \frac{1}{A}[1 - e^{-\alpha[k\pi+\phi-arctg(\alpha)]}] \qquad (7.31)$$

$$F(x_k) = \frac{\frac{R(-1)^k}{\sqrt{1+\alpha^2}}e^{-\alpha[k\pi+\phi-arctg(\alpha)]+W_1}}{W_2} = \frac{W_1}{W_2}[(-1)^k e^{-\alpha(k-1)\pi} + 1] \quad (7.32)$$

where: $k = 1, \cdots n$, n - the number of sign changes of the function:

$$F_s(x) = R(1 - Ax)sin[-\frac{1}{\alpha}ln(1-Ax)] \qquad (7.33)$$

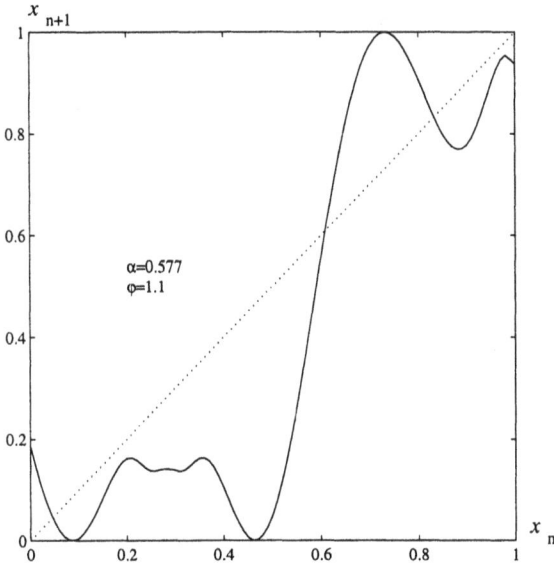

Figure 7.6: An example of the return map Π_R with eleven extrema.

in the unit interval $[0, 1]$ (which depends on α, ϕ and R). In the extreme case $A = \frac{\alpha}{R\sqrt{1+\alpha^2}} = 1$ the map has an infinite number of extrema. In the figure 7.6 we show an example of functions with 11 extrema ($\alpha = 0.577, \phi = 1.1$, $R = 0.5$). General properties of the maps with multiple extrema were considered by many researchers — see eg. Holmes and Whitley[197] and Guckenheimer and Holmes[171]. Most results are related to simple maps eg. cubic map[128,230] or sine map[400,401,431]. Singer theorem[88] (also considered in[431] Guckenheimer-Oster-Ipaktchi theorem) determine the maximum number of attracting periodic orbits which for one-dimensional maps is equal the number of critical points (ie. for example a mapping with two critical points may have at most two attractors). When there are many attractors then the initial condition determines to which of the attractors the solution will actually go.

For the map Π_R we can observe two typical types of bifurcation phenomena. The first type is associated with the changes of stability properties of existing equilibria. The second type is linked with the creation of new equilibria when changing the bifurcation parameter and at the same time changes of stability properties of the existing equilibria. So-called symmetry breaking bifurcation belongs to the latter class[2,171,197]. In this bifurcation a simple equilibrium point (periodic orbit) looses stability (one of the multipliers goes

out from the unit circle through +1) and at the same moment two new stable
equilibria (periodic orbits of the same period as the existing one) are created.
Let us establish the conditions for this kind of bifurcation for the return map
Π_R. Taking into account that $\Pi_R = \Pi_H^2$ and the fact that the fixed point of
Π_R must also be the fixed point or periodic point for Π_H (comp. Fig.7.7). For
the map Π_H symmetry breaking bifurcation takes place for x^* satisfying the
equation:

$$F(x^*) = x^* \tag{7.34}$$

$$\left. \frac{\partial F(x)}{\partial x} \right|_{x = x^*} = -1 \tag{7.35}$$

Symmetry breaking bifurcation takes place for α and ϕ, satisfying the formula:

$$\frac{1-A}{A} = exp\{\alpha arccos(\frac{R(1+e^{\alpha\pi})}{\sqrt{1+\alpha^2}}e^{-\alpha(\pi+\phi-arctg(\alpha))} - \frac{\pi}{2} + \phi arctg\alpha)\}$$

$$\times\{\frac{1}{A} + AR + \sqrt{\frac{e^{2\alpha(\pi+\phi-arctg\alpha)}}{(1+e^{\alpha\pi})^2} - \frac{R^2}{1+\alpha^2}}\} \tag{7.36}$$

Simulation experiments suggest that for all parameter values (α, ϕ) to the right
from this line and below the line of attractor merging CM the mapping always
has two attractors.

To obtain a more detailed picture of the properties of the squeezed spiral
map Π_R a thorough simulation study has been carried out. The computer
experiments revealed several interesting types of behavior. Changing ϕ with
fixed other parameters of the map we observed the displacement of the extrema without change of their number. Figure 7.8 shows typical graphs of the
mapping for changing parameter values and their dependence on ϕ.

Different kind of behavior can be observed for changing α when fixed
ϕ and R; Fig.7.9 shows curves with varying number of extrema and their
position. Creation of new extrema and their displacement changes the stability
properties of the system. Fig.7.10 shows how the observed trajectories change
when changing α with fixed ϕ and R.

To draw the attention to the dependence of the observed trajectories on
the properties of the map all trajectories are shown together with respective
graphs of the map Π_R for the chosen parameter values. The fundamental type
of bifurcation — symmetry breaking (creation of two equilibria) — is linked
with the change of behavior of the graph of the map, its passage above the
tangency at the equilibrium point (as shown in Fig.7.10a,b). Thus two new
equilibria are created corresponding to two new intersections of the graph with
the bisection of the first quadrant. The successive bifurcations are linked only

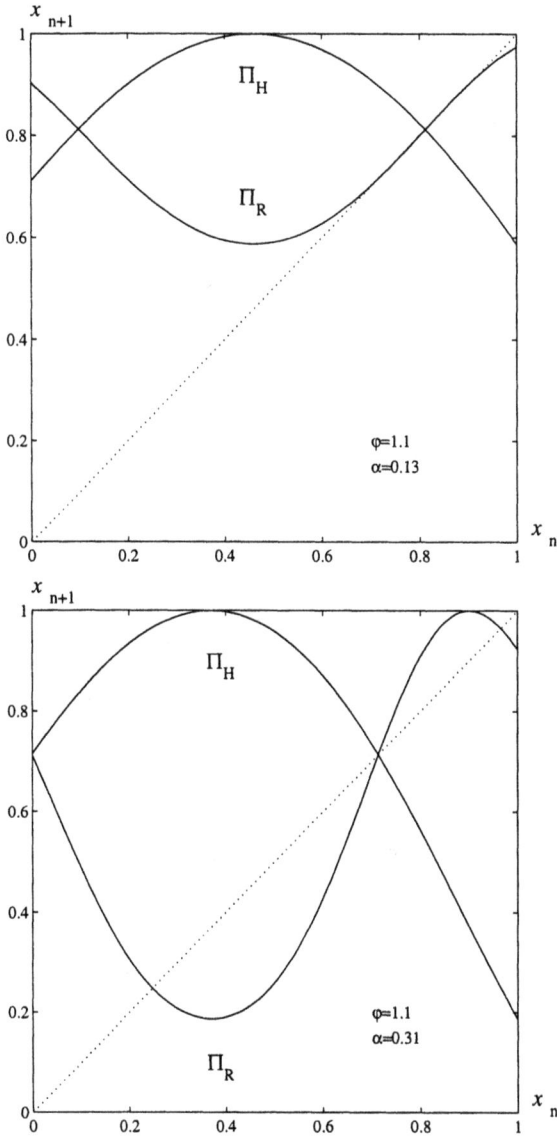

Figure 7.7: Fixed points of the map Π_H and $\Pi_R = \Pi_H^2$ for $\alpha = 0, 13, \phi = 1, 1$ and $R = 0, 5$ (a) ; $\alpha = 0, 31, \phi = 1, 1$ and $R = 0, 5$ (b)

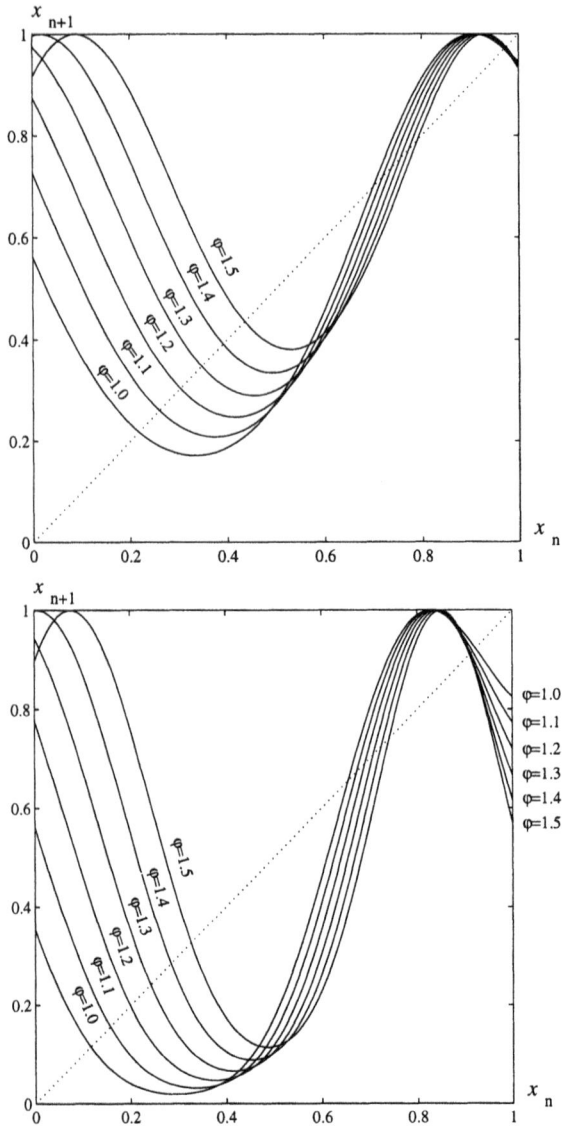

Figure 7.8: Dependence of the map Π_R on the parameter ϕ for fixed $R = 0,5$ and $\alpha = 0,3$ (a); $\alpha = 0,4$ (b); $\alpha = 0,5$ (c)

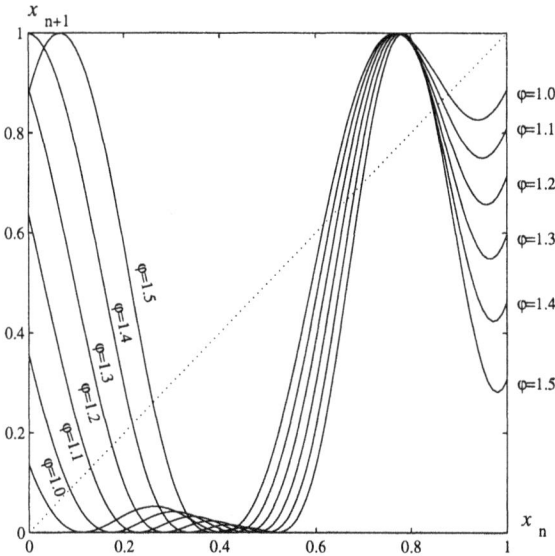

Figure 7.8: Continued

with the changes of stability properties of these equilibria (changes of the slopes of the graph at the equilibria – comp Fig.7.10b-l). The figures 7.10b-f show each two co-existing trajectories. These trajectories are created using different initial conditions. Every attractor has its own basin of attraction. The route leading to chaotic behavior is typically started with the symmetry breaking followed by period-doubling sequence (Feigenbaum[119]) of two orbits created in the symmetry breaking. The route to chaos can be summarized as follows:

- One stable equilibrium point.

- Two stable equilibrium points.

- Period-doubling sequence.

- Two coexisting chaotic attractors.

- Single chaotic attractor.

In the figure 7.10 we show several types of periodic orbits (periodic windows) within the range of existence of chaotic behavior. Analysis of trajectory behavior confirms that the considered system and its associated one-dimensional map is chaotic in the sense of fulfilling the assumptions of the following theorem:

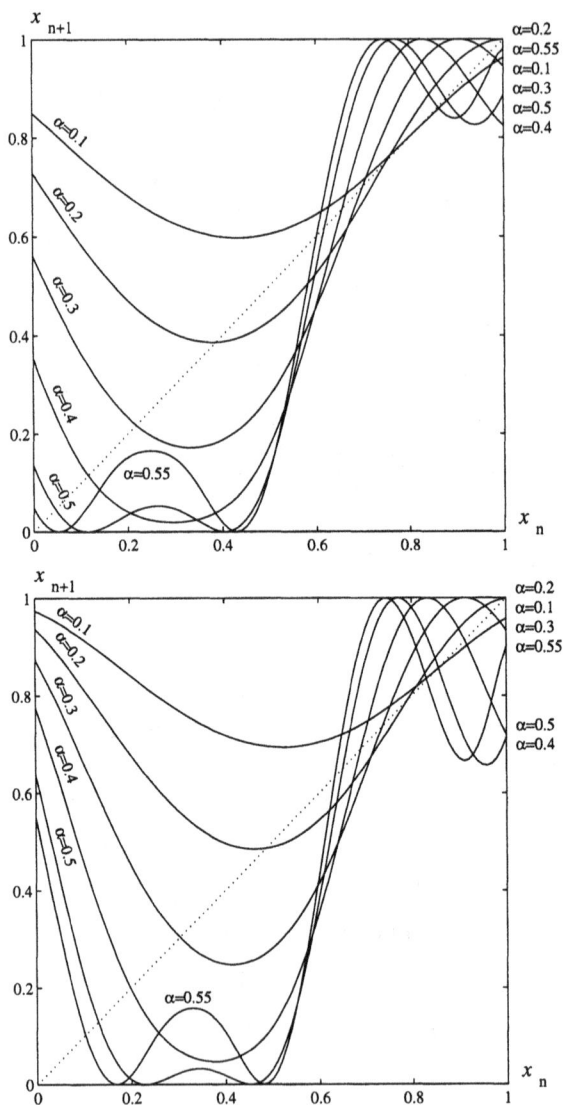

Figure 7.9: Dependence of the map Π_R on parameter α for fixed $R = 0, 5$ and $\phi = 1, 0$ (a) ; $\phi = 1, 2$ (b); $\phi = 1, 55$ (c)

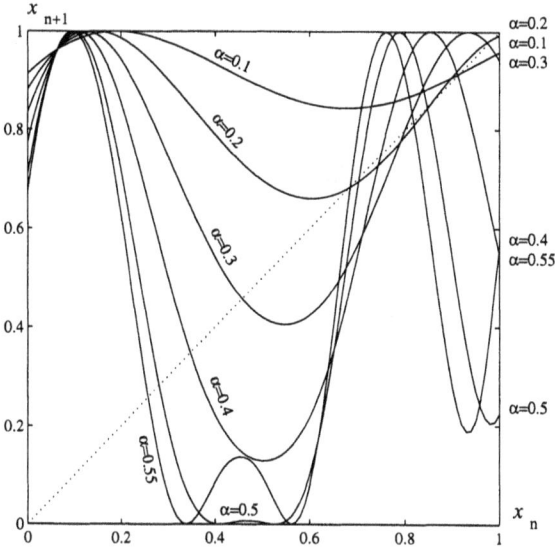

Figure 7.9: Continued

Theorem Li - Yorke[22,265,277]

Let us consider an interval J and its continuous transformation $F : J \to J$.
Let us assume further that there exists a point $a \in J$ for which the successive
iterates of the map F: $b = F(a)$, $c = F^2(a)$, $d = F^3(a)$ satisfy the condition:

$$d \leq a < b < c \text{ or } d \geq a > b > c$$

Then:

- **Thesis 1.** For every $k = 1, 2, \ldots$ there exists in J a point belonging to
  an orbit of period k.

- **Thesis 2.** There exists an uncountable set $S \subset J$ which does not contain
  periodic orbits, satisfying the conditions:

  1. for every $p, q \in S, p \neq q$ we have

$$\limsup_{n \to \infty} |F^n(p) - F^n(q)| > 0 \tag{7.37}$$

  and

$$\liminf_{n \to \infty} |F^n(p) - F^n(q)| = 0 \tag{7.38}$$

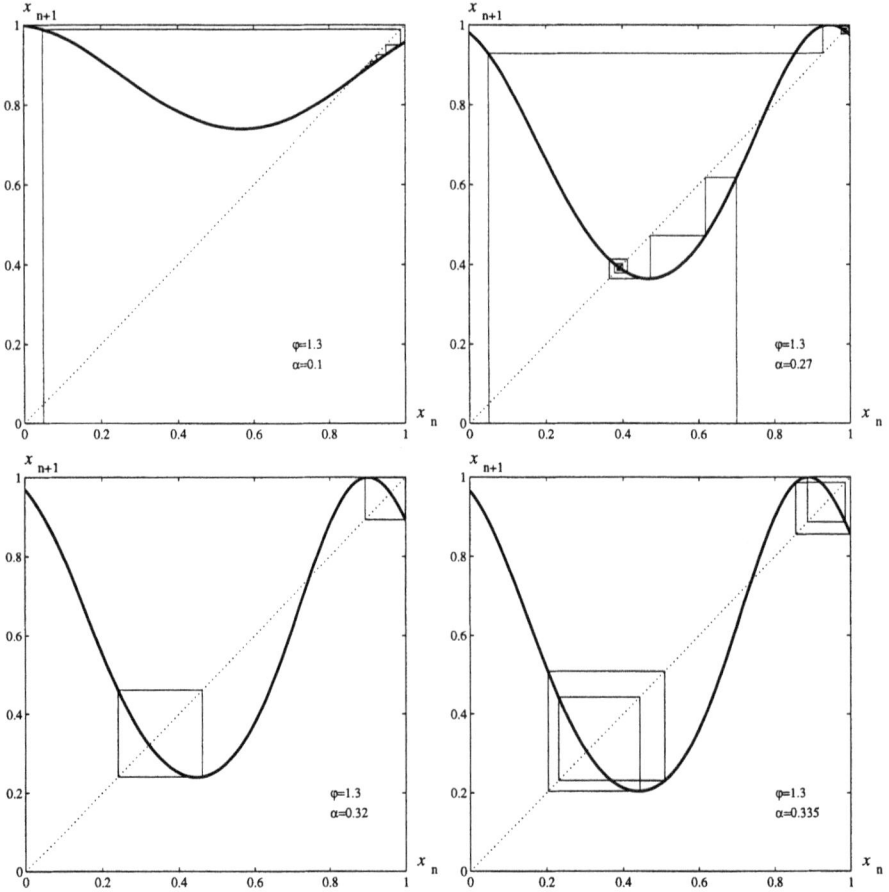

Figure 7.10: Trajectories of the map Π_R for $\alpha = var.$ $\phi = 1, 3$ and $R = 0, 5$. For better understanding we have also indicated the graphs of the map in each case. In the figures (b–f) we have drawn two trajectories corresponding to two coexisting attractors - each observed for a different choice of initial conditions.

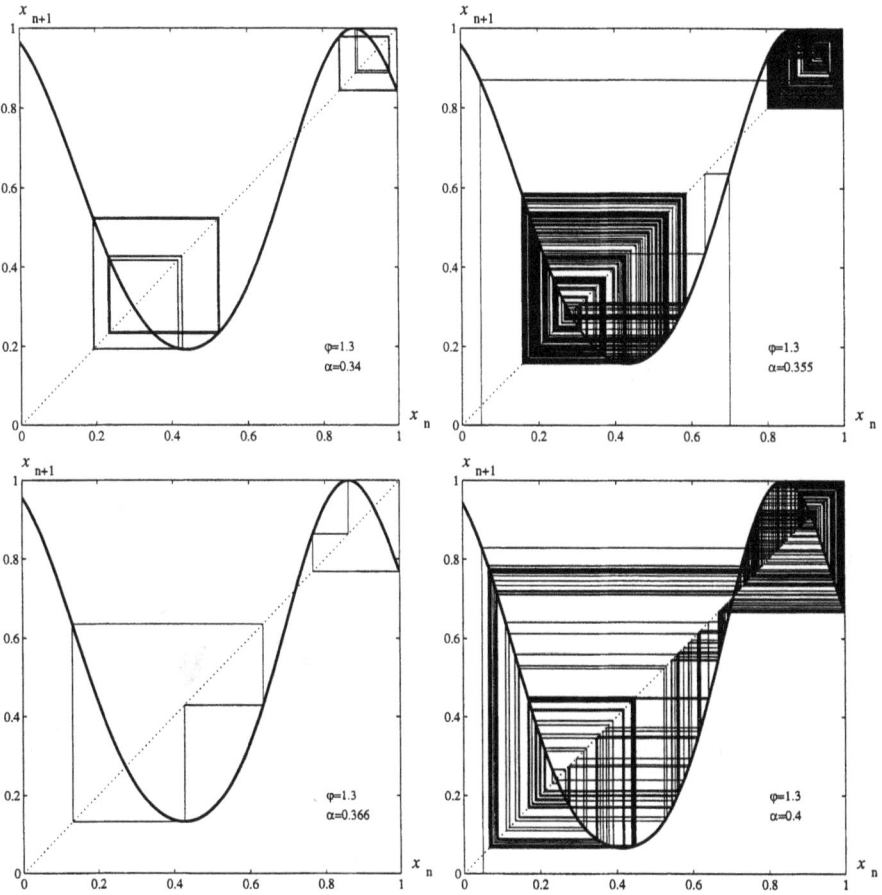

Figure 7.10: Continued

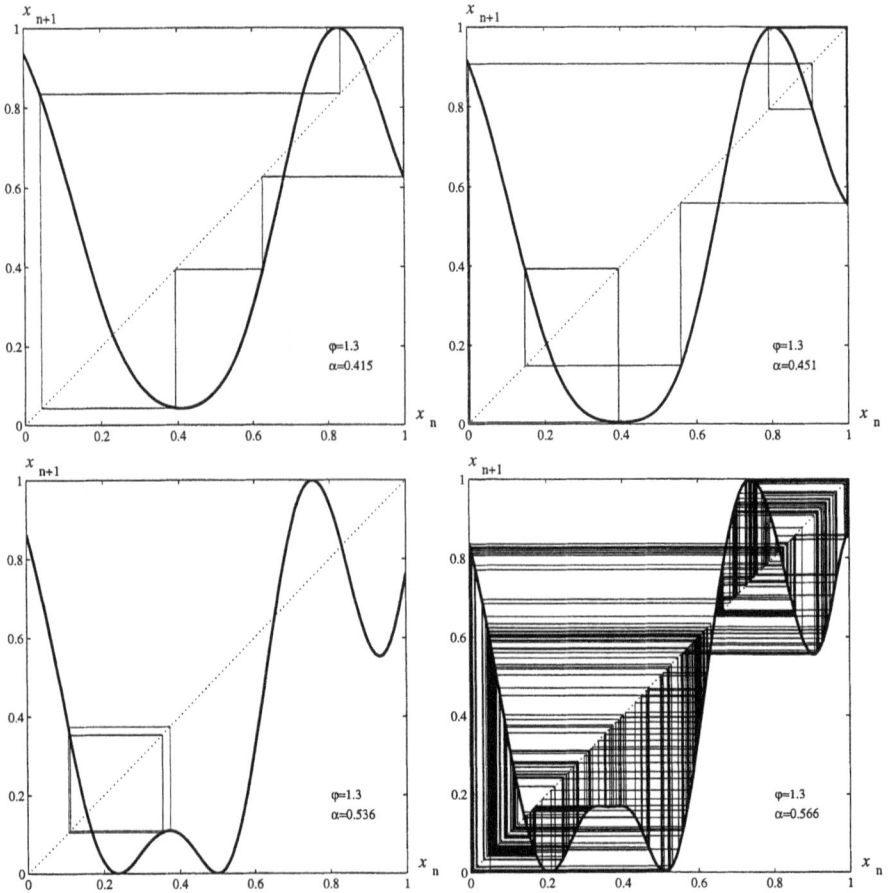

Figure 7.10: Continued

2. for every $p \in S$ and a periodic point $q \in J$ we have

$$\limsup_{n \to \infty} |F^n(p) - F^n(q)| > 0 \qquad (7.39)$$

The above conditions are satisfied in particular when the map F has a period-3 orbit as it was the case of the considered deformed spiral map (Fig.7.10f).

The route from order to chaos is linked with the creation of a "primary bubble" (in the terminology of Bier and Bountis[29]) – a re-merging bifurcation structure of incomplete Feigenbaum sequence of period doublings/period halvings. The development and this "primary bubble" and its development can be seen in the Fig.7.11. The parts of the bifurcation diagrams to the right of the symmetry breaking point (until the attractor merging point Fig.7.11m-s) were constructed in numerical experiments with different choices of initial conditions. Both coexisting branches have exactly the same bifurcation sequences for exactly the same parameter values. This fact becomes obvious when we take into account that $\Pi_R = \Pi_H^2$. The half-return map Π_H does not possess this property of the bifurcation sequences. For small values of ϕ (eg Fig.7.11a) for growing α we observe only one symmetry breaking bifurcation - the diagram has only one pitchfork. For larger ϕ a fully developed bubble ("primary bubble") appears. Further growth of ϕ causes creation of successive bubbles – re-merging bifurcation sequences[29]). There exists a boundary line $\phi = const$ for which the bubble structure is fully developed – the period doubling sequence is fully developed leading to chaos at the point 2^∞. The transition from order to chaos, from the "primary bubble" to the fully developed period doubling/period halving structure is shown in Fig.7.11a, b, c.

In the analyzed mapping the re-merging bifurcation sequences (bubbles) can be found in many cases — the figures 7.11d, e, f show creation of a secondary bubble structure. Bubbles appear also in some of the periodic windows (period doublings/halvings as shown in Fig.7.11g).

An interesting phenomenon is merging of the two coexisting attractors (attractor merging[2,47], complementary band merging[329]). The attractor merging appears when two chaotic attractors are created from coexisting periodic orbits born in a symmetry breaking bifurcation (Fig.7.11k,l etc.). This phenomenon can be explained observing the graphs of the map shown in Fig.7.10 e and g. When changing the bifurcation parameter the slopes of the graph change – the region occupied by each of the coexisting attractors grows approaching the unstable fixed point. There exists a threshold value for which the attractors will touch each-other and later merge (Fig.7.10g). The collision

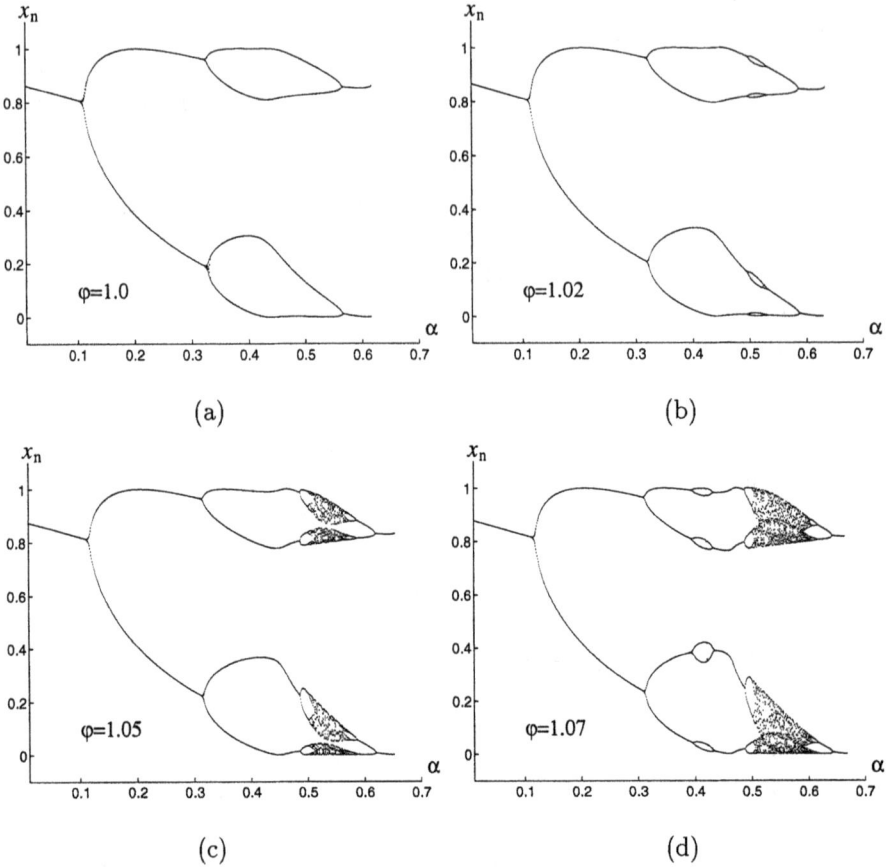

Figure 7.11: Evolution of the bifurcation phenomena observed for changing α values for fixed (growing in successive diagrams) ϕ.

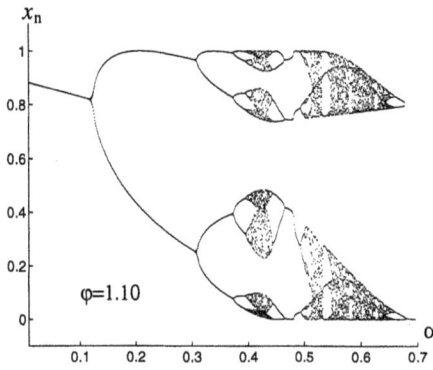

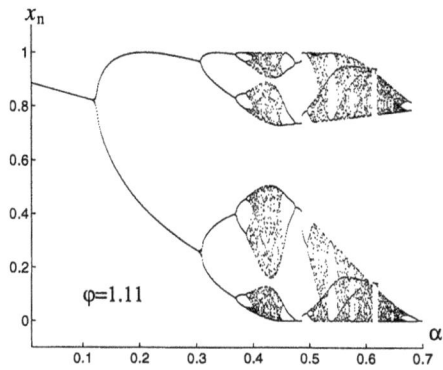

(e) (f)

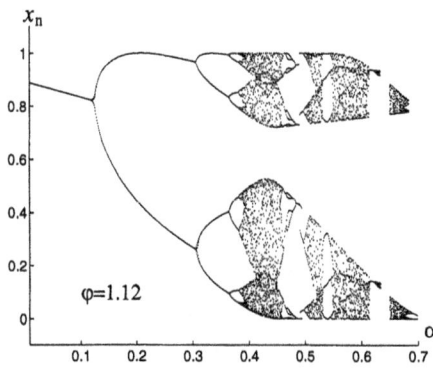

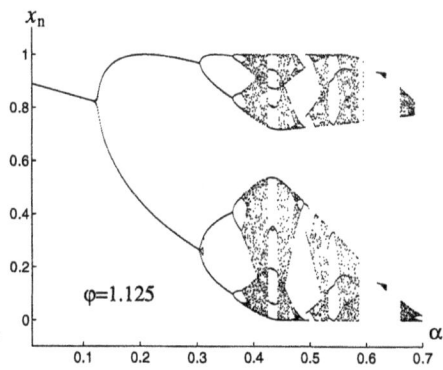

(g) (h)

Figure 7.11: Continued

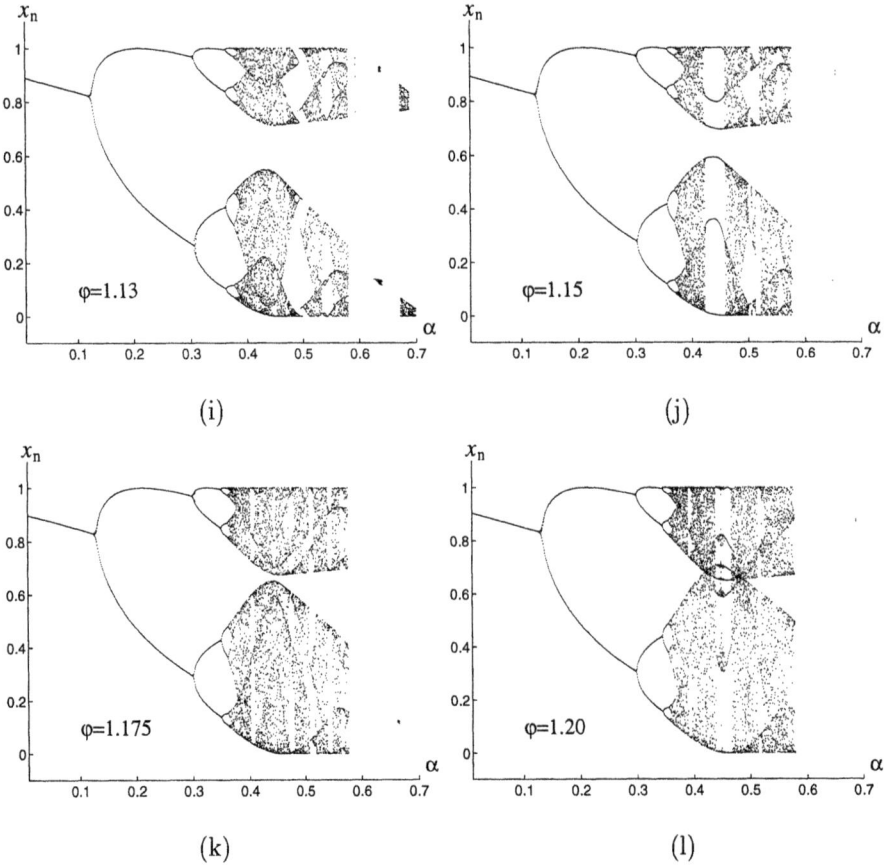

(i)

(j)

(k)

(l)

Figure 7.11: Continued

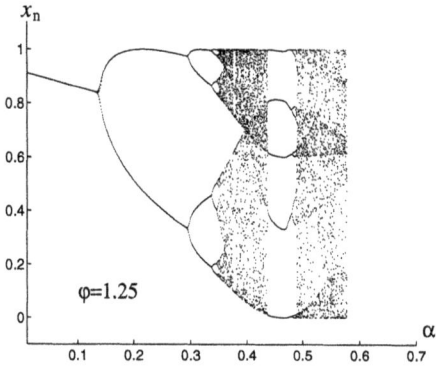

(m)

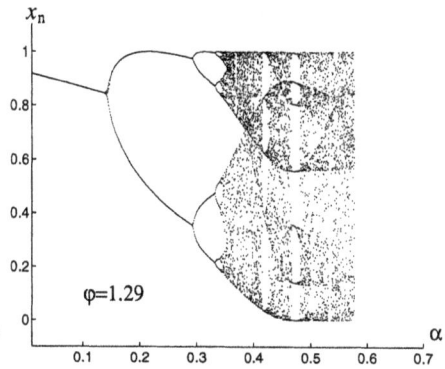

(n)

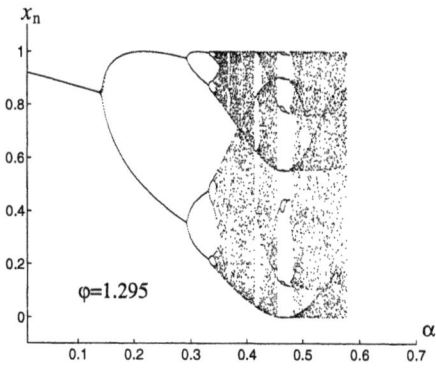

(o)

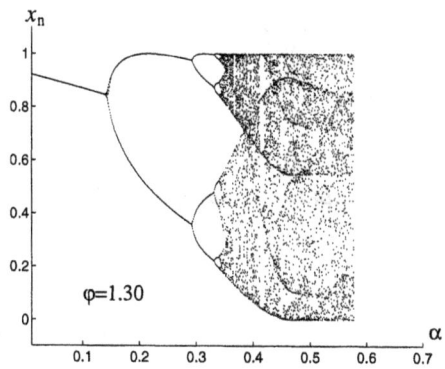

(p)

Figure 7.11: Continued

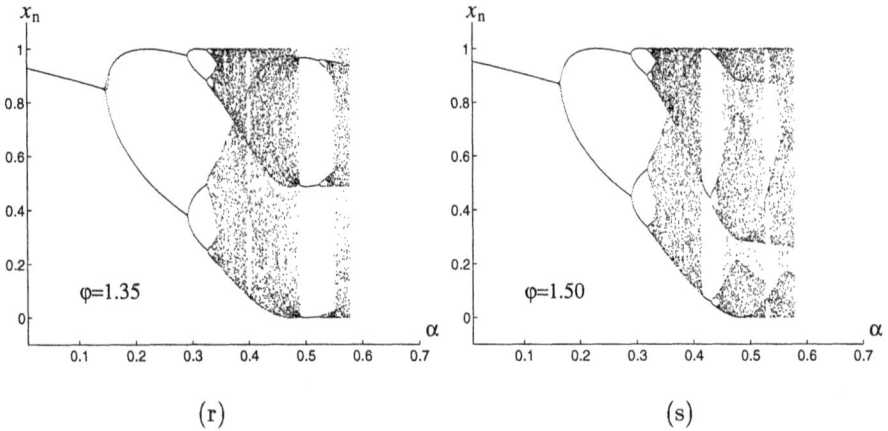

(r)   (s)

Figure 7.11: Continued

of the fixed point with the observed attractor takes place exactly at the point of attractor merging (band merging).

Within the chaotic region we observe very interesting periodic windows (Fig.7.11l-s). Also various crisis-type phenomena can be observed eg. in Fig.7.11g a discontinuity of the diagram is visible – an abrupt change of the size of the orbit. In the Fig.7.11r so-called internal crisis can be observed – jump of the size of the chaotic attractor ("splash").

Let us explain the mechanism leading to attractor disappearance for large values of parameters — note that all bifurcation trees are cut abruptly (comp. Fig.7.11). Let us recall the equation (7.23). It follows from this equation that the map Π_H (Π_R) transforms the unit interval into itself if for fixed R, α satisfies:

$$\alpha \leq \alpha_{gr} = \frac{R}{\sqrt{1-R^2}} \tag{7.40}$$

When passing the limit value $\alpha_{gr} = 0,578$ (for $R = 0,5$) inside the unit interval appear subintervals which are mapped by Π_H (Π_R) outside the unit interval. The mechanism of this kind is shown in Fig.7.12a. For Π_H only the interval $[0; 0,9718]$ is mapped into $[0; 1]$. For Π_R inside this interval there appears a sub-interval mapped outside $[0; 1]$.

Depending on the choice of parameters the size of the sub-interval which is mapped by Π_H outside the unit interval changes. For some ϕ parameter values when passing the limiting value of α_{gr} none of the points of the interval $[0; 1]$ is mapped into its interior – this is the boundary crisis — the attractor

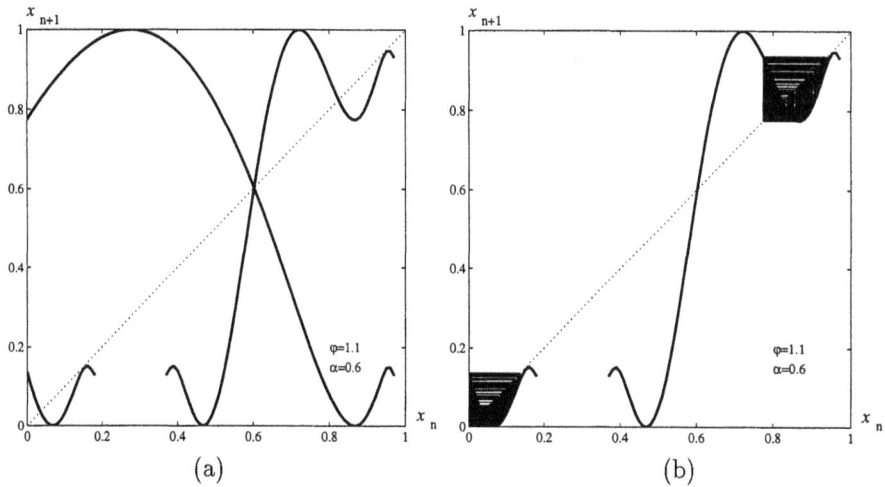

Figure 7.12: Explanation of the crisis phenomenon. For $\alpha \geq 0,578$ $(R = 0,5)$ there exist sub-intervals mapped under Π_H (Π_R) outside the unit interval. When changing the system parameters these sub-intervals can grow until all points are mapped outside the interval where the map is defined – the attractor disappears (a). Selecting the initial conditions we can observe one of the two chaotic attractors or divergent trajectories (b). The figures show results obtained in simulation experiments for $\alpha = 0,6$, $R = 0,5$ and $\phi = 1,1$.

disappears. For some ranges of ϕ there exist subintervals which are mapped into themselves as shown in Fig.7.12b. Choosing initial conditions in a specific way we can observe two trajectories as shown.

A series of figures (7.11) show so-called an abrupt change of attractor size (jump) — as shown by arrows in Fig.7.11g and its magnification Fig.7.18). The Fig.7.13 shows on a background of the respective graphs of the map the changes of periodic orbits for a very small change of the bifurcation parameter α.

It turns out that in a very small α parameter interval the number of the period-two orbits existing in the system doubles — Between α equal $0,485$ and $0,495$ we observe first two co-existing periodic orbits — Fig.7.13a — and later in an abrupt manner two additional orbits of the same period appear — Fig.7.13b. Finally the initially observed orbits disappear as shown in Fig.7.13a. Only the newly created orbits undergo further bifurcations as shown in Fig.7.13c.

This phenomenon is not linked to any crisis-type phenomenon in which two coexisting attractors collide. A very interesting bifurcation phenomenon explains this type of behavior. Let us look in detail at the evolution of the equilibrium points when changing the bifurcation parameter. We observe the equilibria of the second iterate Π_R^2 of Π_R as the period=two orbits of Π_R corresponds to fixed points of Π_R^2). The figure 7.14 shows the graphs of the map Π_R^2 for five different α parameter values. When changing the parameter value we observe first the appearance of two pairs of new unstable equilibria (the graph of Π_R^2 apart from two existing (Fig.7.14a) intersection points with the line $x_{n+1} = x_n$ has four new tangent points as shown in Fig.7.14b. Further change of the bifurcation parameter causes a saddle-node bifurcation of each of these points giving birth to a pair (stable-unstable) of orbits (Fig.7.14c).

The phenomenon described above can be considered as a degenerate kind of the symmetry breaking bifurcation of a periodic orbit. Apart from the single period-two orbit a stable-unstable pair of orbits of the same period is created. They are not born however from the same orbit as it is usually the case in the symmetry breaking bifurcation but from two different orbits lying within some distance from each-other. When changing the bifurcation parameter further we can observe a reversed phenomenon — pairs of the stable equilibria existing from the beginning of the observation (Fig.7.14a) are getting together with the unstable equilibria ("saddle-node bifurcation") — Fig.7.14d), and finally disappear (Fig.7.14e), while the stable points corresponding to new orbits remain. The size of these orbits is different — this is why we observe the hysteresis and jump of the attractor size which in fact is the jump from

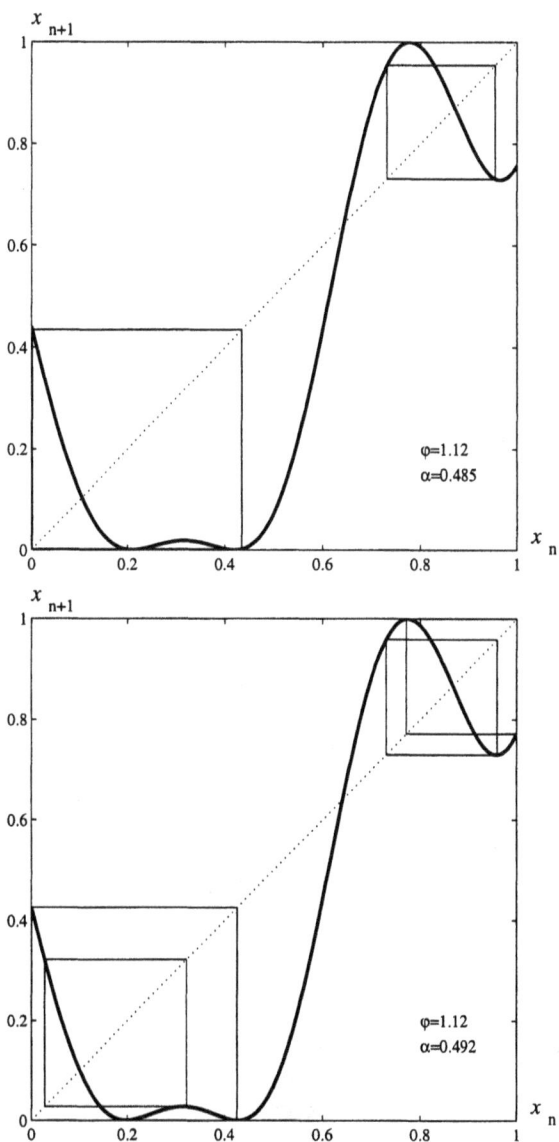

Figure 7.13: Period-two limit cycles, existing for $\alpha = 0,485$, $\alpha = 0,492$ and $\alpha = 0,495$ (for $\phi = 1,12$, $R = 0,5$) together with graphs of the respective graphs of the map Π_R. The observed attractors change in each case.

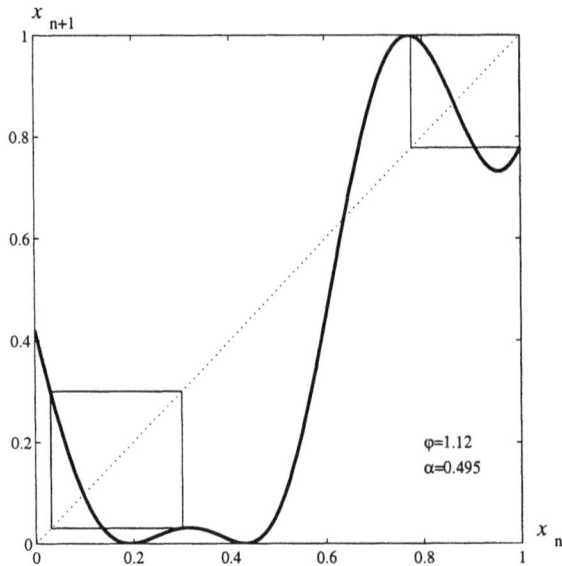

Figure 7.13: Continued

one period-two orbit to another one.

The observed phenomenon can be called "exchange of stability" of two period-two orbits. The orbits existing initially disappear and in their place two new orbits of the same period are created at some distance from the initial ones. At the place of the "jump" really we observe a hysteresis on the branches of the bifurcation tree as shown in Fig.7.15; depending on the direction of changes of the bifurcation parameter the jump is observed at a different value of the parameter α.

7.2.3 Chaos creation mechanism

The analysis of the approximate one-dimensional map enabled us to explain several dynamic phenomena and mechanisms for chaos creation in the original system. We can expect that the horseshoe creation mechanism (stretching and folding) found in the squeezed spiral map is responsible for chaos creation in the original system as well.

To confirm further that the proposed one-dimensional map very well reproduces the properties of the original system we compared the graphs of the return map and the bifurcation diagrams obtained using the approximate map

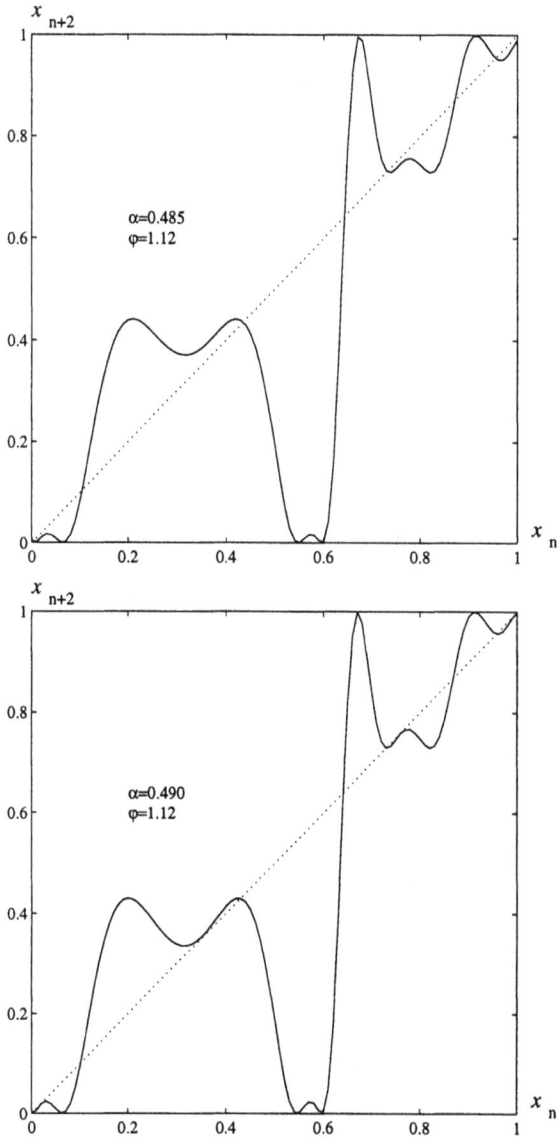

Figure 7.14: Explanation of the "jump phenomenon" — change of the attractor size. The graphs of the map $\Pi_R \circ \Pi_R$ for five values of α — $\alpha = 0,485$, $\alpha = 0,490$, $\alpha = 0,492$, $\alpha = 0,495$ and $\alpha = 0,498$ and fixed $\phi = 1,12$, $R = 0,5$. Full description in text.

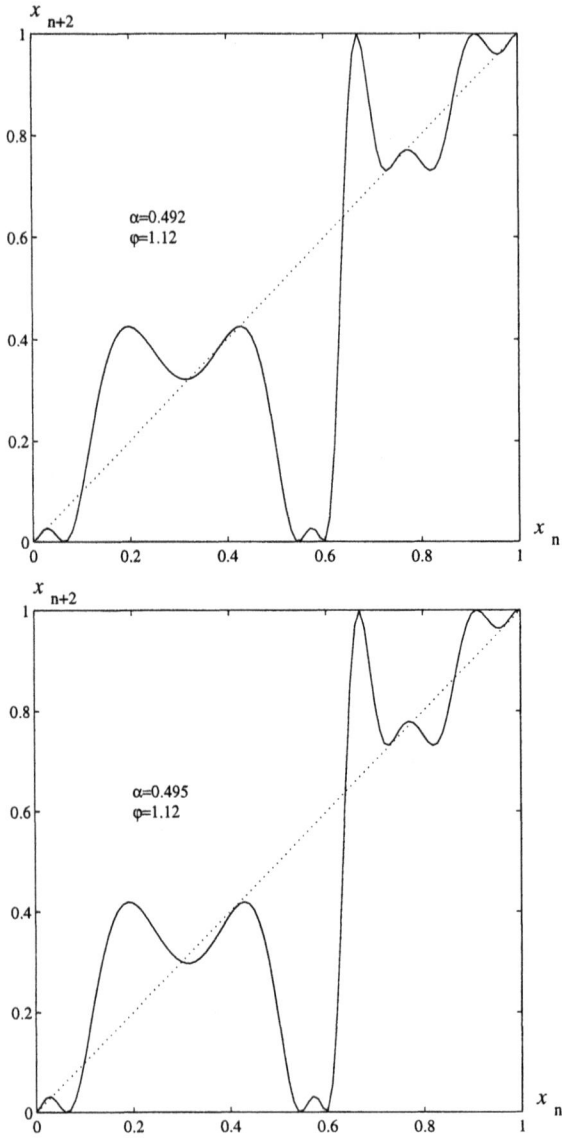

Figure 7.14: Continued

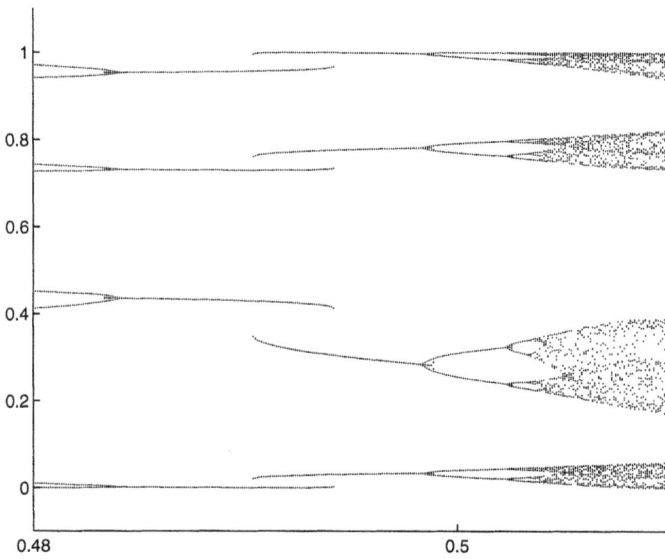

Figure 7.15: Magnified fragment of the bifurcation tree showing the jump phenomenon — abrupt change of the attractor size. The two upper and two lower branches of the diagram represent points belonging to two different orbits — hysteresis of the diagram is clearly visible.

and the original system equations. The results are shown in figures 7.16 and 7.17.

The graphs of the return map obtained via direct numerical integration of the differential equations reveal the existing two-dimensional character of the mapping — some of the branches have multi-layer structure. Comparing with the graph of the one-dimensional map one can see a gap corresponding to the region of discontinuous behavior of the original map.

The agreement of results obtained using the approximate one-dimensional map and integration of the original system equations is very good. The bifurcation sequences are reproduced in great detail confirming existence of a symmetry breaking bifurcation of the periodic orbits created via a Hopf bifurcation (period-1 orbits). The period doubling sequence and "bubbles" are also reproduced in bifurcation diagrams constructed using the approximate one-dimensional map. Bi-directional period doubling sequences ("bubbles") constitute a fundamental route to chaos in this class of systems. Another feature well reproduced by the model is the jump behavior — sudden change of the size of an attractor (periodic or chaotic one) (comp Fig. 7.18). Analysis of properties of the approximate one-dimensional map enables explanation of this phenomenon as well as the disappearing of the attractor via crisis.

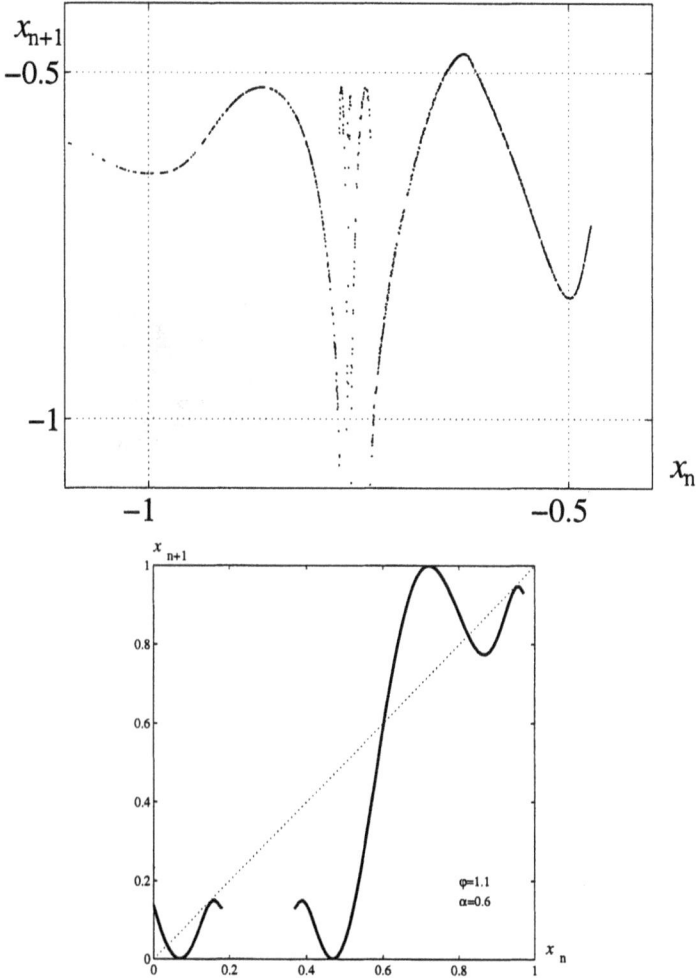

Figure 7.16: Comparison of the graphs of the return map obtained via integration of the original system equations (with $R_1 = R_2 = R_3 = 1$, $C_1 = C_2 = C_3 = 1$, $m_0 = -33,03$, $\sigma_{bp} = 0,2$, $m_1 = 490$) and iterations of the proposed one-dimensional map (for $\phi = 1,10$, $R = 0,5$, $\alpha = 0,6$). The approximate one-dimensional map reflects qualitatively the original behavior in most of the regions.

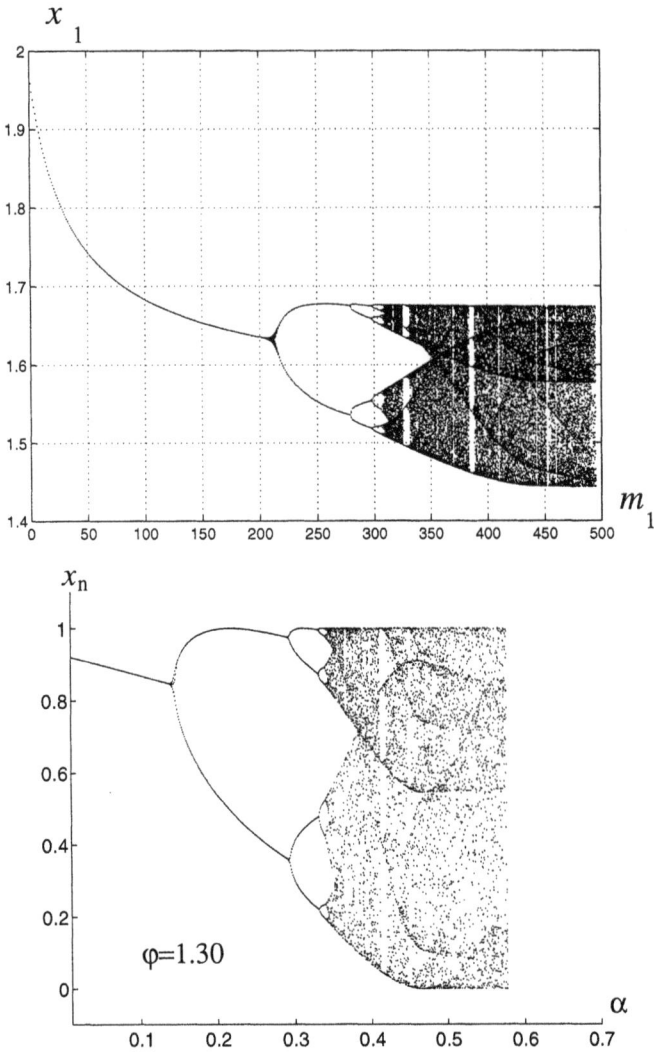

Figure 7.17: Comparison of bifurcation diagrams obtained via direct integration of system equations (with $R_1 = R_2 = R_3 = 1$, $C_2 = C_3 = 1$, $m_0 = -33,03$, $\sigma_{bp} = 0,2$, $m_1 = var$) and by iterations of the proposed one-dimensional map (for $\phi = 1,30$, $R = 0,5$ $\alpha = var.$). The bifurcation sequences are almost identical.

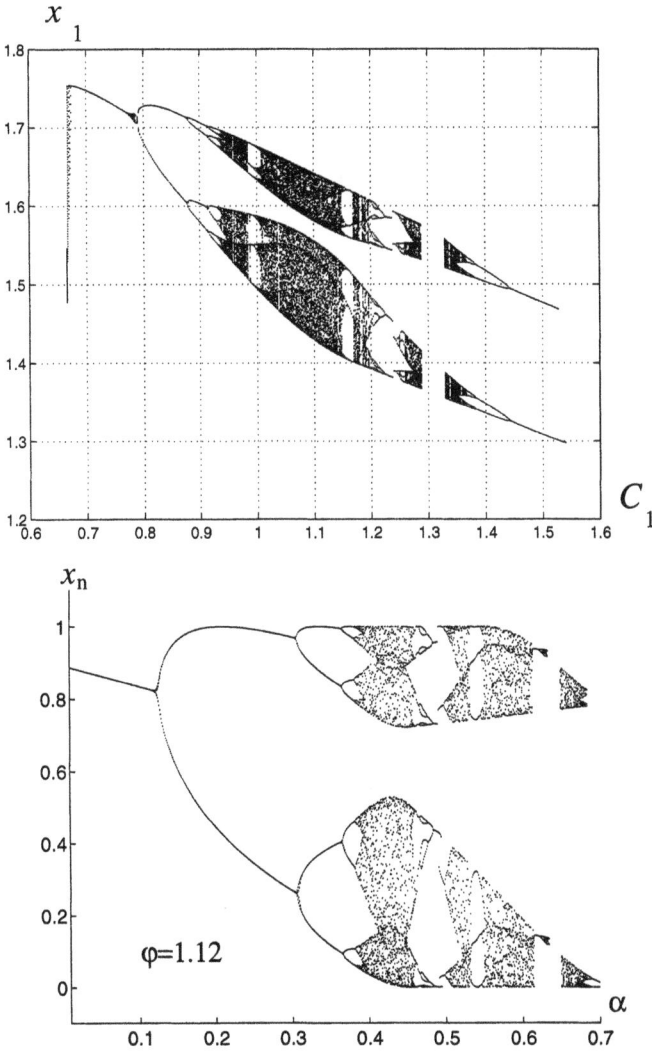

Figure 7.18: Comparison of bifurcation diagrams obtained via direct integration of system equations (with $R_1 = R_2 = R_3 = 1$, $C_2 = C_3 = 1$, $m_0 = -33,03$, $m_1 = 330,$, $\sigma_{bp} = 0,2$) and by iterations of the proposed one-dimensional map (for $\phi = 1,12$, $R = 0,5$). Bifurcation sequences are well reproduced by the model. Even the fine structure of the "bubbles" and jump of the size of the attractor is well reproduced.

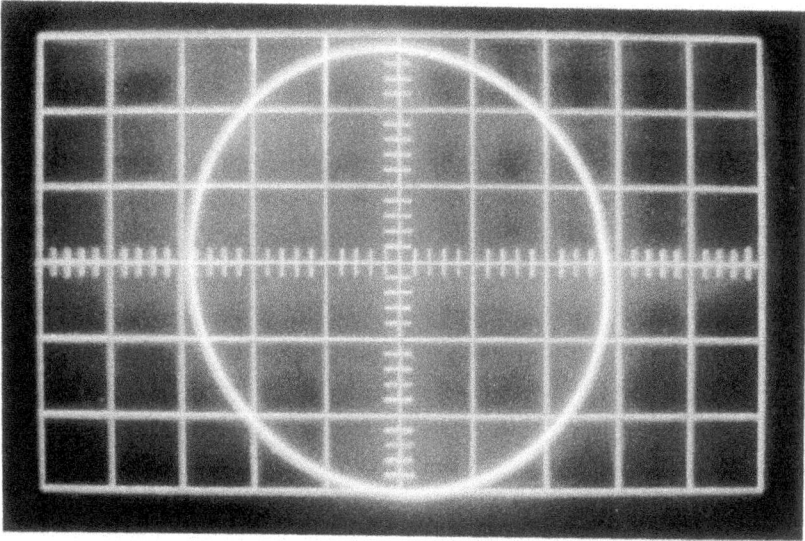

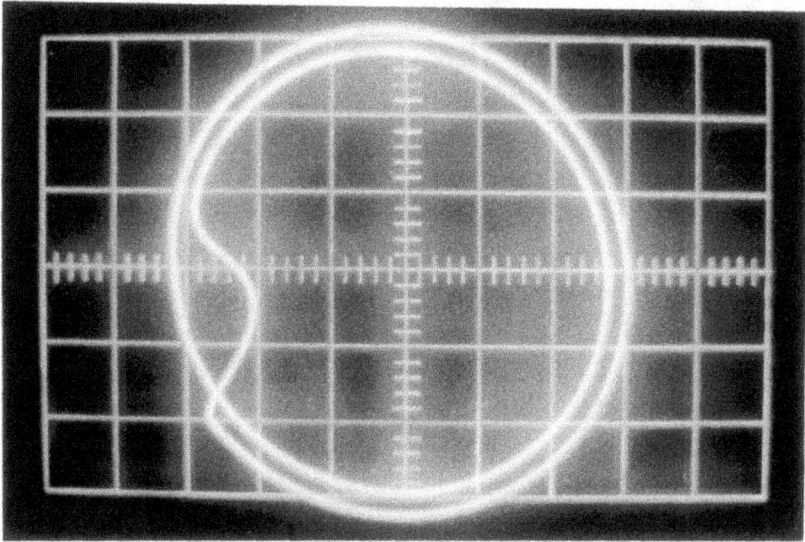

Plate 1. State-space trajectories observed in a laboratory RC-ladder circuit using an XY option on a standard oscilloscope. Changes of qualitative behavior are observed for various values of the slope m_0 of the piecewise linear characteristic. $v_{c1} - v_{C2}$ projections of period-1 and period-2 orbits are shown (horizontal scale 1V/div., vertical scale 0.4V/div.).

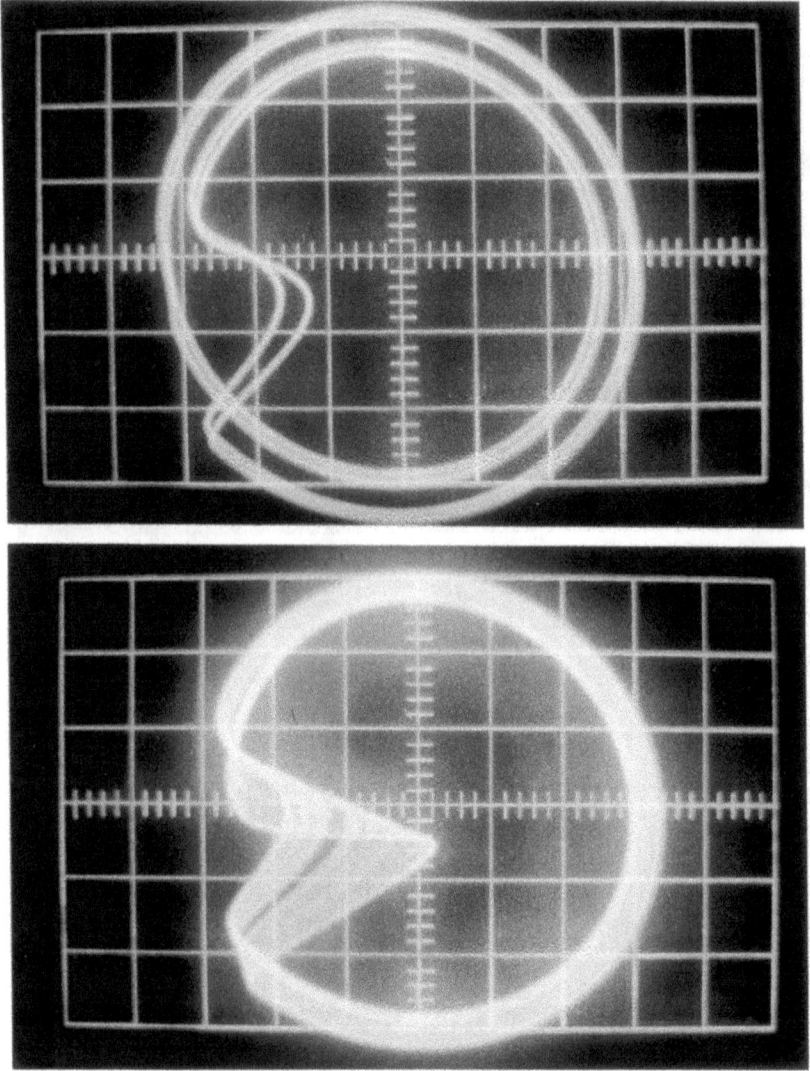

Plate 2. State-space trajectories observed in a laboratory RC-ladder circuit using an XY option on a standard oscilloscope. Changes of qualitative behavior are observed for various values of the slope m_0 of the piecewise linear characteristic. $v_{c1} - v_{C2}$ projections of period-4 orbit and an asymmetrical chaotic Rössler-type trajectory is shown (horizontal scale 1V/div., vertical scale 0.4V/div.).

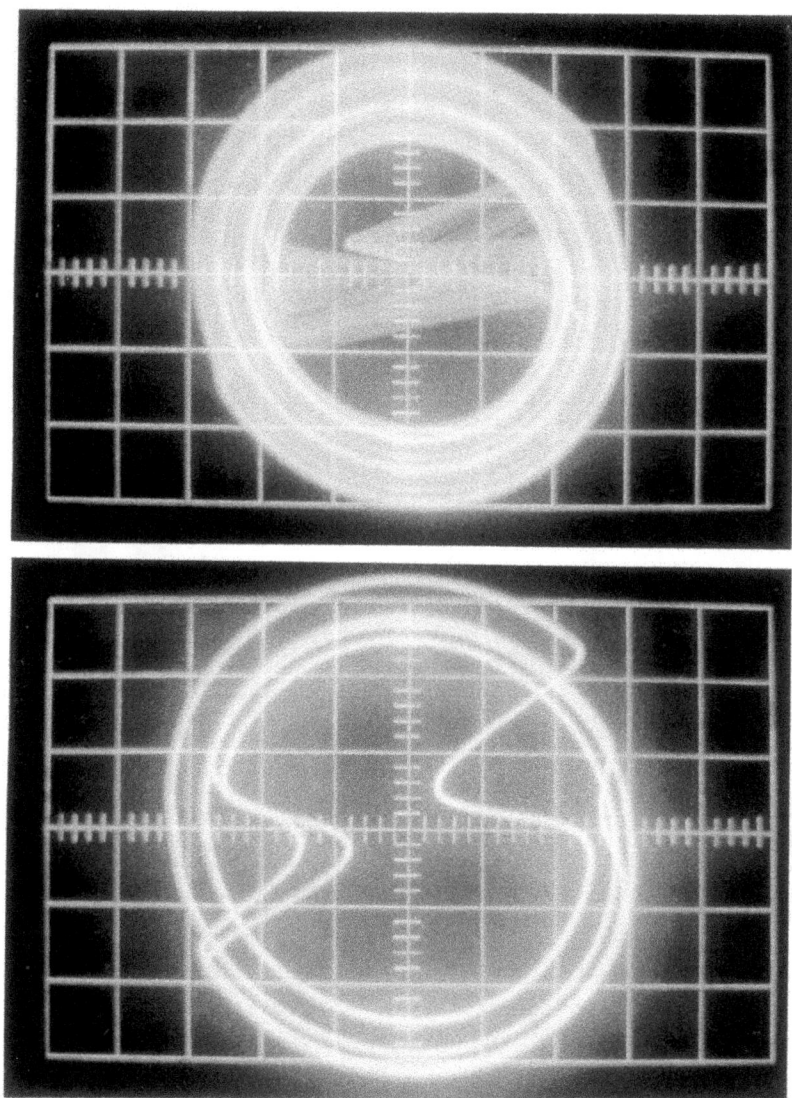

Plate 3. State-space trajectories observed in a laboratory RC-ladder circuit using an XY option on a standard oscilloscope. Changes of qualitative behavior are observed for various values of the slope m_0 of the piecewise linear characteristic. $v_{c1} - v_{C2}$ projections of a Rössler-type chaotic orbit with center symmetry and one of periodic windows within the chaos range are shown (horizontal scale 1V/div., vertical scale 0.4V/div.).

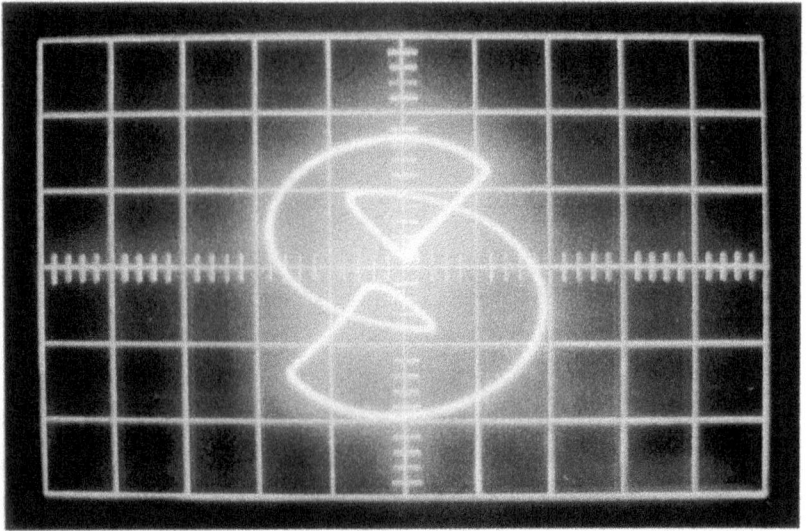

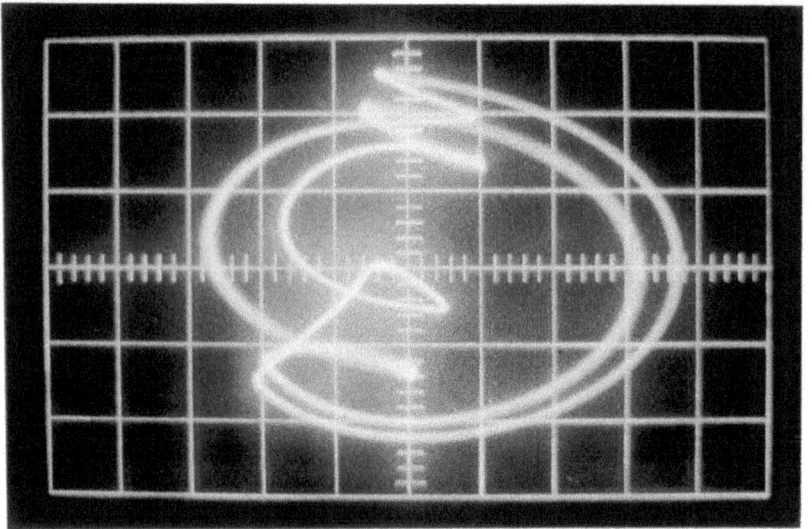

Plate 4. When using a dual type of nonlinear characteristic different types of attractors are observed in the RC-ladder oscillator. Universal route to chaos via period doublings is still observed – $v_{c1} - v_{C2}$ projections of period-1 and period-2 orbits (horizontal scale 1V/div., vertical scale 0.4V/div.).

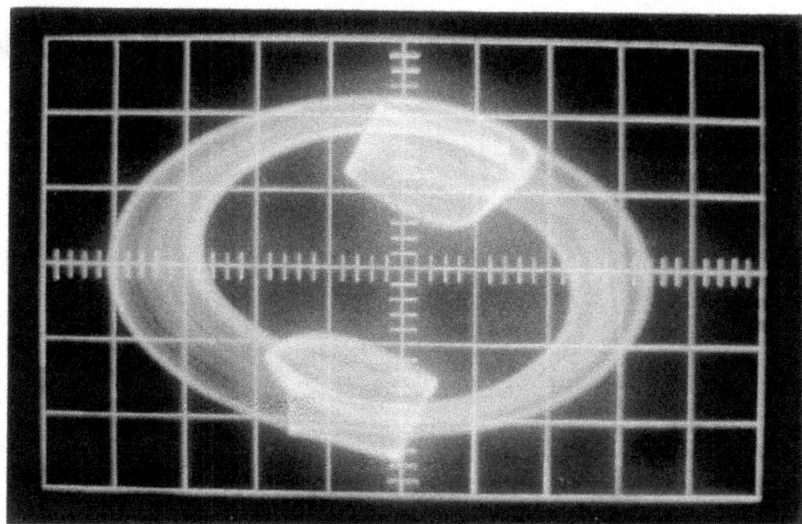

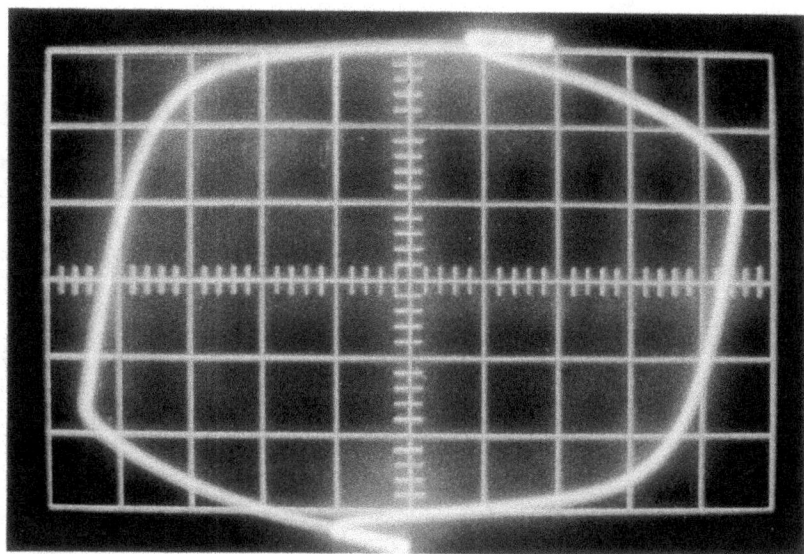

Plate 5. Chaotic orbit observed in the RC-ladder generator with a dual nonlinear characteristic. When passing the chaos range one can observe a large amplitude periodic orbit surrounding all orbits observed in previous experiments (comp. Plate 4.). $v_{c1} - v_{C2}$ projections of attractors are shown (horizontal scale 1V/div., vertical scale 0.4V/div.).

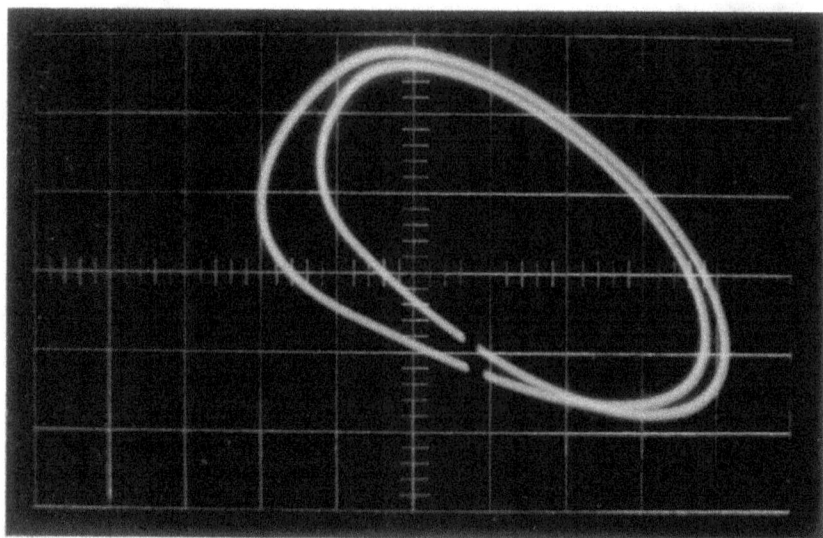

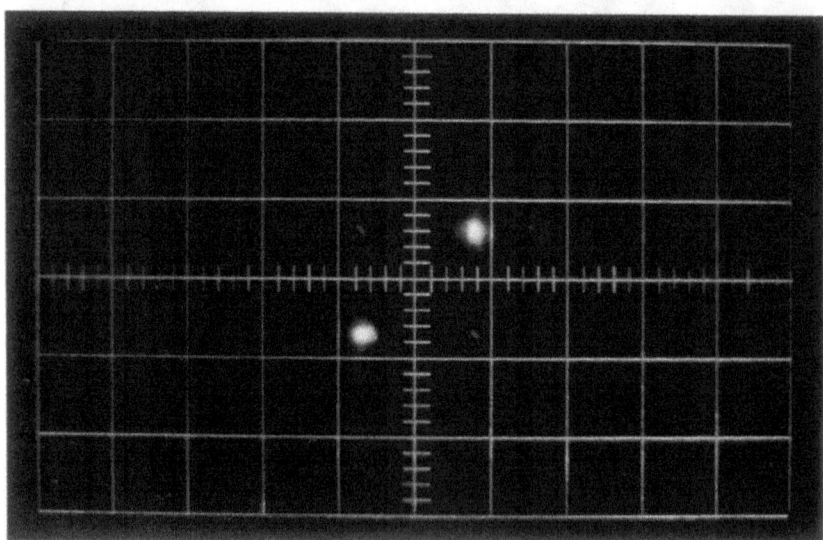

Plate 6. Period-2 orbit (horizontal scale 1V/div.,vertical scale 0.5V/div.) observed in Chua's circuit and corresponding Poincaré map composed of two points obtained electronically (horizontal scale 40mV/div.,vertical scale 0.1V/div.)

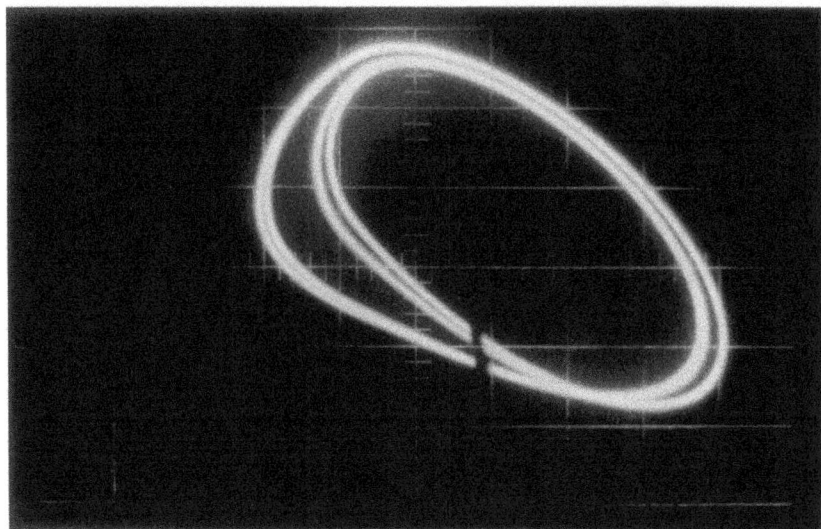

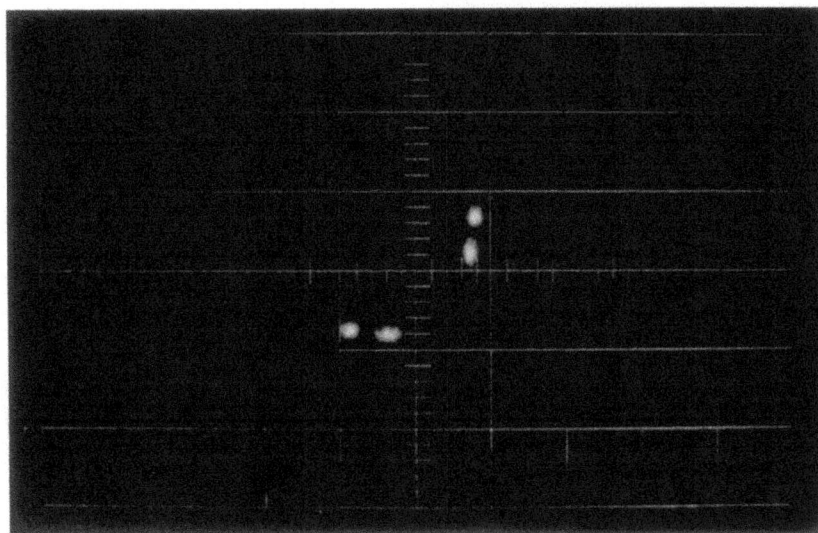

Plate 7. Period-4 orbit (horizontal scale 1V/div.,vertical scale 0.5V/div.) observed in Chua's circuit and corresponding Poincaré map composed of four points obtained electronically (horizontal scale 40mV/div.,vertical scale 0.1V/div.)

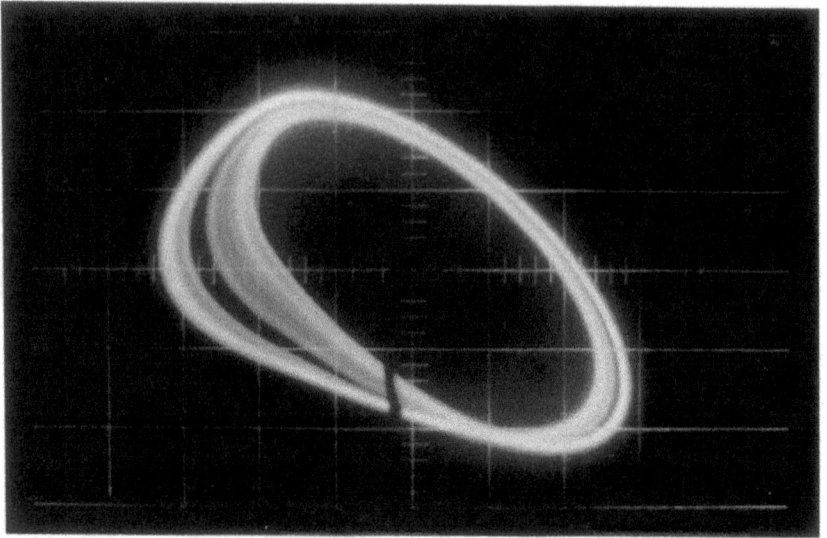

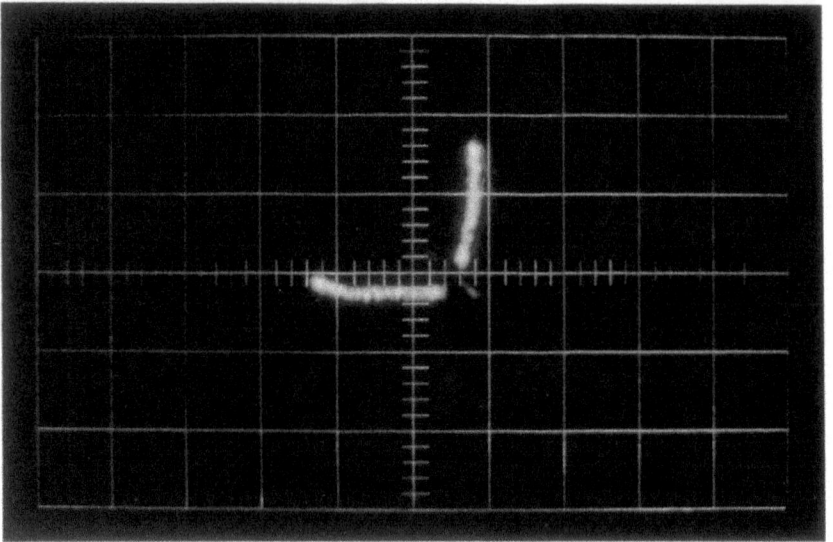

Plate 8. Chaotic orbit of a ribbon type (horizontal scale 1V/div.,vertical scale 0.5V/div.) observed in Chua's circuit and corresponding Poincaré map composed of two sections obtained electronically (horizontal scale 40mV/div.,vertical scale 0.1V/div.).

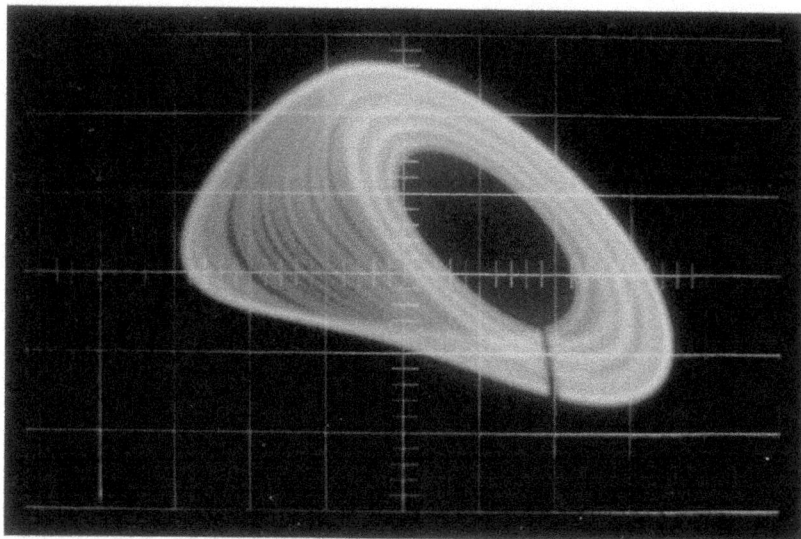

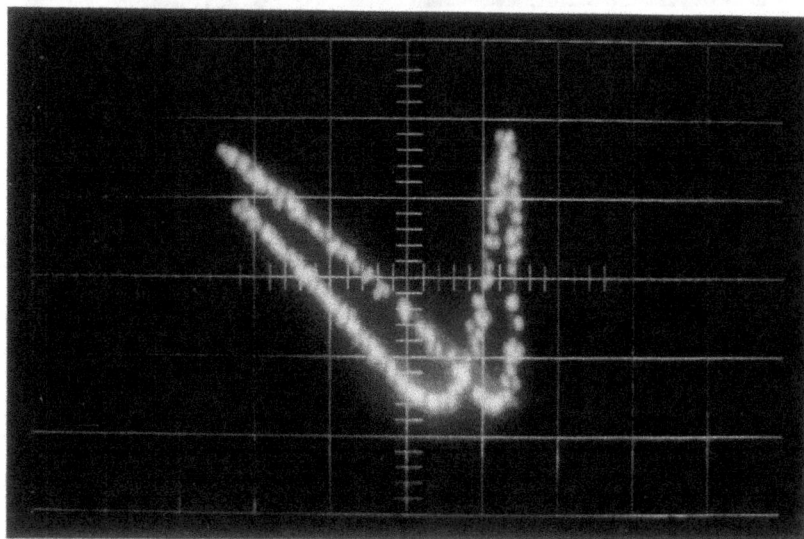

Plate 9. Rössler-type spiral attractor (horizontal scale 1V/div.,vertical scale 0.5V/div.) observed in Chua's circuit and corresponding Poincaré map obtained electronically (horizontal scale 40mV/div.,vertical scale 0.1V/div.)

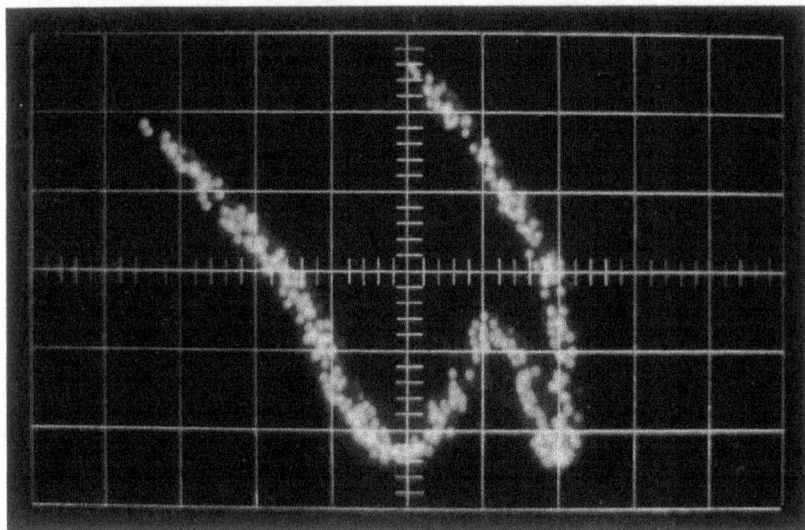

Plate 10. Rössler-type spiral attractor (horizontal scale 1V/div.,vertical scale 0.5V/div.) observed in Chua's circuit and corresponding Poincaré map obtained electronically (horizontal scale 40mV/div.,vertical scale 0.1V/div.). The position of the section plane has been changed in comparison with Plate 9. - this results in a different shape of the Poincaré map.

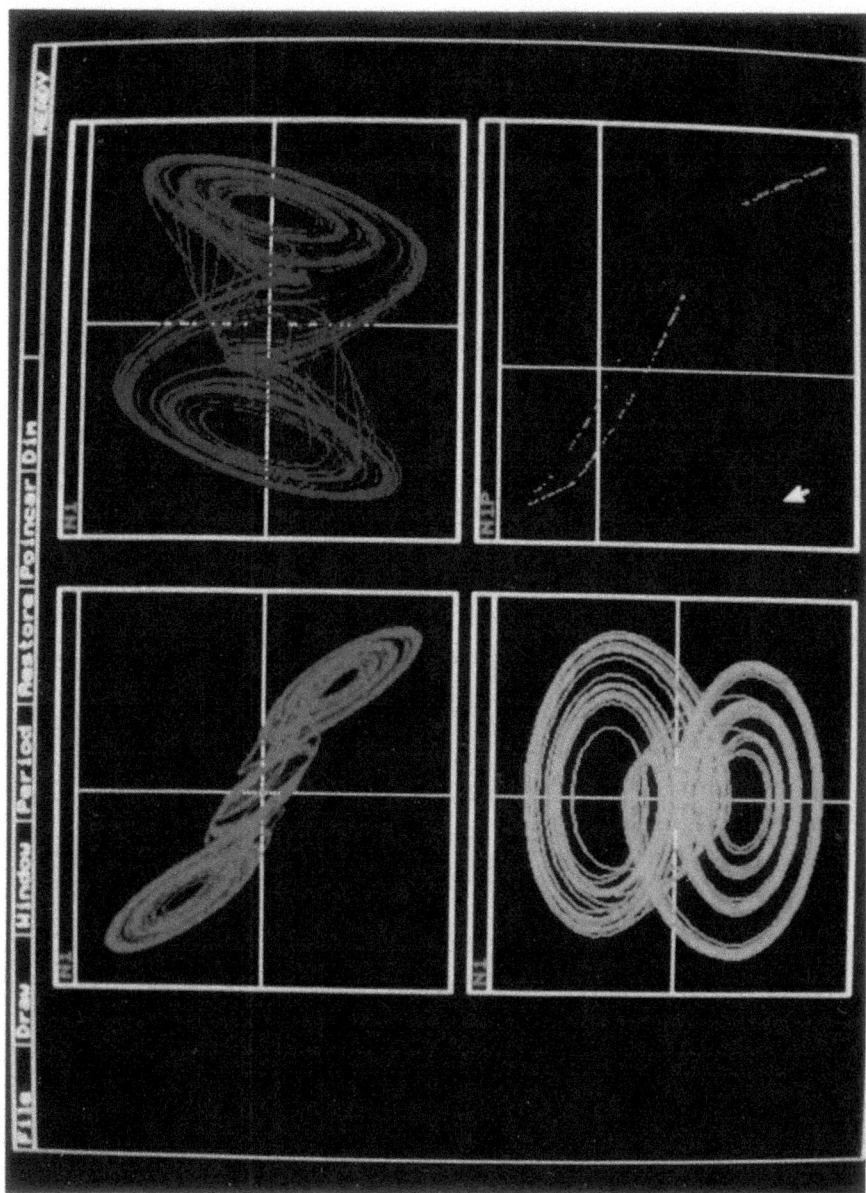

Plate 11. Analysis of the double scroll attractor using computer methods — three two-dimensional projections of the attractor and a Poincaré map are shown.

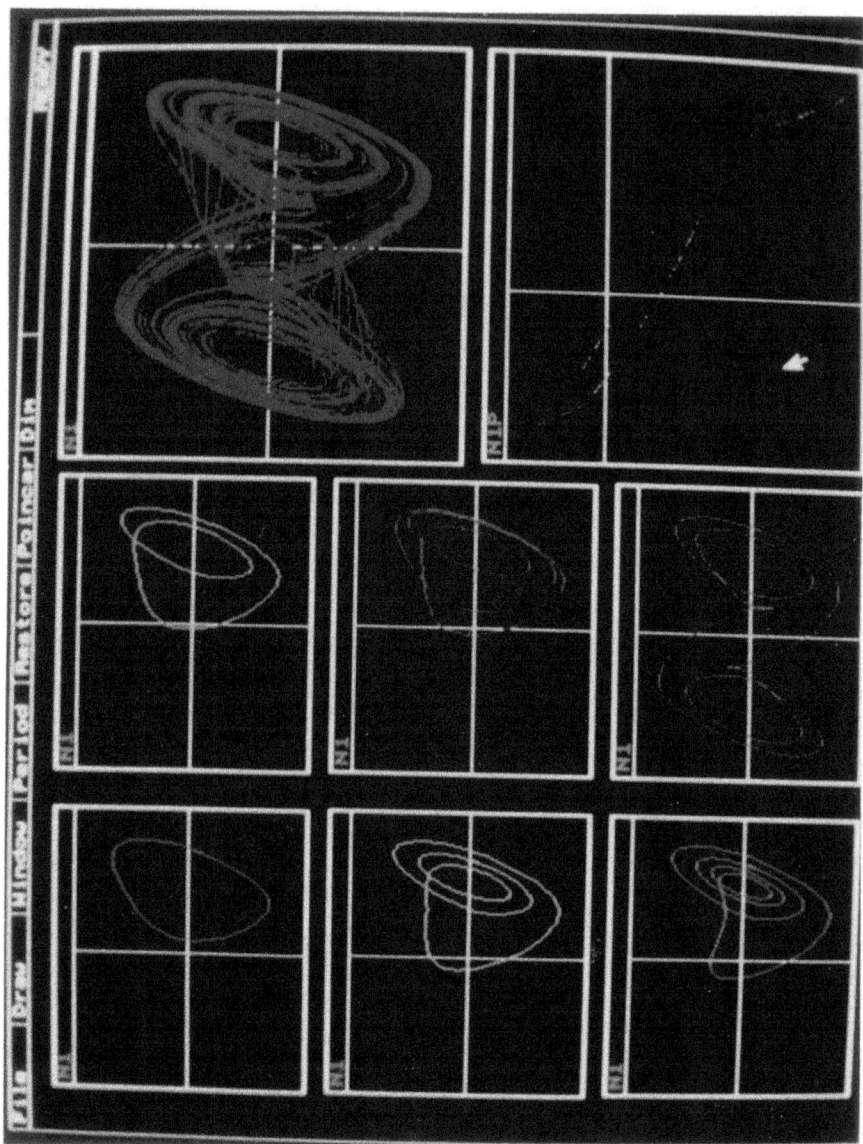

Plate 12. Analysis of the double scroll attractor using computer methods — Some of the unstable periodic orbits found using one-dimensional time series are shown.

Plate 13. Global color-coded (a, b)-parameter bifurcation diagram for the second-order digital filter with saturation arithmetic. Each color code (except black) denotes a region of existence of orbits of period p ($p < 16$), where $p = 1$ - gray (one fixed point - light gray, two fixed points - dark gray), $p = 2$ - blue, in the region marked light green both fixed point and period- two orbits coexist, $p = 3$ - light blue, $p = 4$ - light red, $p = 5$ - magenta, $p = 6$ - yellow, $p = 7$ - white, $p = 8$ - blue , $p = 9$ - green, $p = 10$ - cyan, $p = 11$ - red, $p = 12$ - magenta, $p = 13$ - brown. High period orbits $p \leq 50$ are coded as $p \bmod (15) + 7$.

Plate 14. Details of the "circular" regions and fine structure of Arnold tongues within the largest "circular" region: "sausage" structures.

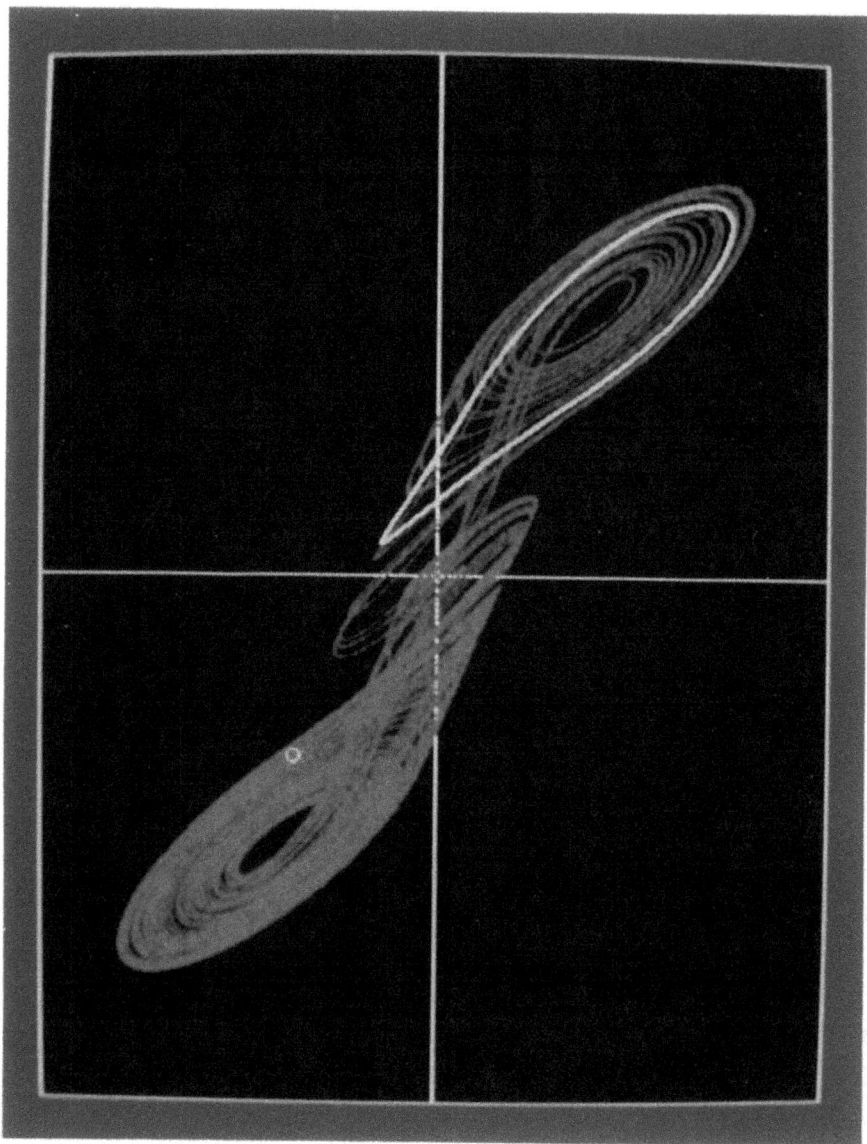

Plate 15. Stabilization of period-one orbit in Chua's circuit using OGY technique. The controlled orbit is shown in yellow. The red trajectory represents transient behavior before the trajectory entered the region where control could be applied.

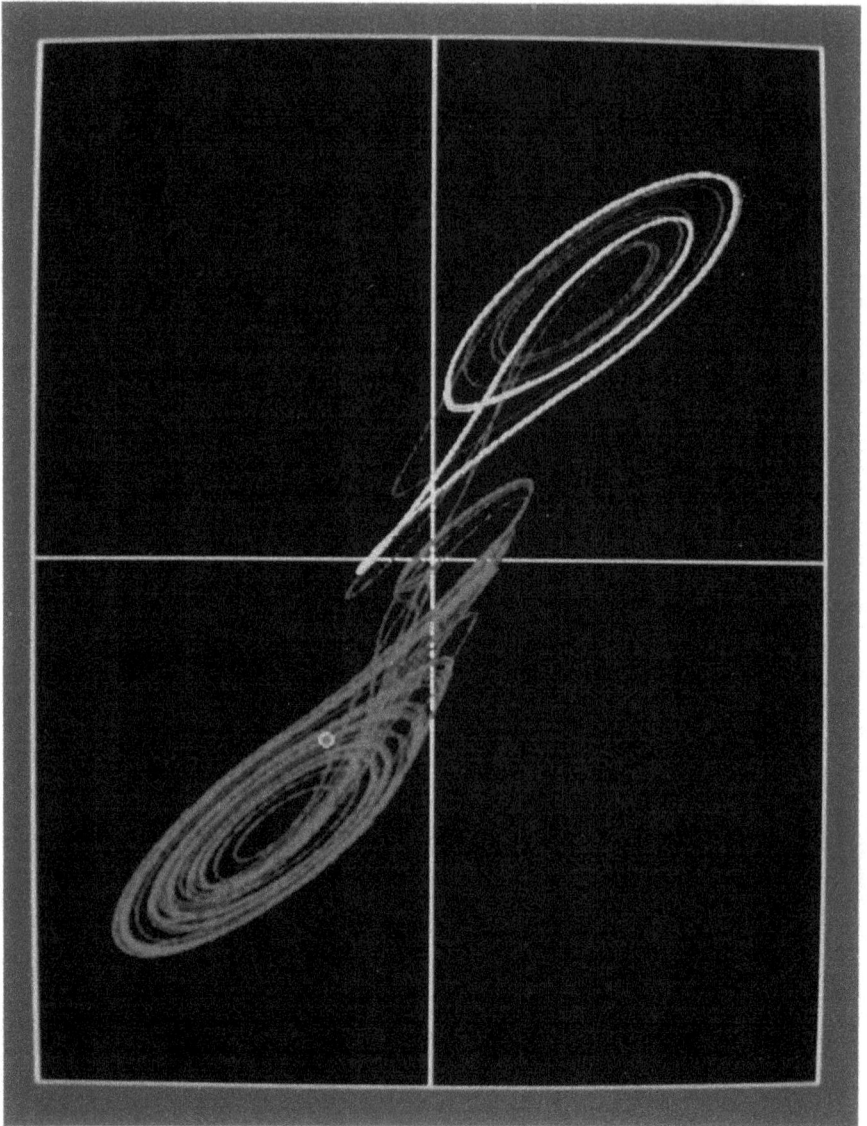

Plate 16. Stabilization of period-two orbit in Chua's circuit using OGY technique. The controlled orbit is shown in yellow. The red trajectory represents transient behavior before the trajectory entered the region where control could be applied.

CHAPTER 8

CONJECTURE ON EXISTENCE OF CHAOTIC OSCILLATIONS

From the circuit analysis or design points of view the methodologies presented in this book are quite difficult to apply. Every single circuit represents a problem on its own and except from simulation experiments and laboratory experiment there is little common approaches to their analysis possible. Always gathering of these results is very time consuming and their interpretation is a matter of personal experience, certain "feeling" and often some luck.

Having analyzed a large number of examples we have come to a conclusion that many circuits and systems exhibiting complicated behavior and chaos share some common properties which although cannot serve as a criterion of existence of chaos but give a good indication that such type of behavior is highly probable. These common properties are possible to be put into a form which is easily verifiable on the basis of stability analysis of a linear case and verification of the position of the nonlinearity with respect to the linear stability sector.

8.1 Conjecture on chaos generation

Let the dynamics of the system (circuit) under consideration be described by so-called Lure systems:

$$\dot{x}(t) = Ax(t) + BF[C^T x(t)] \tag{8.1}$$

where : $\mathbf{x}(t), \mathbf{B}, \mathbf{C} \in \mathbb{R}^n$, $\mathbf{A}$ - matrix $n \times n$, $F : R \to R$ locally Lipschitz function, $F(0) = 0$. Let the pairs: $(\mathbf{A}, \mathbf{B})$ - controllable, $(\mathbf{A}, \mathbf{C})$ - observable. The transfer function $G(s) = \mathbf{C}^T(s\mathbf{I} - \mathbf{A})^{-1}\mathbf{B}$ has no poles at zero.

Let us consider next a linear system associated with (8.1), in which the nonlinear function $F(\sigma)$ has been substituted by a linear one: $F(\sigma) = K\sigma$:

$$\dot{\mathbf{x}}(t) = \mathbf{A}\mathbf{x}(t) + \mathbf{B}K[\mathbf{C}^T\mathbf{x}(t)] \tag{8.2}$$

Let us assume that there exists at least one nonzero interval $[K_1, K_2]$, such that the system (8.2) is stable for every $K \in [K_1, K_2]$ (we say that there exist a non-zero stability sector (Hurwitz sector) for the system with linear feedback).

Let us further assume that the slope of the nonlinear characteristic F near the equilibrium points defined by the solutions of the equation:

$$\mathbf{y} + F(\mathbf{y})G(0) = 0 \qquad (8.3)$$

does not belong to the interval $[K_1, K_2]$ and whit growing distance from the equilibria (ie. for growing σ) the slope changes in such a way that the characteristic intersects the linear stability sector.

Then it is possible to adjust the slopes of the nonlinear characteristic in such a way that for fixed parameters of the linear part of the system defined by $(\mathbf{A}, \mathbf{B}, \mathbf{C})$, the system will generate chaotic oscillations[320,328].

Typical characteristics satisfying the conditions of the above conjecture are shown in Fig.8.1.

Comments:

- There exists a link between the shape of the nonlinear characteristic and the existence of homoclinic orbits in the system. It is however impossible to give this relationship in terms of mathematical closed-form formulas even for piecewise-linear systems.

The conjecture was confirmed for many different kinds of chaotic systems. It has been also extended for the classes of circuits with switches [7] and some systems containing transmission lines.

8.1.1   Verification of the conjecture – examples

Chua's circuit

The schematic diagram of Chua's circuit is recalled in Fig.8.2. The dynamics of the circuit is described by equations of the form:

$$C_1 \frac{dv_{C_1}}{dt} = G(v_{C_2} - v_{C_1}) - g(v_{C_1}) \qquad (8.4)$$

$$C_2 \frac{dv_{C_2}}{dt} = G(v_{C_1} - v_{C_2}) + i_L \qquad (8.5)$$

$$L \frac{di_L}{dt} = -v_{C_2} \qquad (8.6)$$

Depending on the choice of circuit parameters and slopes of the piecewise-linear function $g(v)$ this system is known to generate a large number of chaotic orbits with distinct properties. The figure 8.3 shows typical trajectories observed in this circuit.

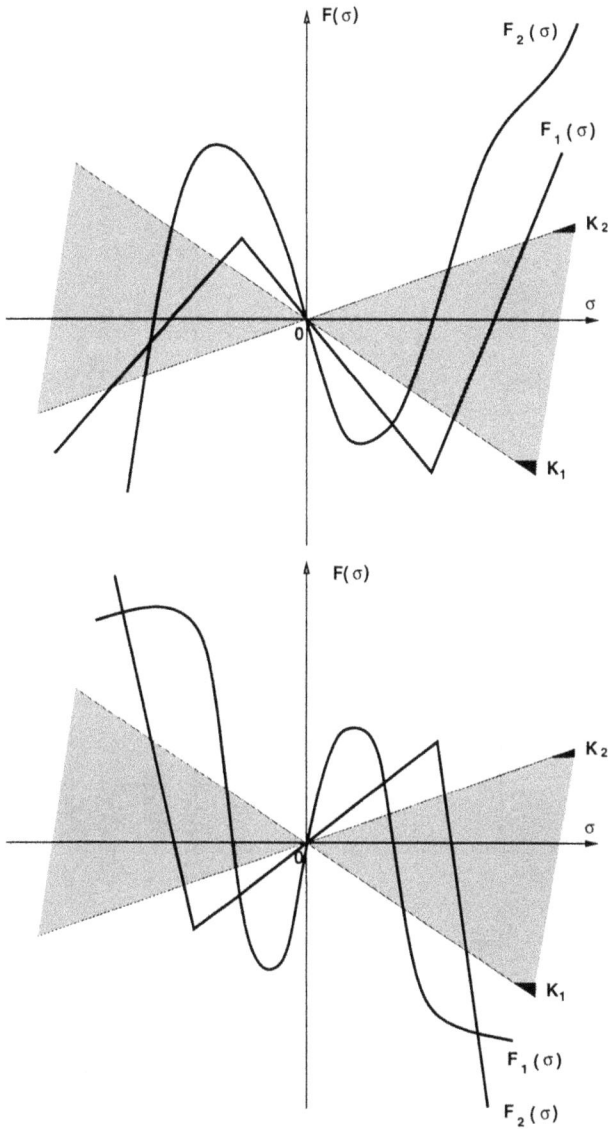

Figure 8.1: Typical nonlinear characteristics satisfying the conjecture on chaos generation. Position is shown with respect to the Hurwitz sectors of the linearized system.

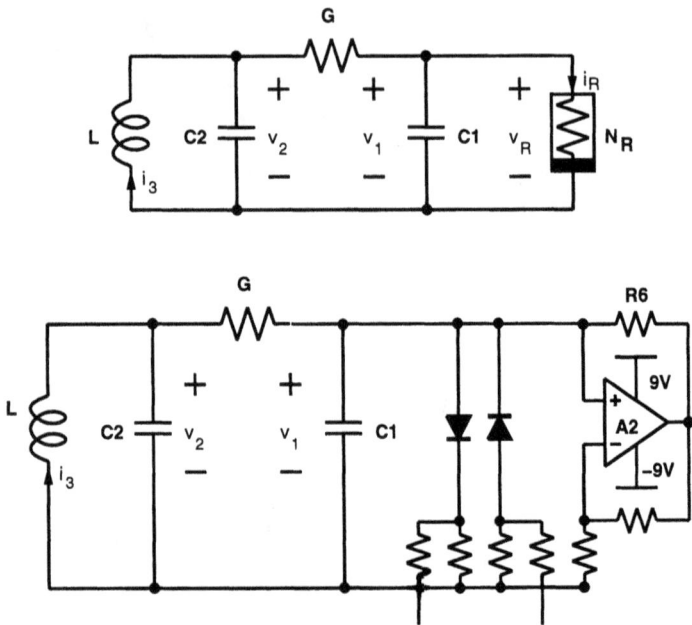

Figure 8.2: Circuit diagram of Chua's circuit and its implementation using single operational amplifier.

- The Figure 8.3a shows trajectories observed for $G = 0,7$, $C_1 = 1/8.5$, $C_2 = 1$, $m_0 = -0,5$, $m_1 = -0,8$, $v_b = 1$. This kind of orbits were called by Rössler [385] as "spiral attractor" (the trajectories lie on a geometric structure resembling a Möbius band and encircle a fixed point).

- Figure 8.3b shows a trajectory observed for $G = 0,7$, $C_1 = 1/8,8066$, $C_2 = 1$, $m_0 = -0,5$, $m_1 = -0,8$, $v_b = 1$. In Rössler's classification this is a screw-type attractor".

- Figure 8.3c shows a trajectory observed for $G = 0,7$, $C_1 = 1/9$, $C_2 = 1$, $m_0 = -0,5$, $m_1 = -0,8$, $v_b = 1$. The observed attractor is called "double scroll" [73,279,282,283,284] and is one of the best studied examples known to date. The geometric structure seems to be created by gluing together of two "screw-type" attractors lying symmetrically against the origin in the state space.

- Figure 8.3d shows yet another type of attractor observed in Chua's circuit for $G = 0,751$, $C_1 = 1/9$, $C_2 = 1$, $m_0 = -0,5$, $m_1 = -0,8$, $v_b = 1$. It is sometimes referred to as "ribbon-type" attractor[283].

- Qualitatively different type of behavior is observed for the choice of parameters: $G = 0,539, L = -0,3005, C_1 = -0,0647, C_2 = 0,3180, m_0 = -0,5013, m_1 = -1,3475$ (two elements have negative values!!) and the choice of nonlinear characteristic as in the Fig.8.5. The observed attractor was called the "double hook" 8.3e - "double hook"[25] - because of the shape of the graph of the Poincaré map in this case resembling two fish-hooks[25]).

- When choosing the nonlinear characteristic as in Fig.8.6, and adjusting the parameters: $G = 0,7, L = 1/7, C_1 = 1/9, C_2 = 1, m_0 = -0,75, m_1 = -0,44$ (so-called dual Chua's circuit[363]) we observe again an attractor of the known type - Rössler band[385]) shown in Fig.8.3f.

In all considered cases the nonlinear (piecewise-linear) characteristics satisfy the conditions mentioned in the formulated conjecture - they intersect the Hurwitz sectors of the associated linear feedback system. It is interesting to notice that depending on the choice of parameters of the linear part the system can possess a single stability sector or multiple stability sectors[323]. In every case in this circuit there are three unstable fixed points.

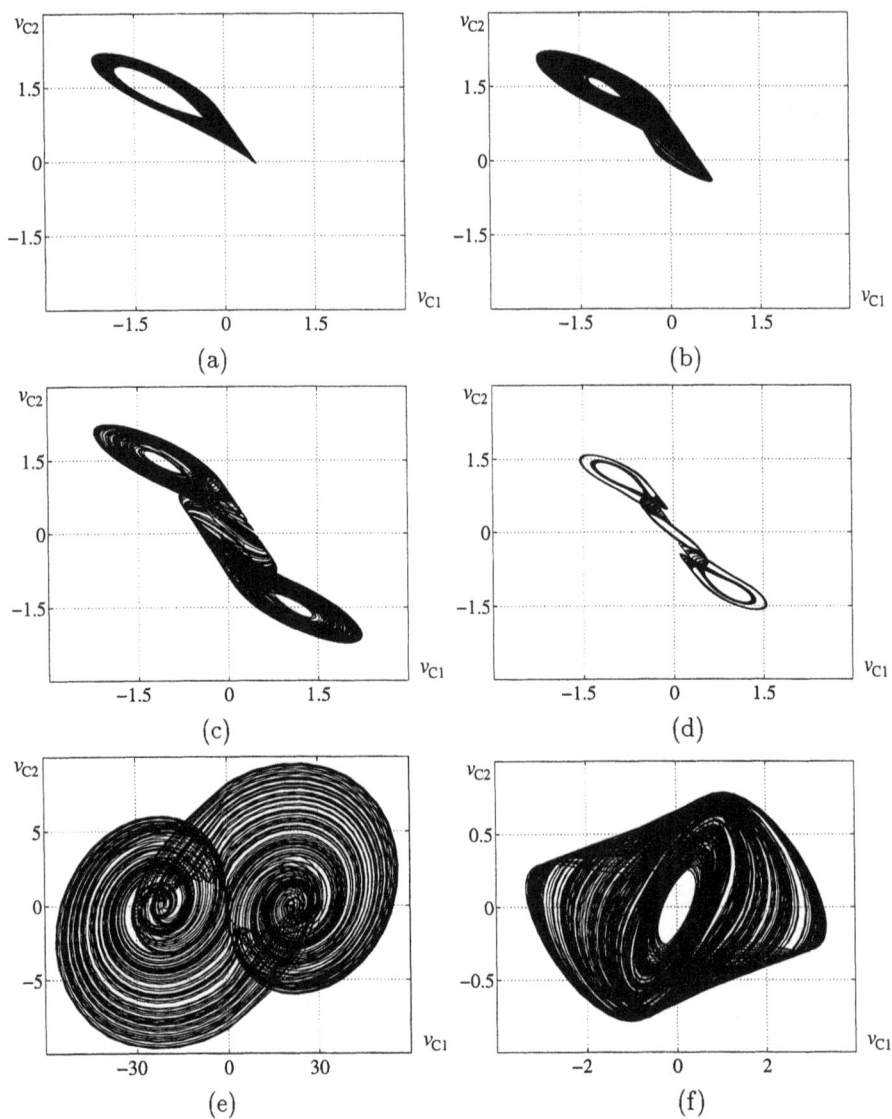

Figure 8.3: Typical attractors observed in Chua's circuit (projections on a chosen plane). Description in the text.

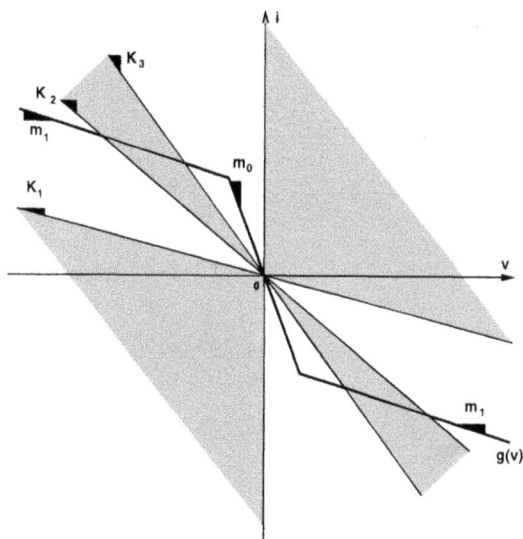

Figure 8.4: Characteristic of the nonlinearity of Chua's circuit with respect to stability sectors of the linearized system. For characteristics of this shape we observe various chaotic attractors eg. as the one shown in Fig.8.3a-d.

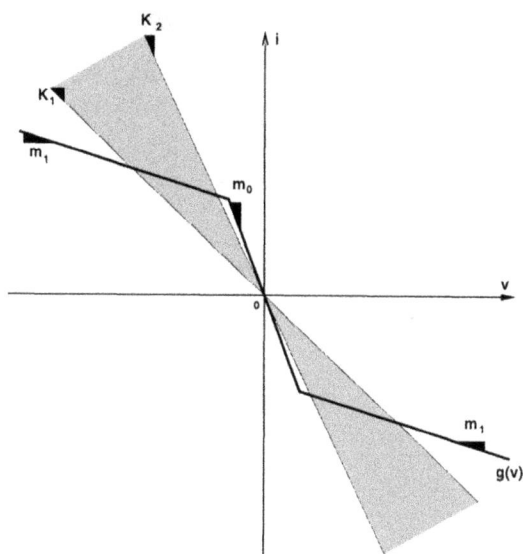

Figure 8.5: Position of the nonlinear characteristic in the Chua's circuit in the case of generation of the "double hook" attractor – see Fig.8.3e).

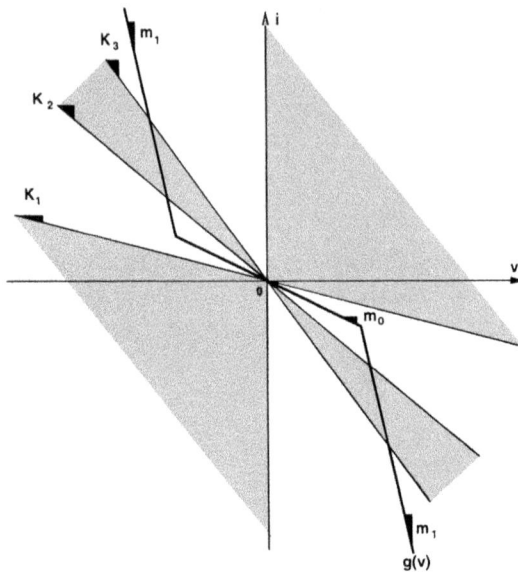

Figure 8.6: Position of the nonlinear characteristic with respect to the stability sector in the case of generation of the Rössler band in dual Chua's circuit (see Fig.8.3f).

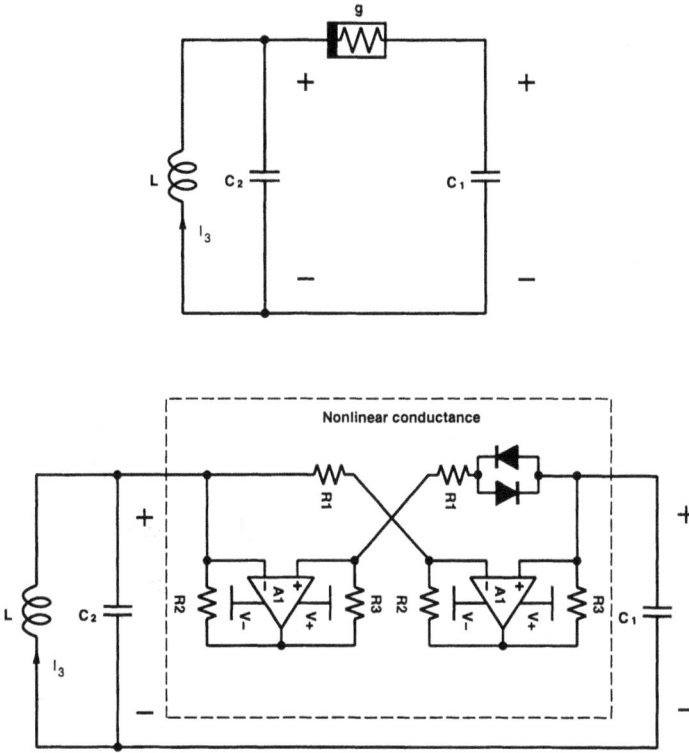

Figure 8.7: Diagram of the "folded torus" circuit and its implementation using operational amplifiers.

"Folded torus" circuit

Analyzing the chaos generation mechanisms in circuits one of possible routes to chaos was confirmed to be torus breakdown – this kind of mechanism has been observed eg. in so-called "folded torus" circuit shown in Fig.8.7. The route to chaos in this circuit does not involve the standard mechanism of period doublings. Changing parameters we observe first a creation of a single periodic orbit then its bifurcation into a torus (quasiperiodic orbit) and further destruction of the torus structure into a chaotic attractor. Fig.8.8 shows the projections of the toroidal (quasiperiodic) attractor (a) and the resulting "folded torus" chaotic attractor created via torus destruction (b). The dynam-

(a)

(b)

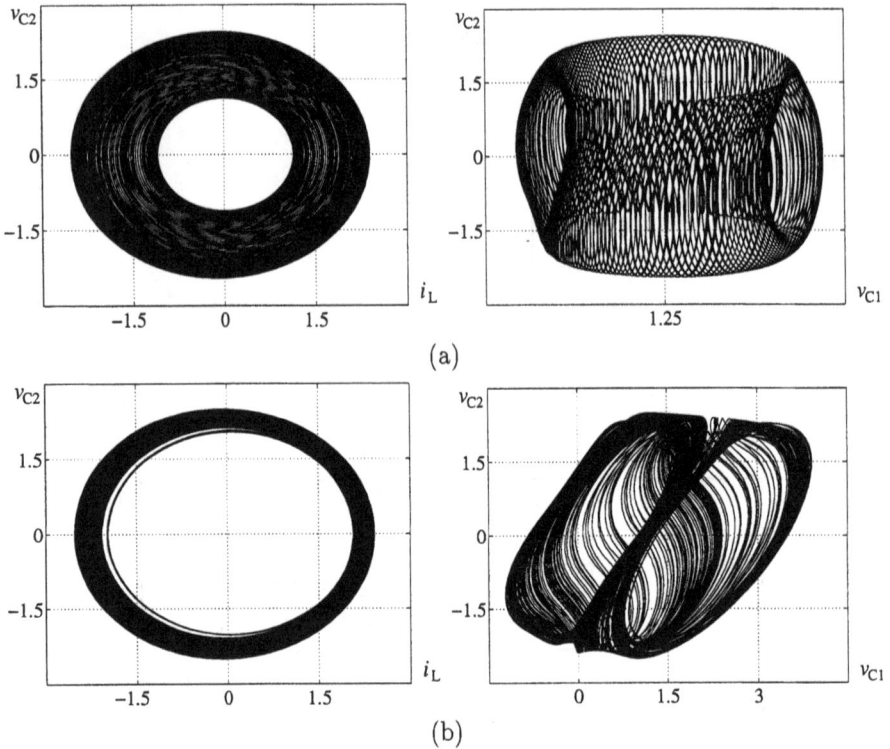

Figure 8.8: Trajectories observed in the "folded torus" circuit. Toroidal attractor (a) observed for $m_0/C_2 = 0,07$, $m_1/C_2 = 0,1$, $1/LC_2 = 1$ and $C_2/C_1 = 2,0$. Chaotic attractor (b), observed for $m_0/C_2 = 0,07$, $m_1/C_2 = 0,1$, $1/LC_2 = 1$ and $C_2/C_1 = 15,0$.

ics of this circuit is described by equation of the form:

$$C_1 \frac{dv_{C_1}}{dt} = g(v_{C_2} - v_{C_1}) \tag{8.7}$$

$$C_2 \frac{dv_{C_2}}{dt} = g(v_{C_1} - v_{C_2}) - i_L \tag{8.8}$$

$$L \frac{di_L}{dt} = v_{C_2} \tag{8.9}$$

It is interesting to analyze the stability of the linearized circuit in this case. We obtain slightly disturbing result – we have a stability sector of zero area – namely the line with slope $k = 0$. The nonlinear characteristic obviously intersects this line – Fig.8.9. We have to classify this example as a pathological

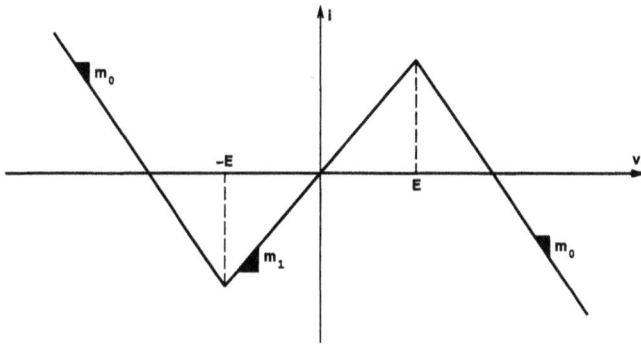

Figure 8.9: Characteristic of the nonlinear resistor in the "folded torus" circuit; the Hurwitz sector is zero in this case (liné with slope $k = 0$).

one despite the fact that the conjecture is also satisfied.

RC-ladder network with nonlinear feedback

This example has been the object of a thorough study in the previous chapters. The circuit diagram is shown in Fig.5.4 (equations (4.1)-(4.5)). Let us investigate the position of the characteristic of the nonlinear controlled source with respect to the stability sector of the linearized circuit. Chaotic oscillations were observed in the system for choices of the characteristics as shown in Fig.8.10a. As in the case of several other examples this nonlinear characteristic intersects the stability sector of the linearized system.

In this circuit chaotic oscillation have been observed also for other choices of nonlinear characteristics ("dual" characteristics) eg. the characteristic f_2 shown in Fig.8.10b [296,322] (similar to the dual Chua's circuit described above).

The conjecture is also confirmed for the cases of higher order RC-ladders containing more then three RC sub-circuits. It was confirmed that circuits of this kind up to 10-th order can generate chaotic oscillation when the nonlinear characteristic intersects the linear stability sector and the slopes are adjusted in a suitable way[296,322,325].

Single diode chaos generator

Inaba and Mori[207] and Saito[392] proposed a number of of circuits generating chaos containing only one nonlinear element – a diode. Let us analyze the applicability of our conjecture in this case. The circuit diagram of the generator

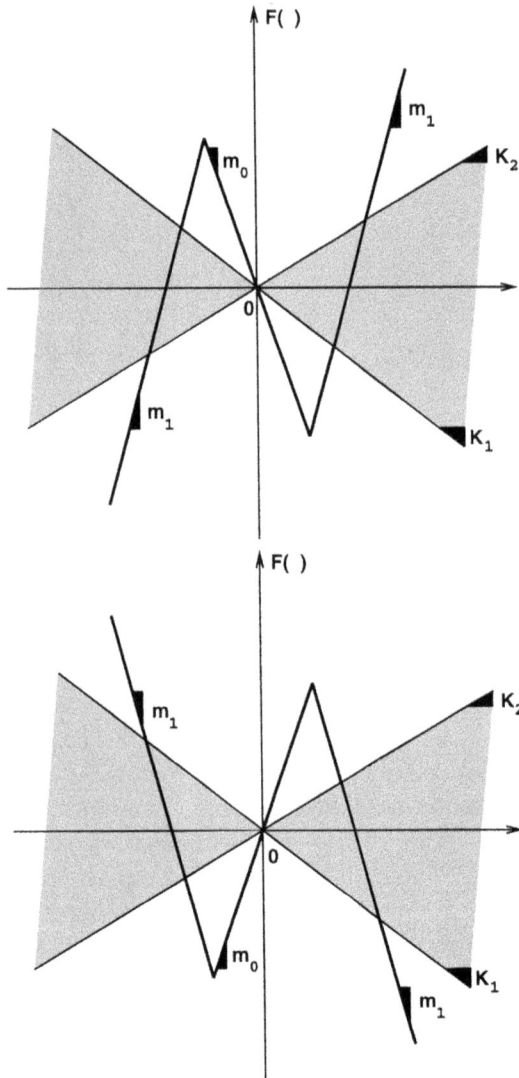

Figure 8.10: Characteristics of the nonlinear controlled source guaranteeing existence of chaotic oscillations in the RC-ladder circuit. The characteristic intersects the stability sector of the linearized circuit.

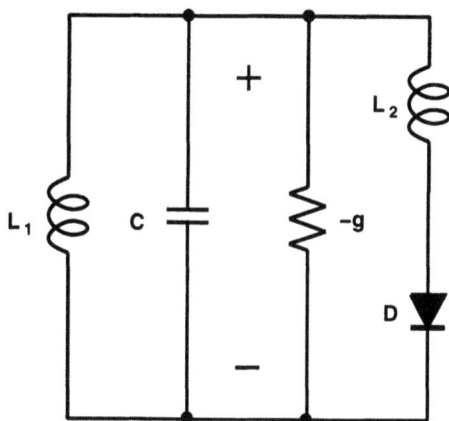

Figure 8.11: Circuit diagram of the chaos generator with single diode.

proposed by Inaba and Mori[207] is shown in Fig.8.11. The dynamics of this circuit is described by state equations of the form:

$$C\frac{dv}{dt} = -i - i_d + gv \tag{8.10}$$

$$L_1\frac{di}{dt} = v \tag{8.11}$$

$$L_2\frac{di_d}{dt} = v - f(i_d) \tag{8.12}$$

Chaotic oscillations have been observed in the circuit for the cases when the characteristic (idealized) of the diode were as shown in Fig.8.12. We should note here that in this case the nonlinear characteristic is non-symmetric – the characteristic intersects however the linear stability sector. In this case the circuit has only one equilibrium point 0 which changes its stability properties when the slope of the nonlinearity switches. All conditions of our conjecture are satisfied.

The two-dimensional projections of a typical chaotic attractor observed in this circuit for $C = 0,0153\mu F$, $L_1 = 909mH$, $L_2 = 100,4mH$ are shown in Fig.8.13.

Saito studied the behavior of this circuit using analytical methods and proved existence of chaos in proper mathematical sense in terms of existence of an infinite number of unstable periodic orbits and Smale horseshoe. Thus our conjecture finds also full mathematical confirmation in this case.

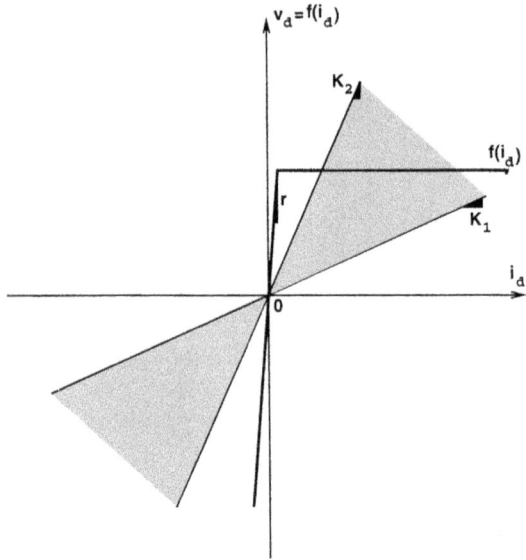

Figure 8.12: Characteristic of the diode versus the linear stability sector for the diode chaos generator.

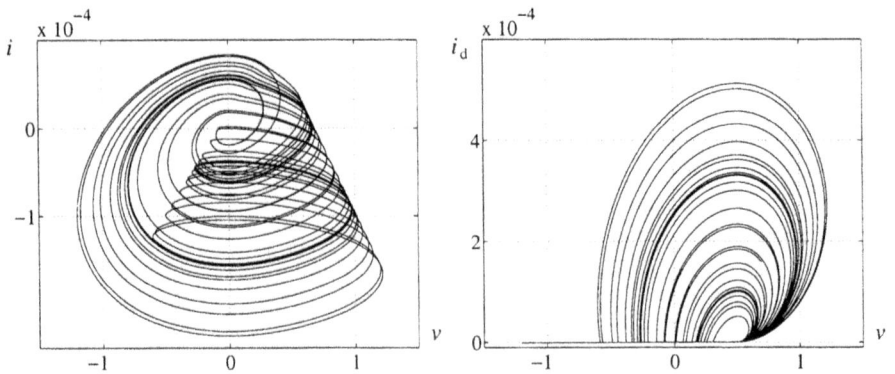

Figure 8.13: Chaotic attractor observed in the diode circuit proposed Inaba and Mori for $C = 0,0153\mu F$, $L_1 = 909mH$, $L_2 = 100,4mH$.

CHAPTER 9

COMPLEX BEHAVIOR IN DIGITAL FILTERS

In this chapter we will present some results confirming possibilities of complex behavior in extremely simple second-order digital filter structures.

Two types of such behavior have been confirmed by simulation experiments and rigorous mathematical analysis: 1). chaotic behavior defined as the existence of aperiodic, bounded trajectories displaying sensitive dependence to the initial states. Such trajectories often form very complex, self-similar patterns in the state-space. 2). an abundance of oscillatory solutions and final state sensitivity with respect to system parameters. This type of behavior can be identified by an extremely complicated structure of the parameter space (existence of Arnold tongues) and fine structure of changes of dynamic behavior when varying filter parameters (devil's staircase). The first type of behavior has been discovered in filter sections employing 2's complement adder overflow characteristics and confirmed via symbolic dynamics technique and modern mathematical approaches. Properties of the second type have been found in extensive numerical experiments carried out for filter sections employing saturation arithmetic and in which a full confirmation via the mathematical analysis of an associated one-dimensional model of the system has been made.

9.1 Digital Filtering and Complex Behavior in Digital Systems

In recent years, with advances in semiconductor device, integrated- circuit, and computer technology, digital filtering techniques have made their way to a wide class of practical applications. The idea of digital signal processing is itself not new and dates back to the first attempts to simulate analog systems with the aid of a digital computer, however, only the progress in high-speed logic devices, large-capacity memories and sophisticated digital signal processors (often application- specific) made the applications feasible in terms of reasonable size and cost. There are two more reasons for the ever growing interest in applications of digital signal processing in general and digital filtering in particular. First, use of digital representations in processing and transmission of signals is more efficient in terms of accuracy, noise immunity, stability etc. Second, digital systems can be easily implemented with software on a general purpose computer, therefore they are easy to build, test, modify and adapt.

Some of the important applications of digital filters such as speech and image processing, digital communications including telephony, telemetry and transmission, satellite communication, instrumentation, radar- sonar systems, geoscience, biomedicine, automatic control and robotics, are already included in the standard textbooks on digital signal processing[9,36,54,351,375]. Recent papers[214] and[93] give a good insight into up-to-date applications in instrumentation and medicine.

New applications in consumer electronics such as digital tape- recorders (DAT), compact- disk (CD), digital TV including high-definition television (HDTV)[8,216], video[317], cellular radio-telephony can be spotted in everyday life, in our homes and offices.

With so many applications in mind we can understand why the research in the area of digital filters has been so important during the last decade and why is it a vital problem to understand fully the functioning of digital systems.

9.2   Digital filtering

Digital signal processing and in particular digital filtering are basically transformations of the input data in the form of a sequence of numbers (integer or real valued) into another data set representing sequence of numbers on the system (filter) output. This transformation is carried out following a prescribed rule (algorithm) and implemented either in software or hardware form. The term digital means that all signals are discrete in time and their amplitude is also discretized, i.e.. the amplitude can take only a finite number of so-called representable values within the filter dynamic range, defined as the range of admissible signal amplitudes. When operating in normal conditions it is desired that the transformations of signals be linear. The digital encoding of signals and operations performed implies however that in reality linearity can be achieved only to a certain degree. In the literature the influence of *nonlinearities* introduced in a real world (hardware) implementations of digital systems is referred to as *finite word-length effects*[9,53]. Nonlinearities often cause stability problems, locking into a false operating point and various kinds of oscillatory behavior.

While oscillatory behavior in digital filters has been studied already in the seventies[297,299,87,107,456,457] it is only recently that more complicated behavior has been accounted for[52,75,133,334,396,397]. In typical applications these are unwanted exceptional phenomena. The analysis of these phenomena, however, poses not only a challenge for researchers, but their full understanding could perhaps open new areas for unconventional applications.

9.2.1   What is considered as complex behavior in digital systems?

What are the features which enable us to recognize complex behaviors? As already mentioned in the initial chapters of this book there is no commonly accepted definition for such phenomena. We can classify system orbits following chosen criterion and thus try to understand the meaning of these notions. In the case of discrete-time systems the trajectories can be divided into four classes depending on their asymptotic behavior as k (time) $\to \infty$:

- constant trajectories (limit set consists of a point)

- periodic orbits (limit set consists of a finite number of points visited periodically)

- quasi-periodic orbits (limit set consists of an n-torus or a closed curve in the two- dimensional case)

- recurrent trajectories (limit set consists of a Cantor-like set called a strange attractor)[a]

Following such a classification, the chaotic orbits belong to the last group: they are bounded, unstable and possess a limit set having a Cantor-like structure often called a strange attractor.

The adjective *complex* is used here to describe not only chaotic behavior of the output signals but also the structure of the parameter or initial condition spaces, which even in the case of non-chaotic attractors could composed of intermingled regions of extremely complicated shapes having a fractal dimension.

Complex behavior can be recognized in the system (and in a digital filter in particular) by the presence of one of the following characteristic features:

- **Sensitive dependence on initial conditions**
  Despite the fact that two trajectories start from initial conditions lying very close to each other they will eventually separate as time goes on and will be totally different (but they remain bounded). This property is very important from the practical point of view. The inherent property

---

[a] A Cantor set is a nonempty set having neither internal nor isolated points

of any digital system is that the initial state can not be specified with infinite precision. Such real world imperfections include measurements taken with some error ε, errors in A /D and D /A conversion, actual computations affected by the finite word-length: rounding and truncation errors in operations done in the processor, etc. Consequently, below some fixed level we are not able to distinguish between two different initial conditions.

- Referring to the trajectory classification it is possible to distinguish different types of solutions on the basis of their phase- portraits or time evolutions (e.g. the output signal from the filter).
  (a) to the first category belong all trajectories which converge towards a point on the phase portrait or a constant value in the time evolution;
  (b) second type of trajectories correspond to a finite set of points on the phase portrait;
  (c) phase portraits of the quasiperiodic orbits behave in a "regular" way but fill densely an $(n-1)$ - dimensional torus (eg. a closed curve in the case of a second-order filter);
  (d) solutions that behave in a highly irregular way in the observation time interval, forming apparently self-similar fractal patterns which we define to be "practically chaotic" in this paper. (Thus periodic solutions with periods longer then the available observation time are "practically chaotic"[268]).

- detection of one of the known universal "routes" to chaos when varying system parameters (e.g.existence of a period doubling sequence).

- **Fractal basin boundaries**
  In nonlinear dynamical systems possessing more then one attractor often uncertainty in the initial conditions leads to uncertainty in the final asymptotic state (it should be stressed that this property differs from the sensitivity to initial conditions in which case the asymptotic state was unique - but chaotic). The sets of points belonging to the boundaries between the domains of attraction of each of the attractors have a fractal dimension. In other words in any neighborhood of an arbitrarily chosen initial condition there are always points that will be attracted to a different set then the chosen point. In measurements this means that even though the attractors are simple (asymptotic behavior is not chaotic) the experimental results will appear random due to uncertainty in the initial conditions.

- **Fractal structure of the parameter plane**

  In experiments we sometimes find extremely complicated patterns in the bifurcation parameter space, where regions of existence of trajectories of a distinct type are inter-wound, often forming self-similar structures. In this case, there exist domains in the parameter space where it is no longer possible to specify the type of behavior due to the finite precision in our specification of system parameters.

9.3 Finite Word-length Effects in Digital Filters

In the design of a digital filter the usual requirement is to perform the desired signal transformations in a linear way - most design procedures[9,33,36,352,380] treat the filter as a discrete time linear system and do not take into account the digital encoding until the implementation stage. Despite the fact that deviations from linear behavior in principle can be made arbitrarily small by choosing sufficiently long binary representations, in practice it is necessary to consider the filter as a nonlinear system and account for the effects of nonlinearities introduced by finite word-length in the analysis of its performance. In practice, the nonlinear effects are usually included in the filter model as an independent nonlinear, memoryless (i.e. $f : R \to R$) block inserted into a linear discrete structure. It should be noted, however, that not all finite word-length effects could be treated in such a way[53].

9.3.1 Sources of nonlinearities in digital filters

There are three basic sources of nonlinear effects in a real world digital filter:
(a) finite word-length of the signals being processed,
(b) finite word-length of the filter coefficients,
(c) nonlinear effects introduced by addition and multiplication operations within the filter.

All these categories can be divided into two groups: quantization and overflow, treated in modeling and analysis as decoupled, independent effects. As mentioned above typical filter synthesis procedures are based on the theory of linear discrete systems and no nonlinear effects are considered until the consideration of the actual hardware or software implementation. At the implementation level as a first step one has to choose the representation of numbers in the filter - in typical applications this representation is defined by the properties of the analog-to-digital converter at the input of the filter and choice of registers within the filter. Once the representation is assigned, the set of representable numbers is fixed. At the second step one has to define the arithmetic operations. There are two basic choices of representations:

(a). fixed point arithmetic, defined by:

- dynamic range and quantization step,

- number of bits,

- type of representation of negative numbers (signed magnitude [b] one's complement[c] or two's complement[d]),

- quantization rule,

- overflow scheme.

(b). floating point arithmetic, defined by:

- number of bits of exponents,

- number of bits of mantissa,

- representation of negative numbers (mantissa and exponents),

- quantization rule,

- overflow scheme.

---

[b]In the signed magnitude arithmetic a binary coded fractional number $N = \pm\Sigma^0_{i=-m}b_i2^i = \pm0.b_{-1}b_{-2}...b_{-m}$, is represented as:

$$N_0 = \begin{cases} 0.b_{-1}b_{-2}...b_{-m} & \text{for } N \geq 0 \\ 1.b_{-1}b_{-2}...b_{-m} & \text{for } N \leq 0 \end{cases}$$

i.e. the first bit (sign bit) is equal 0 for positive numbers, and is set to 1 for negative numbers

[c]In the 1's complement arithmetic a binary- coded fractional number $N = \pm\Sigma^0_{i=-m}b_i2^i = \pm0.b_{-1}b_{-2}...b_{-m}$, is represented as

$$N_1 = \begin{cases} N & \text{for } N \geq 0 \\ 2 - 2^{-L} - |N| & \text{for } N \leq 0 \end{cases}$$

where L is the number of bit locations in the register to the right of the binary point. The 1's complement of a negative number can be found by representing the number by $L+1$ bits and then changing 0s into 1s and 1s into 0s.

[d]In the 2's complement arithmetic a binary- coded fractional number $N = \pm\Sigma^0_{i=-m}b_i2^i = \pm0.b_{-1}b_{-2}...b_{-m}$, is represented as

$$N_1 = \begin{cases} N & \text{for } N \geq 0 \\ 2 - |N| & \text{for } N \leq 0 \end{cases}$$

9.3.2 Quantization rules

Since the set of representable values is finite due to finite register (word) length it is necessary to give the rules for treating the signals which Three types of quantization schemes are currently used in digital signal processing:
(a) rounding,
(b) magnitude truncation,
(c) value truncation.
In the case (a) the numbers are rounded to the nearest machine- representable number (e.g. in the case of fixed point arithmetic the function F_Q assigns to a real number x the product of an integer k times the quantization step q:
$F_Q(x) = kq$ when $(k - \frac{1}{2})q \leq x < (k + \frac{1}{2})q)$ while in cases (b) and (c) all bits that cannot be accommodated in the register are truncated (i.e. in the case of the fixed- point representation in the case (b)

$$F_Q(x) = \begin{cases} kq & \text{for } x > 0 \text{ and } (k-1)q \leq x < (k+1)q \\ -kq & \text{for } x \leq 0 \text{ and } (k-1)q \leq |x| < (k+1)q \end{cases}$$ and in the case

(c) $F_Q(x) = kq$ when $(k - 1)q \leq x < (k + 1)q)$. Due to the application of the quantization rules the characteristics F_Q of the filter become *multi-step-functions* with the number of steps and their width depending on the word-length (number of bits), dynamic range and choice of fixed- or floating- point arithmetic.

9.3.3 Overflow schemes

Since the addition and multiplication operations in general can produce over-flows i.e. results of these operations can fall outside of the dynamic range of the filter, there are three three common types of actions that could be taken to set the filter back to normal operation. There are three typical types of overflow characteristics F_O:
(a) saturation,
(b) zeroing,
(c) two's complement (modulo 2).
These characteristics are defined in Fig.9.1.

Finally the resulting characteristic of the filter is a combination of both the quantization and the overflow nonlinear characteristics, and its specific properties are defined by filter implementation specifications:

$$F(\sigma) = F_O[F_Q(\sigma)] \tag{9.1}$$

Nonlinear characteristics introduced cause in general two types of parasitic dynamic behaviors: quantization (small amplitude) oscillations and overflow

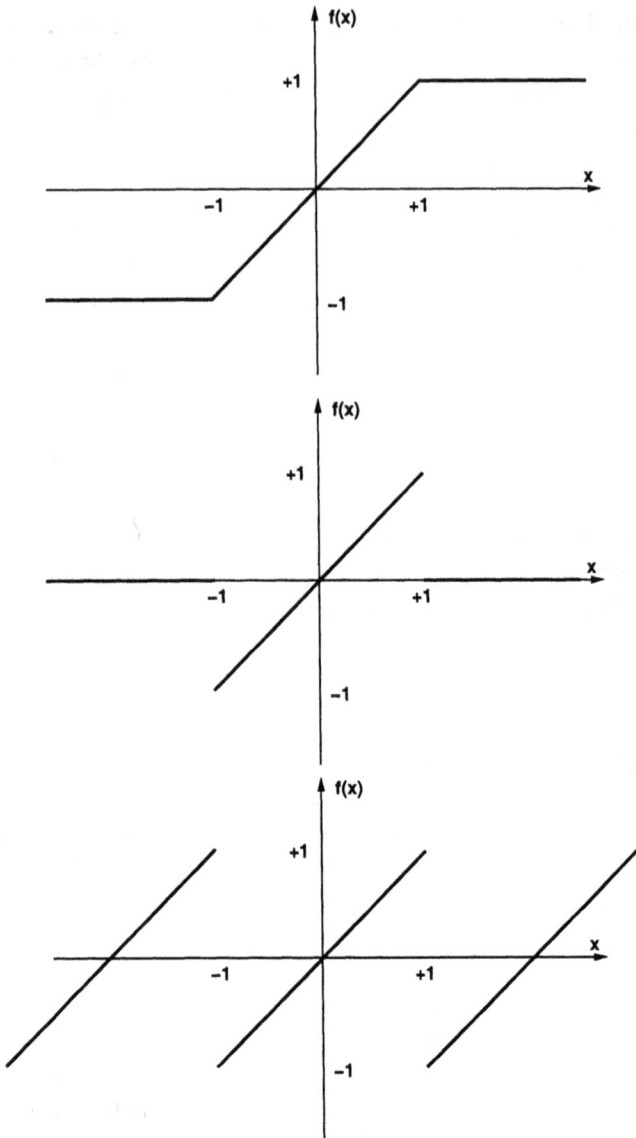

Figure 9.1: Typical characteristics F_O used for overflow correction (a). saturation; (b). zeroing; (c). two's complement (modulo 2)..

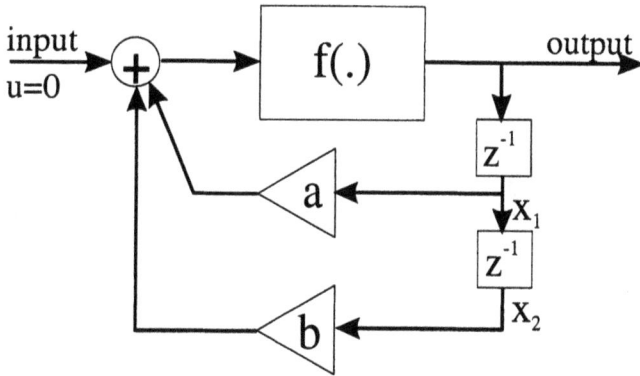

Figure 9.2: Considered direct form realization of digital filters.

(large amplitude) oscillations. In this paper we will consider the second type of behavior only. In particular we will consider second- and third-order filter structures realized in the *direct form* [e]shown in Fig.9.2 whose dynamics can be described by a state equation of the form (discrete map):

$$x(k+1) = f[x(k), u(k)] \qquad (9.2)$$

with $x(k) \in R^2$ or R^3, $u(k) \in R$. The nonlinear function f takes into account the overflow only. Considering linear operation of the filter only (no overflow nonlinearity) we can easily establish stability regions in the parameter space corresponding to various combinations of eigenvalues. The partition of the parameter space is shown in Fig.9.3.

9.4 Chaotic Behavior in Digital Filter Sections With 2's Complement Arithmetic

The simplest digital filter structure producing apparently aperiodic solutions under zero input conditions has been described and analyzed by L.O.Chua and his collaborators (see eg.[75,77,78,268]) and several other authors[35,94,95]. It was a second order digital filter realized in the direct form using a 2's complement arithmetic for the overflow rule (modulo 2 nonlinear characteristic), and operating at the boundary of the region of asymptotic stability. Its dynamics is

---

[e]By *direct form*, we mean a structure (realization) of the filter corresponding directly to the state equations (or transfer function). In contrast, a non-direct (e.g. cascade, parallel, ladder, wave etc.) form means that for such a realization some kind of factorization, expansion into partial fractions, or some other transformation of the equations (transfer function) is needed.

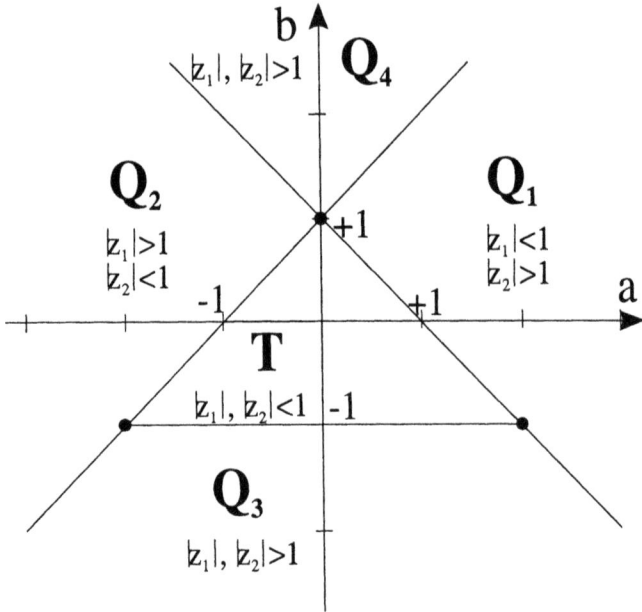

Figure 9.3: Partition of the parameter plane into sets T, Q_1, Q_2, Q_3, Q_4, for $(a, b) \in Q_1 \cup Q_2$ one eigenvalue lies inside and one outside the unit circle, for $(a, b) \in Q_3 \cup Q_4$ both eigenvalues lie outside the unit circle, for $(a, b) \in T$ both eigenvalues lie inside the unit circle and the linear system is asymptotically stable.

governed by the following second- order system of discrete time equations:

$$\begin{bmatrix} x_1(k+1) \\ x_2(k+1) \end{bmatrix} = \begin{bmatrix} x_2(k) \\ F[-x_2(k) + ax_1(k)] \end{bmatrix} = f[x(k)] \qquad (9.3)$$

with a modulo (2's complement) characteristic for the function F (Fig.9.1c). The system (9.3) can be considered as a map $f : I^2 \rightarrow I^2$ (where $I^2 = \{(x_1, x_2) : -1 \leq x_1 < 1, -1 \leq x_2 < 1\}$). It is possible to show that the map f is one-to-one and onto (see[75]). Let $f^k(x)$ denote the k-th iterate from the initial point x. The trajectory of system (9.3) starting from $x(0) = x$ is defined as:

$$\Gamma(x) = \{x(k) = f^k(x), k \geq 0\} \qquad (9.4)$$

9.4.1 Simulation results - self similar patterns of trajectories

Three types of trajectories have been discovered in the system (9.3) during simulation experiments:
(a) Trajectories filling an ellipse. This is the case for all

$$x(0) \in \Omega_L = \{(x_1(0), x_2(0)) : [\frac{[x_1(0) + x_2(0)]^2}{2 + a} + \frac{[x_1(0) - x_2(0)]^2}{2 - a}]^{\frac{1}{2}} < 1\}$$

(9.5)

when there is no overflow and system behaves in a linear way.
(b) Trajectories which jump among finite number of ellipses,
(c) Trajectories evolving in an apparently random way among an infinite number of ellipsis but creating fractal patterns. A typical trajectory of this kind is shown in Fig.9.4. One can clearly recognize the existence of a large central elliptical region, two smaller ellipsis positioned symmetrically, further six, ten, eighteen, ... etc. smaller ellipses. There are more and more ellipsis repeatedly generated when the number of iterates becomes larger and larger. An interesting fact is that for the same value of $a = 0.5$ the trajectory pattern depends on the initial condition chosen. For example the initial condition taken from inside the blank ellipses produce eventually periodic trajectories traveling among a finite number of ellipses of the same size. There exist however regions in the initial condition space which generate apparently aperiodic trajectories visiting different types of ellipses (Fig.9.4) and in the limiting case, defined by $k \to \infty$, an infinite number of them. Magnification of portions of such a trajectory reveals self-similar patterns.

9.4.2 Analysis via symbolic dynamics

Let us introduce the matrices:

$$A = \begin{bmatrix} 0 & 1 \\ -1 & a \end{bmatrix}, b = \begin{bmatrix} 0 \\ 2 \end{bmatrix}, \overline{A} = A^{-1} = \begin{bmatrix} a & -1 \\ 1 & 0 \end{bmatrix}, \overline{b} = \begin{bmatrix} 2 \\ 0 \end{bmatrix}$$

(9.6)

It has been shown in[75] that:

$$f(x) = Ax + bs$$

(9.7)

where

$$s_k = \begin{cases} -1, & \text{if } -x_1(k) + ax_2(k) \geq 1 \\ 1, & \text{if } -x_1(k) + ax_2(k) < -1 \\ 0, & \text{otherwise} \end{cases}$$

(9.8)

and the $k - th$ iterate can be expressed as:

$$f^k(x) = Af^{k-1}(x) + bs_{k-1}$$

(9.9)

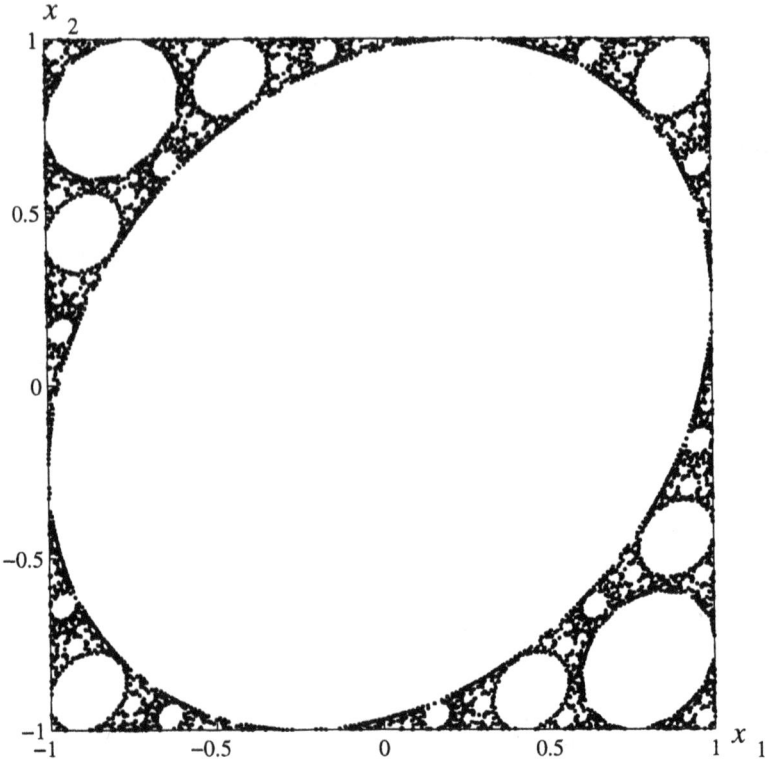

Figure 9.4: Self-similar pattern generated by a trajectory of the second-order digital filter with 2'complement arithmetic for $a = 0.5$, $b = -1$ and $x(0) = [0.61356, -0.61356]^T$

Thus we can define a map

$$S : I^2 \to \Sigma = \{\bar{s} = (s_0 s_1 s_2...) : s_k = -1, 0, 1, k = 0, 1, 2, ...\} \qquad (9.10)$$

with s_k defined by (9.9) and characterize the system behavior in terms of symbol sequences $\bar{s}$; namely the s_k tells us in which of the three regions:

$$\begin{align}
I_{-1} &= (x_1, x_2) : -x_1 + ax_2 \geq 1 \qquad (9.11)\\
I_1 &= (x_1, x_2) : -x_1 + ax_2 < -1 \text{ or} \qquad (9.12)\\
I_0 &= I^2 \setminus (I_{-1} \cup I_1) \qquad (9.13)
\end{align}$$

the $k - th$ iterate $f^k(x)$ lies. Having defined Σ_f as $\Sigma_f = S(I^2)$, we say that the sequence $\bar{s}$ is admissible if $\bar{s} \in \Sigma_f$. A number of theorems are available for checking whether a given sequence $\bar{s}$ is admissible or not[75,78,134]. In[134] we found the following admissible sequences:

$k = 1$ 0
$k = 2$ 1-1
$k = 3$ 00-1 and
            001
$k = 10$ 0000-100001
$k = 18$ 00-100-11-110010011-11-1
$k = 27$ 00-100-100-100-100-11-11001001-11-1 and
            001001001001001-11-100-100-11-11
$k = 38$ 00-1000010000-100-11-110010000-100001001-11-1
$k = 42$ 00-1000010000-1000010000-1000010000-100-11-11-11-1 and
            0010000-1000010000-1000010000-100001001-11-11-11
$k = 64$ 00-100-100-100-100-11-110010000-100001001-11-1
            00-100-100-100-100-11-11001001-11-1 and
            001001001001-11-100-1000010000-100-11-11
            001001001001-11-100-100-11-11
$k = 76$ 00-10000000001001-11-100-1000010000-100-11-11
            001000000000-100-11-110010000-100001001-11-1
$k = 134$ 00-10000000001001-11-100-10000000001001-11-100-1
            000010000-1000010000-100-11-11
            001000000000-100-11-11001000000000-100-11-11001
            0000-1000010000-100001001-11-1

To understand the mechanisms of the filter behavior let us introduce next the following partition of the set Σ[75]:

(a) $\Sigma_\alpha = \{\bar{s} = (s_0 s_1 s_2...) \quad : \quad \bar{s}$ is periodic, i.e.

$$s_k = s_{k+K} \text{ for some } K \text{ and } \forall k \geq 0\} \quad (9.14)$$

(b) $\Sigma_\beta = \{\bar{s} = (s_0 s_1 s_2...) \quad : \quad \bar{s}$ is eventually periodic, i.e. $\bar{s} \notin \Sigma_\alpha,$

$$\text{but } \sigma^p(\bar{s}) \in \Sigma_\alpha \text{ for some } p > 0\} \qquad (9.15)$$

where σ^p denotes a p-shift of the symbol sequence $(\sigma((s_0 s_1 s_2...)) = (s_1 s_2...)))$.

$$\text{(c) } \Sigma_\gamma = \Sigma \setminus (\Sigma_\alpha \cup \Sigma_\beta) \qquad (9.16)$$

Corresponding to such a partition I^2 can also be divided into three subsets:

$$I_\alpha = S^{-1}(\Sigma_\alpha \cap \Sigma_f) \qquad (9.17)$$

$$I_\beta = S^{-1}(\Sigma_\beta \cap \Sigma_f) \qquad (9.18)$$

$$I_\gamma = S^{-1}(\Sigma_\gamma \cap \Sigma_f) \qquad (9.19)$$

Properties of these sets were studied in[75] and can be summarized as follows:

Theorem 9.1 I_α *is the union of elliptical regions and can be expressed as:*

$$I_\alpha = \bigcup_{s_{0,K-1} \in \Sigma_\alpha \cap \Sigma_f} \Pi(\rho_m, z_0) - M \qquad (9.20)$$

where: $s_{0,K-1}$ *denotes a K-periodic sequence,* $\Pi(\rho_m, z_0)$ *denotes an ellipse centered at z_0, which is defined by the solution of the equations:*

$$z_1 \quad = \quad Az_0 + bs_0 \qquad (9.21)$$

$$\vdots \qquad (9.22)$$

$$z_{K-1} \quad = \quad Az_{K-2} + bs_{K-2} \qquad (9.23)$$

$$z_0 \quad = \quad Az_{K-1} + bs_{K-1} \qquad (9.24)$$

$$(9.25)$$

with the maximum diameter ρ_m among all ρ such that:

$$\Pi(\rho, z_i) \subseteq \bar{I}^2 \ , \ i = 0, 1, 2...K-1 \qquad (9.26)$$

and

$$M = \bigcup_{s_{0,K-1} \in \Sigma_\alpha \cap \Sigma_f} M_{s_{0,K-1}} \qquad (9.27)$$

where $M_{s_{0,K-1}} = \{x \in I^2 : G_{l \cdot K}(x) = x_j, l = 0, 1, 2..., j = 1, 2...J\}$ *are dense countable subsets of $45°$ inclined ellipses $\Omega(\rho_m, z_0)$ (Here, $G_0(x) = x$, $G_{k+1}(x) = AG_k(x) + bs_k$, $k = 0, 1, ...$ for any $x \in \Pi(\rho_m, z_0)$ and $x_j, j = 1, ...J$ denote the tangent points of Π in ∂I_1).*

Theorem 9.2 *I_β consists of all pre-images of points y which are on the boundary of the elliptical region $\Pi(\rho_m, z_0)$ and satisfy:*

$$I_\beta = \bigcup_{s_{0,K-1} \in \Sigma_\alpha \cap \Sigma_f} N_{s_{0,K-1}} \qquad (9.28)$$

with

$$N_{s_{0,K-1}} = \bigcup_{p=1}^{\infty} f^{-p}[\{y \in I^2 : G_{-1} \in \Omega(\rho_m, z_{K-1}) \cap \partial I_1 \qquad (9.29)$$

where $G_{-1}(y) = \overline{A}y + \overline{b}s_{K-1}$.

9.4.3 Properties of the set I_γ

The set I_γ can be obtained by successive removal of the elliptical regions $I_{s_{0,K-1}}$ and the sets $N_{s_{0,K-1}}$ following the admissible periodic sequences found before. It is conjectured that the trajectories starting from I_γ are visiting all of the boundaries of the ellipses thus creating chaotic dynamics and forming fractal patterns. The dynamic behavior of these trajectories is, however, still not fully understood and requires further analysis.

9.4.4 Analysis using modern mathematical tools

Measure-theoretic theory

In this section we shall consider the probabilistic (measure-theoretic) approach in the analysis of chaotic system.
An important concept in studying measure preserving dynamical systems is mixing property. Mixing means that a set of initial conditions of nonzero measure will eventually spread over the whole phase space as the system evolves.

Definition 9.1 *Let (M, Ω, μ) be a normalized measure space, and $G : M \to M$ a measure-preserving transformation $(\mu(G^{-1}(A)) = \mu(A)$ for all $A \in \Omega)$. G is called mixing if*

$$\lim_{n \to \infty} \mu(A \cap G^{-n}(B)) = \mu(A)\mu(B) \quad \forall A, B \in \Omega$$

Another concept is exactness. For exact systems a set of initial conditions of nonzero measure will eventually fill the whole phase space.

(a)

a=4.   b=−1.

(b)

a=3.   b=−1.1

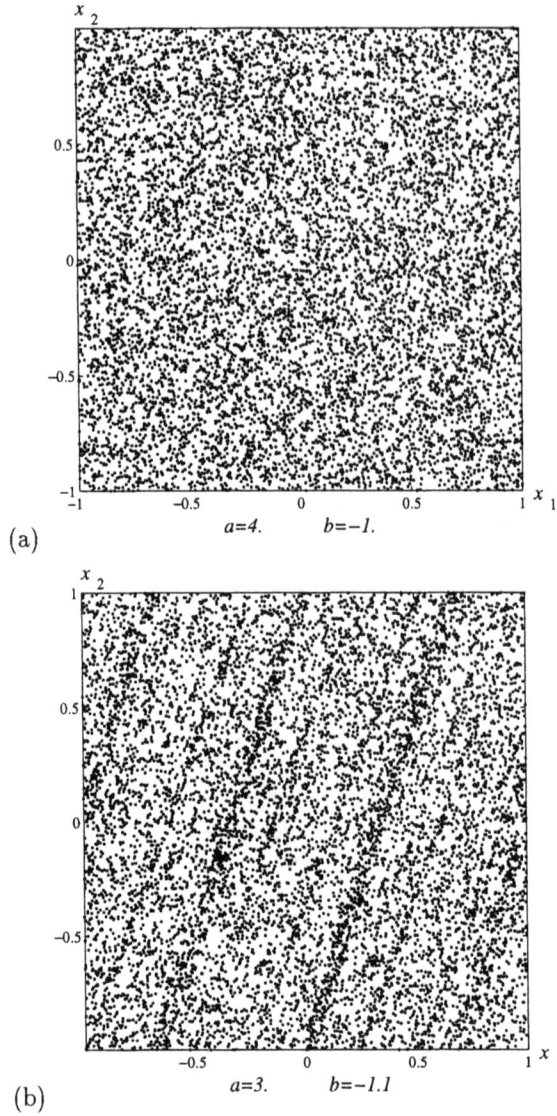

Figure 9.5: The trajectory of 10000 points for (a) $a = 4, b = -1$, (b) $a = 3, b = -1.1$

Definition 9.2 *Let (M, Ω, μ) be a normalized measure space, and $G : M \to M$ a measure-preserving transformation such that for all $A \in \Omega$, $G(A) \in \Omega$. G is called exact if*

$$\lim_{n \to \infty} \mu(G^n(A)) = 1 \quad \text{for every } A \in \Omega, \mu(A) > 0$$

We apply the measure-theoretic theory to the considered digital filter. The measure we will use for the map **F** will be the Borel measure.

Lemma 9.3 *Let $b \neq 0$ be an integer. Then the map **F** is measure-preserving.*

Theorem 9.4 *If $b = -1$, $a > 2$ and a is an integer, then **F** is mixing.*

Theorem 9.5 *If a, b are integers, such that $b \neq a+1$, $b \neq -a+1$ and $b \neq 0, \pm 1$, then the map **F** is exact.*

It can be proved that exactness implies mixing. The converse is not necessarily true; the mixing map **F** for $b = -1$ and a an integer larger than 2, is not an exact map.

Fig. 9.5 shows four examples of complex trajectories. Of particular interest is Fig. 9.5(a), which shows uniform distribution of points in the state space – such a filter structure could be possibly used as a noise generator. In Fig. 9.5(b) the uniform distribution is no longer valid. One could see stripes that are more frequently visited. Fig. 9.6(a) and (b) show examples when the trajectory visits fragments of the state space. Thus one can see that changing parameters we can influence the statistical properties of the generated sequences.

9.4.5 Higher-order filters

In this section we will show some simulation results confirming the existence of chaotic trajectories also in a third-order digital filter with 2's complement arithmetic. Chaos and fractals in this kind of filters (with zero-input and no quantization effects) were first reported in[77]. Dynamic behavior of the direct-form realization of such a filter is governed by a system of non-linear discrete equations:

$$\begin{bmatrix} x_1(k+1) \\ x_2(k+1) \\ x_3(k+1) \end{bmatrix} = \begin{bmatrix} x_2(k) \\ x_3(k) \\ F[cx_1(k) + bx_2(k) + ax_3(k)] \end{bmatrix} + \begin{bmatrix} 0 \\ 0 \\ 1 \end{bmatrix} u(k) \quad (9.30)$$

where $F(.)$ - denotes a 2's complement overflow nonlinear characteristic.

Below we show typical results of experiments carried out with this system in the case of fixed coefficient values $a = 1.5$, $b = -1.5$, $c = 1$ and eight different

(a) $a=2.$ $b=-1.5$

(b) $a=0.4$ $b=-1.05$

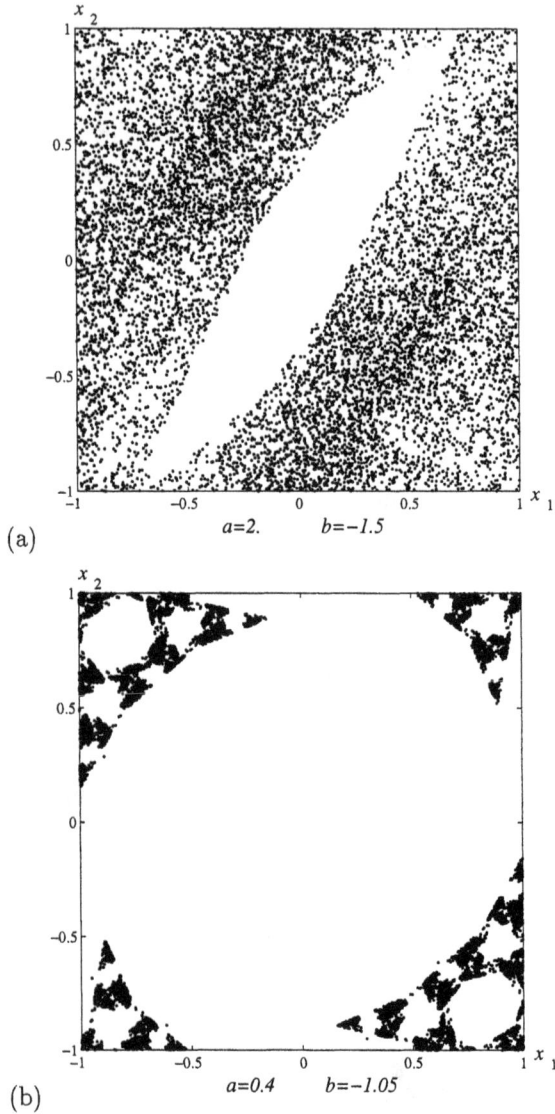

Figure 9.6: The trajectory of 10000 points for (a) $a = 2, b = -1.5$, (b) $a = 0.4, b = -1.05$ — typical behavior of mixing maps.

initial conditions:

$$
\begin{aligned}
\text{trajectory 1 } x(0) &= [0.2, 0.7, -0.94]^T \\
\text{trajectory 2 } x(0) &= [0.8, -0.4, -0.2031]^T \\
\text{trajectory 3 } x(0) &= [0.5, -0.6, 0.6548]^T \\
\text{trajectory 4 } x(0) &= [-0.9, 0.4, -0.4837]^T \\
\text{trajectory 5 } x(0) &= [0.2, 0.76, -0.9999]^T \\
\text{trajectory 6 } x(0) &= [0.5455, -0.1, -0.99]^T \\
\text{trajectory 7 } x(0) &= [0.5, -0.99, 0.15]^T \\
\text{trajectory 8 } x(0) &= [0.2, -0.356, 0.155]^T
\end{aligned}
$$

Figures 9.7-9.14 show the two-dimensional projections on the $x_1 - x_2$ plane of trajectories observed in the third-order filter in these four cases. All trajectories show a very complex geometrical structure of various fractal patterns. Behavior of this three-dimensional system could be analyzed in a similar way as described for the two-dimensional case but the symbolic characterization involves now seven symbols[77]. Use of symbolic characterization enables us also in this case to uncover the complex structure of the invariant set of filter trajectories. It is conjectured that the trajectories visit an infinite sequence of boundaries of elliptical regions contained on several planar sheets in the three- dimensional space[77], thereby generating chaotic dynamics and fractal patterns.

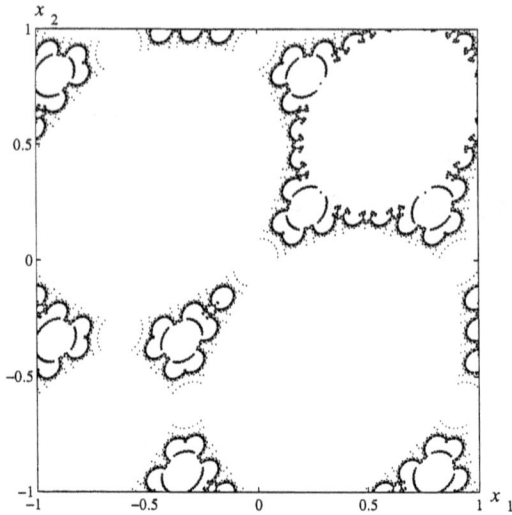

Figure 9.7: Trajectory of the third-order digital filter observed for $a = 0.5$, $b = -1$, $c = 1$ and $x(0) = [0.2, 0.7, -0.94]^T$.

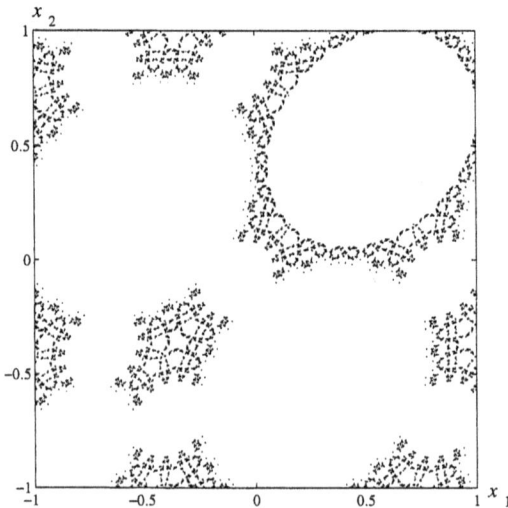

Figure 9.8: Trajectory of the third-order digital filter observed for $a = 0.5$, $b = -1$, $c = 1$ and $x(0) = [0.8, -0.4, -0.2031]^T$.

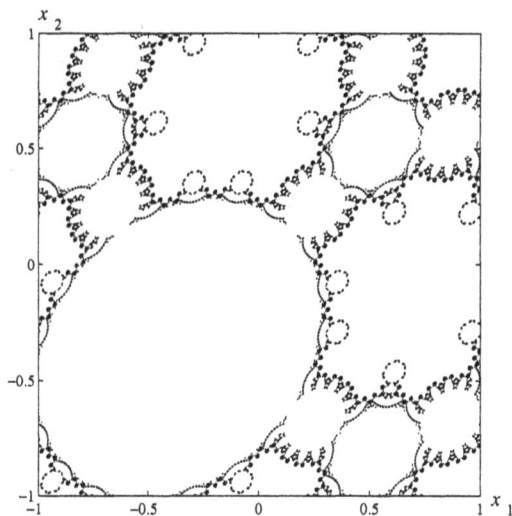

Figure 9.9: Trajectory of the third-order digital filter observed for $a = 0.5$, $b = -1$, $c = 1$ and $x(0) = [0.5, -0.6, 0.6548]^T$.

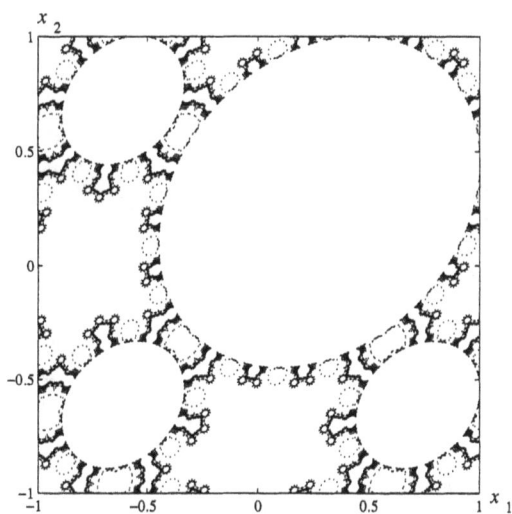

Figure 9.10: Trajectory of the third-order digital filter observed for $a = 0.5$, $b = -1$, $c = 1$ and $x(0) = [-0.9, 0.4, -0.4837]^T$.

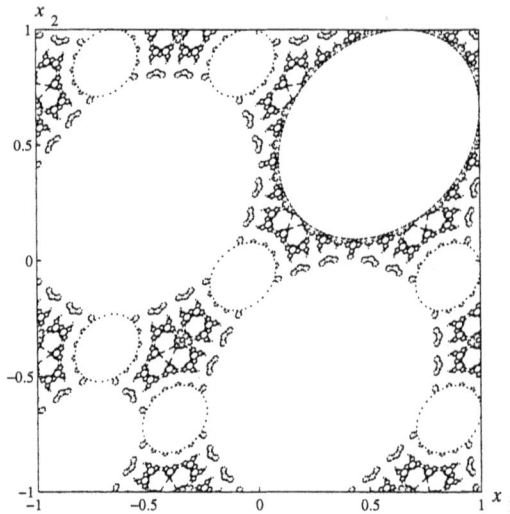

Figure 9.11: Trajectory of the third-order digital filter observed for $a = 0.5$, $b = -1$, $c = 1$ and $x(0) = [0.2, 0.76, -0.9999]^T$.

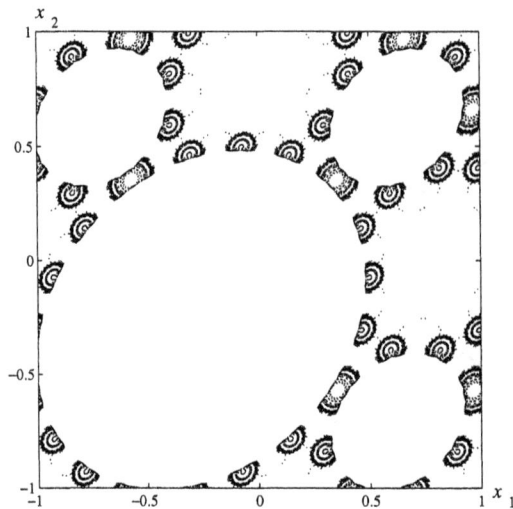

Figure 9.12: Trajectory of the third-order digital filter observed for $a = 0.5$, $b = -1$, $c = 1$ and $x(0) = [0.5455, -0.1, -0.99]^T$.

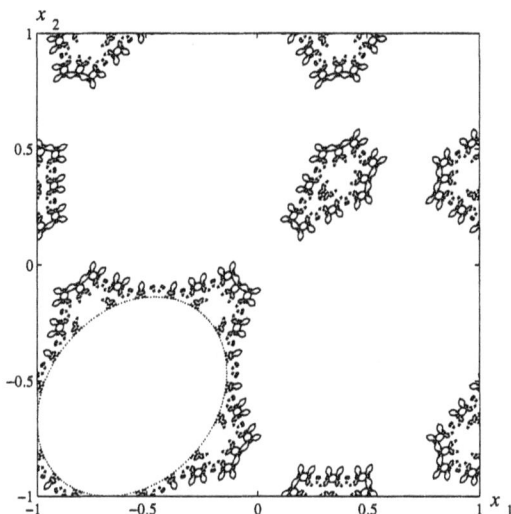

Figure 9.13: Trajectory of the third-order digital filter observed for $a = 0.5$, $b = -1$, $c = 1$ and $x(0) = [0.5, -0.99, 0.15]^T$.

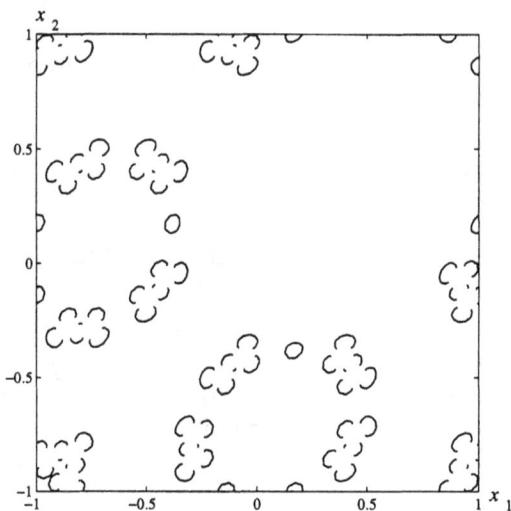

Figure 9.14: Trajectory of the third-order digital filter observed for $a = 0.5$, $b = -1$, $c = 1$ and $x(0) = [0.2, -0.356, 0.155]^T$.

9.5　Final State Sensitivity in Digital Filters with Saturation Arithmetic

Different type of dynamic behavior has been found in the case of a second-order filter with a *saturation - type* adder overflow characteristic. The dynamics of this system (with zero input) are described by equations of the form:

$$\left[\begin{array}{c} x_1(k+1) \\ x_2(k+1) \end{array} \right] = \left[\begin{array}{c} x_2(k) \\ F[bx_1(k) + ax_2(k)] \end{array} \right] = f[x(k)] \qquad (9.31)$$

where : $a, b \in R$, $F : R \to R$, $F(\sigma) = \sigma$ for $|\sigma| < 1$, $F(\sigma) = 1$ for $\sigma \geq 1$, $F(\sigma) = -1$ for $\sigma \leq -1$ as shown in Fig.9.1(a).

Dynamic behavior and in particular self-sustained overflow oscillations in this class of systems have been studied by a number of authors (see eg.[298] and references therein).

A result of particular interest obtained via the Lyapunov function technique can be found in[456] who has shown that the nonlinear system under consideration is asymptotically stable for all parameters a and b within the linear stability triangle $(F(\sigma) = \sigma)$ ie. $b > -1$, $b < a + 1$ and $b < -a + 1$ (this triangle is indicated in light gray in Plate 13.

During simulation experiments in this system we did not find any trajectories forming fractal patterns as it was the case in the filter with 2's complement characteristic. The dynamics of this systems appears, however, to be extremely reach and interesting types of complex behavior have been observed.

9.5.1　*Parametric analysis of system's behavior: Arnold tongues and devil's staircase*

Using the recently developed software package KRAKFIL[135] we constructed a two parameter bifurcation diagram for the above filter with a saturation arithmetic. Plate 13 shows color-coded pictures of regions of existence of orbits of different rotation numbers[274] on the $a - b$ parameter plane. The portrait of the parameter space shown in the figure (a) has been constructed in the case when we limited the maximum number of iterates for each set of parameters a, b to 2000. Outside the linear light gray stability triangle one can clearly see the Arnold tongue structure - the principal "tongues" 1/2 and 0/1 being the half- planes $b > a + 1$ and $b > -a + 1$ respectively. The quadrant delimited by $b < a - 1$ and $b < -a - 1$ constitutes the 1/4 tongue. The 1/3 tongue has the borders:

$$b = \frac{a^2 + 1}{a - 1} \text{ and}$$

$$a = \frac{b^2 + 1}{b - 1}$$

which asymptotically approach, respectively, the borders of the 1/2 and the 1/4 tongue. In a similar way the curves delimiting the 1/6 tongue:

$$b = -\frac{a^2 + 1}{a + 1} \text{ and}$$

$$a = -\frac{b^2 + 1}{b + 1}$$

approach asymptotically the lines delimiting the 0/1 and 1/4 region, respectively. It is easy to notice the existence of an infinite number of (shrinking by 1/2) half- circular regions (see Plate 14) in which we were not able to calculate the rotation numbers - there were no periodic orbits of period less then 2000. Note that all tongues visualized in Plate 14 have their origin at these half- circle lines. We were able to find the equations defining the largest two circular regions:

$$b(1 + b) = -a(1 + a) \text{ and}$$
$$b(1 + b) = a(1 - a)$$

Numerical experiments are presently in progress to find out what kind of behavior could be observed when the parameters a and b are chosen within the half-circular regions. The first figure in Plate 14 shows in more detail the structure of the half- circular regions and the second one shows the detailed structure of the largest circular region $\Omega = \{(a, b) : b(1 + b) > a(1 - a), b < -1, a > 0\}$ discovered in the earlier experiment. This time we continued the experiment for 50000 iterates for each chosen a and b. In these cases we again discovered regions of existence of periodic orbits. However, the convergence of the trajectories is very weak. Further extension of the observation interval above 50000 iterates reveals even finer structures of the Arnold's tongues. One can easily notice the existence of so- called "sausage structures" and apparently an infinite number of curves where the successive "sausages" representing solutions of the same rotation number touch each- other. The exhibited pattern has apparently a self- similar infinite structure, repeated in every half- circular region.

The structure is so fine that below a given accuracy level ε for the filter parameters it is not possible to specify what will be the observed system behavior. The system displays final state sensitivity, which we define as the sensitive dependence of the observed final state on the variations of the filter coefficients.

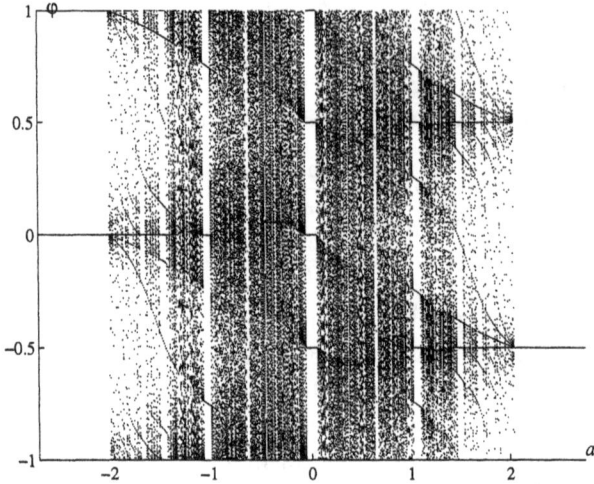

Figure 9.15: Typical bifurcation diagram obtained for varying a (for fixed $b = -1.05$) showing an abundance of various regimes when varying the bifurcation parameter.

Using our software package KRAKFIL[135] we carried out also one-parameter bifurcation analysis. The Arnold tongue diagram in fact could be constructed with finite (low) precision and we were not able to find any confirmation of the existence of aperiodic (chaotic) orbits. In some cases we continued the experiments up to 200000 iterates still not finding any periodicity in the observed trajectory.

To have some more insight into the underlying mechanisms of system behavior we constructed some one-parameter bifurcation diagrams[133]. These reveal interesting bifurcation patterns and confirm the coexistence of more than one periodic orbits (of equal period) in some regions of the $a - b$ plane[339]. We conjecture that there is always a single stable - unstable pair of orbits of (the same) even period, while the orbits of odd period always constitute stable - unstable twins (i.e. depending on the initial condition chosen we observe in our experiments one of the two stable orbits of odd period, existing in the system). The bifurcation phenomena of the periodic orbits in this system are still not fully understood and require further study.

9.5.2 *Circle-map theory and its applications*

Let us recall some important definitions and theorems from the circle-map theory.

Let $f : S^1 \longmapsto S^1$ be a continuous map of S^1 to itself. The map $\Pi :$ $\mathbb{R} \longmapsto S^1$ defined by $\Pi(t) = (cos(2\pi t), sin(2\pi t))$ is continuous and onto. Let $F : \mathbb{R} \longmapsto: \mathbb{R}$ be a *lift* of f, i.e. F is continuous, $\Pi \circ F = f \circ \Pi$, and for each $x \in \mathbb{R}$, $F(x + 1) = F(x) + k$, where k is an integer constant. The integer k is unique for a given continuous map f and is called *the degree of f*. In this paper we are interested in degree-one maps only. An important concept in the study of degree-one maps is the rotation number. Let f be a degree-one map, and let F be a lift of f. If $x \in \mathbb{R}$, then the *rotation number* of x under F is defined by:

$$\rho_F(x) = \lim_{n \to \infty} \frac{F^n(x) - x}{n}. \tag{9.32}$$

We say that a map of a circle is *non-decreasing* if its lift is non-decreasing.
Proposition 1 ([428]) *If f is a non-decreasing degree-one map, F is a lift of f, then $\rho_F(x)$ exists for every $x \in \mathbb{R}$ and does not depend on initial point x. $\rho_F(x)$ is rational iff f has a periodic point.*
Thus for non-decreasing degree-one maps we can define the rotation number in the following way:

$$\rho_f = \rho_F(x) \; (mod \; 1),$$

where x is an arbitrary real value and F is an arbitrary lift of f.

We say that a point x is *non-wandering* if $\forall U$-neighborhood of x $\exists n > 0 : f^n(U) \cap U \neq \emptyset$. We say that a point x is *chain recurrent* if $\forall \varepsilon > 0$ $\exists n > 0$ $\exists x_0, \ldots, x_n : d(f(x_i), x_{i+1}) < \varepsilon$ and $x = x_0 = x_n$. Using the concept of non-wandering and chain-recurrent points we were able to prove the following two lemmas.

Lemma 9.6 *If f is a non-decreasing degree-one map and f is not injective, then the rotation number of f is rational.*

Lemma 9.7 *If f is a non-decreasing degree-one circle map and the rotation number of f is rational then for every $x \in S^1$ the limit set $\omega(x)$ is periodic.*

We consider the case $(a, b) \in Q_3 = \{(a, b) : b < -1, b < a + 1, b < -a + 1\}$. For other cases the behavior of the filter is well-known, the only possible limit sets are period-1 and period-2 orbits[138]. In[138] it was shown that for $(a, b) \in Q_3$ there exists an absolutely convex polygon with invariant boundary $\mathbf{W}^\infty$. It was proved that every non-trivial trajectory in finite time enters the set $\mathbf{W}^\infty$ and remains in it. Thus we can reduce our study to the analysis of one-dimensional map of $\mathbf{W}^\infty$ into itself. As $\mathbf{W}^\infty$ is homeomorphic to a circle we can define

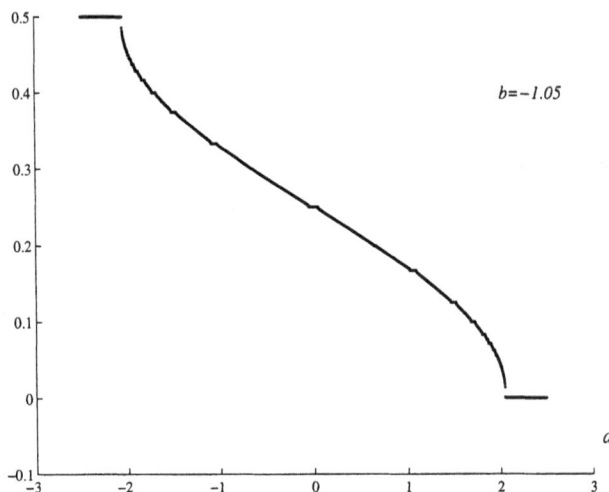

Figure 9.16: The rotation number as a function of parameter a.

the rotation number of $\Phi := \mathbf{F}|\mathbf{W}^\infty : \mathbf{W}^\infty \longmapsto \mathbf{W}^\infty$. The map Φ is weakly monotone which implies (Proposition 1) the existence of a unique rotation number for each pair (a, b), denoted by $\rho_{\mathbf{F}}$.

Theorem 9.8 *If* $(a, b) \in Q_3$, $x \neq O$, ρ *is the rotation number of* Φ, *then*

1. *If* Φ *is not a homeomorphism then* ρ *is rational.*

2. *If* ρ *is rational* ($\rho = p/q$) *then the limit set of* x *is a period-q orbit contained in* $\mathbf{W}^\infty$.

3. *If* ρ *is irrational then the limit set of* x *is dense in* $\mathbf{W}^\infty$.

From the above theorem it follows that no chaotic behavior is possible when we use the saturation nonlinearity for the overflow rule. In the region Q_3 the behavior of the filter strongly depends on the circuit parameter values. A very small change of filter parameters can cause the qualitative change of circuit behavior (change of the period of the periodic orbit or change of the type of the orbit, from periodic to the aperiodic one).

Rotation number is a monotonic continuous function with plateaus of finite width at every rational value[23]. Zooming-in this picture shows repeating finer structures of this kind (self- similarity). Changes of rotation numbers obey the

Farey's rule: between two cycles of rotation numbers $\frac{p}{q}$ and $\frac{r}{s}$ there is always an orbit with rotation number $\frac{p+r}{q+s}$.

Now we will discuss the structure of the Arnold tongues.

Definition 9.3 *We say that point (a, b) belongs to the ω-Arnold tongue denoted by A_ω if there exists an orbit with rotation number $\omega \in \mathbb{R}$ for $\mathbf{F}$ with parameters (a, b):*

$$A_\omega = \{(a, b) : \exists x \neq O : \rho_{\mathbf{F}}(x) = \omega\}. \tag{9.33}$$

From Proposition 1 it follows that $(a, b) \in A_{p/q}$ implies the existence of a periodic orbit with period q and rotation number p/q. It is also clear that Arnold tongues with different ω are disjoint. For any real ω the ω-Arnold tongue is closed and path-wise connected[274].

In Fig. 9.16 we present the rotation number as a function of parameter a. One can easily see the devil's staircase structure. It can be proved that if we change the parameter a for a given value of b then the rotation number changes in a weakly monotonic way. It can also be proved that if ω is irrational then the interior of A_ω is empty. It is clear that all Arnold tongues have nonempty intersection with the interval $T_3 = \{(a, b) : b = -1, a \in [-2, 2]\}$ - the bottom side of the triangle T. Namely $A_\omega \cap T_3 = \{(2\cos(2\pi\omega), -1)\}$. In Fig. 9.17 the structure of Arnold tongues on parameter plane is shown.

It is clear that for such parameter choices the structure does not have the filtering properties but can generate oscillations of any chosen period.

9.5.3 *Invariant sets and limit sets of trajectories. Analysis of system dynamics via a one-dimensional map*

To give a justification for the results described in previous sections we will present now an in-depth analysis of the behavior of system trajectories. As already mentioned above at the beginning of this section the considered system is asymptotically stable for all parameters a and b within region $S = \{(a, b) \in R : b > -1, b < a + 1, b < -a + 1\}$. Let us divide the part of the $a - b$ parameter plane outside the stability triangle into several subsets and analyze the trajectory behavior for the parameters in each of them[335]:

$Q_{11} = \{(a, b) : b \leq -1 + a, b > 1 - a\}$
$Q_{12} = \{(a, b) : |b - a| < 1, a \geq 1\}$
$Q_{21} = \{(a, b) : b \leq -1 - a, b > 1 + a\}$
$Q_{22} = \{(a, b) : |b + a| < 1, a \leq -1\}$
$Q_3 = \{(a, b) : a + b < -1, b - a < -1\}$
$Q_4 = \{(a, b) : a + b > 1, b - a > 1\}$

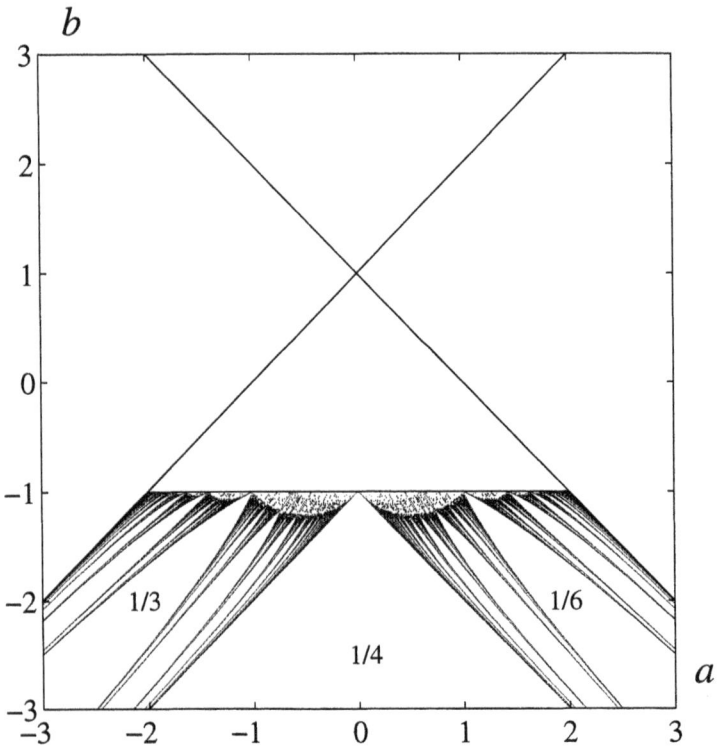

Figure 9.17: The ranges of parameters (a, b) with a given rotation number, points (a, b) lying inside the half-circular regions correspond to homeomorphic Φ (possible quasi-periodic limit sets), below these regions the map Φ is not homeomorphic and the rotation number is rational (periodic limit sets).

$Q_{51} = \{(a,b) : |b+a| < 1, b < -1, b(b+1) + a(a-1) \geq 0, b(1+a) + 1 + a^2 \leq 0\}$
$Q_{61} = \{(a,b) : |b-a| < 1, b < -1, b(b+1) + a(a+1) \geq 0, b(1-a) + 1 + a^2 \leq 0\}$
$Q_{52} = \{(a,b) : b < -1, |a+b| < 1, b(1+a) + 1 + a^2 > 0\}$,
$Q_{53} = \{(a,b) : b < -1, b(b+1) + a(a-1) < 0\}$,
$Q_{62} = \{(a,b) : b < -1, |b-a| < 1, b(1-a) + 1 + a^2 > 0\}$ or
$Q_{63} = \{(a,b) : b < -1, b(b+1) + a(a+1) < 0\}$
$P_1 = \{(a,b) : b = 1 - a, a > 2\}$; $P_2 = \{(a,b) : b = 1 + a, a < -2\}$;
$P_3 = \{(a,b) : b = -1 - a, a > 0\}$; $P_4 = \{(a,b) : b = -1 + a, a < 0\}$;
$P_5 = \{(a,b) : b = 1 + a, a > 0\}$; $P_6 = \{(a,b) : b = 1 - a, a < 0\}$;

Singular cases

Let us next introduce the set I_0^2 of all initial conditions satisfying:
(a) $z_1 x_2(0) + b x_1(0) = 0$ when $b < 1 - a$ and $b > 1 + a$, or $z_2 x_2(0) + b x_1(0) = 0$,
when $b > 1 - a$ and $b < 1 + a$; in which cases the trajectories tend to zero;
(b) $x_2(0) = -x_1(0) = x$ when $b = 1 + a$; in which case there exists an infinite
number of period-two trajectories $(x, -x) \rightarrow (-x, x)$;
(c) $x_2(0) = x_1(0) = x$ when $b = 1 - a$; in which case there exist an infinite
number of fixed points (x,x),
where : $z_1 = \frac{a - \sqrt{a^2 + 4b}}{2}$, $z_2 = \frac{a + \sqrt{a^2 + 4b}}{2}$

Invariant sets

Our study is based on an elementary observation summarized in lemma 1
below.

Lemma 9.9 *For any choices of parameters and initial conditions the trajectories after two iterations enter the set:*

$$I^2 = \{(x_1, x_2) \in R : |x_1| \leq 1, |x_2| \leq 1\} \qquad (9.34)$$

Thus all system trajectories are uniformly bounded and I^2 is an invariant
set. A natural question to ask is: What is the trajectory behavior within
this set? The basic properties of the invariant set are summarized in the two
lemmas given below.

Lemma 9.10 *For any a, b outside the absolute stability triangle and $x(0) \neq 0$
there $\exists k \geq 0$ such that $f^k(x) \in \overline{AB} \cup \overline{CD}$, where $A(1,1)$, $B(1,-1)$, $C(-1,-1)$,
and $D(-1, 1)$ denote the four corners of the square $\Omega = \{x \in R^2 : x_1, x_2 \in [-1, 1]\}$.*

This means that for every trajectory must eventually hit one of the sides AB
or CD of the square I^2

Lemma 9.11 *The images of the square Ω under f satisfy:*

$$f^n(\Omega) \supset f^{n+1}(\Omega) \tag{9.35}$$

and $f^n(\Omega)$ is an absolutely convex polygon (i.e. convex and symmetric with respect to the origin).

From the above stated lemmas it follows the main result summarized in the following theorem:

Theorem 9.12 *For $x(0) \neq 0$ all trajectories eventually enter the set $W_\Omega = \partial f^\infty(\Omega)$. W_Ω is a closed, absolutely convex, nonempty invariant set, which as a product of polygons is also a polygon and is homeomorphic to a circle.*

Thus the study of dynamics of our system can be reduced to analysis of a one-dimensional map of a polygon into itself which is.

In some parameter regions it is possible to give more precise characterization of the invariant sets. These can be summarized as follows :

Lemma 9.13 *We consider five cases:*
a). For $(a,b) \in Q_{11} \cup Q_{12} \cup P_1 \cup P_2$ there exists an invariant set consisting of two points.
b). For $(a,b) \in Q_{12} \cup Q_{22}$ there exists an invariant set composed of two disjoint broken lines.
c). For $(a,b) \in Q_3 \cup Q_4 \cup P_3 \cup P_4 \cup P_5 \cup P_6$ there exists an invariant set composed of the four sides of the square I^2.
d). For $(a,b) \in Q_{51} \cup Q_{61}$ there exists an invariant set which is a hexagon.
e). For $(a,b) \in Q_{52} \cup Q_{62} \cup Q_{53} \cup Q_{63}$, there exists an invariant set which is a more complicated polygon with an even number of sides.

Some details of the structure of the regions in the parameter space with different invariant sets are depicted in Fig.9.18. The numbers denote the number of sides in the invariant polygons in each of these regions.

Limit sets of system trajectories

Let us next give a characterization of the limit sets in each of the invariant sets described above. The following lemmas specify the trajectory behavior in the limit $k \to \infty$ in the above-mentioned four cases.

Lemma 9.14 *For $(a,b) \in Q_1 = Q_{11} \cup Q_{12} \cup Q_{13}$ there are two cases :*
(a). $x(0) \in I_0^2 \Rightarrow$ all trajectories tend to the origin,
(b). $x(0) \notin I_0^2 \Rightarrow$ trajectories tend to one of the equilibrium points $(1,1)$ or $(-1,-1)$.

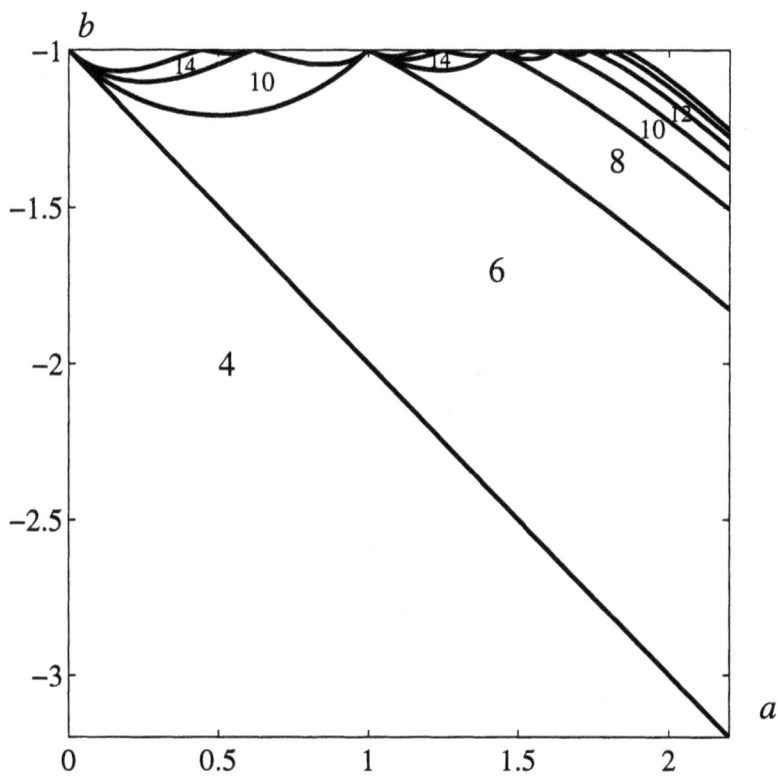

Figure 9.18: Regions of existence of invariant polygons in the $a - b$ parameter space.

This region is colored dark grey in the Plate 13.

Lemma 9.15 *For $(a,b) \in Q_2 = Q_{21} \cup Q_{22} \cup Q_{23}$ there are two cases :*
(a). $x(0) \in I_0^2 \Rightarrow$ all trajectories tend to the origin,
(b). $x(0) \notin I_0^2 \Rightarrow$ all trajectories tend to a period-2 limit cycle $(1,-1) \rightarrow (-1,1)$.

This region is colored sky blue in Plate 13.

Lemma 9.16 *For $(a,b) \in Q_3$ there are two cases :*
(a). $x(0) = (0,0) \Rightarrow$ trajectory is trivial, constant
(b). $x(0) \neq 0 \Rightarrow$ there exist two period-four limit cycles :
$(1,1) - (1,-1) - (-1,-1) - (-1,1)$ *which is stable and*
$(1, \frac{a}{b+1}) - (\frac{a}{b+1}, -1) - (-1, -\frac{a}{b+1}) - (-\frac{a}{b+1}, 1)$ *which is unstable.*

This region is colored light red in Plate 13.

Lemma 9.17 *For $(a,b) \in Q_4$ there are two cases :*
(a). $x(0) = (0,0) \Rightarrow$ trajectory is trivial, constant
(b). $x(0) \neq 0 \Rightarrow$ there exist two stable fixed points :
$(1,1)$ *and* $(-1,-1)$ *and three period-two orbits :*
$(1,-1) - (-1,1)$ *which is stable, and*
$(1, -\frac{a}{1-b}) - (-\frac{a}{1-b}, 1), (-1, \frac{a}{1-b}) - (\frac{a}{1-b}, -1)$ *both unstable.*

This case correspond to the parameters a and b within the region marked light green Plate 13.

Lemmas 5 - 8 give rigorous confirmation of a part of numerical results presented in the color-coded bifurcation diagram.

Lemma 9.18 *For $(a,b) \in Q_{51} \cup Q_{61}$ after a finite number of iterations the trajectory enters the invariant set which is a hexagon:*
$(1,1),(1,b+a),(\frac{-1-b}{a},-1),(-1,-1),\ (1,-b-a),(\frac{1+b}{a},1)$ *for $(a,b) \in Q_{51}$ or*
$(1,-1),(-\frac{1+b}{a},-1),(-1,b-a),(-1,1),(\frac{1+b}{a},1),\ (1,-b+a)$ *for $(a,b) \in Q_{61}$.*
For any parameter choice of $(a,b) \in Q_{51} \cup Q_{61}$ the system trajectories map the invariant hexagon onto itself. Moreover this map is homeomorphic to a map of a circle onto itself and is non-decreasing.

A more general result can be stated as follows:

Theorem 9.19 *For $(a,b) \in Q_4 \cup Q_5 \cup Q_6$ the invariant polygon W_Ω is mapped into itself and $h : W_\Omega \rightarrow W_\Omega$ is a weakly monotone map homeomorphic to a map of a circle. This implies the existence of a unique rotation number for each pair (a,b)*[274].

Corollary 9.20 *The system under consideration can possess periodic or quasi-periodic solutions only.*

In the cases included in theorem 4 the limit sets of the trajectories are composed of a finite number of points in the case of an orbit with a rational rotation number and of a countable number of points in the case of an orbit with an irrational rotation number[274].

Typical graphs of maps of invariant polygons are piecewise-linear with "flat" regions as shown in Fig.9.19. For better understanding in each case we included a simple sketch showing how the mapping is constructed.

9.6   Conclusions

This paper presents an exposition of complex dynamic behaviors encountered in extremely simple structures of digital filters. Two types of nonlinear digital filters have been considered: 1). filters with 2's complement adder overflow nonlinearity, 2). filters with saturation - type adder overflow characteristic.

In the first case the presence of chaotic trajectories generating fractal patterns in the state- space has been confirmed via simulation experiments and analyzed using symbolic dynamics technique.

In the second case no chaotic behavior of filter trajectories has been found; however, the dynamics of the system is extremely complex displaying final state sensitivity to changes of system coefficients, fine Arnold tongues structure of the parameter space and devil's staircase route of changes of orbit rotation numbers in one-parameter bifurcation experiments. Analytical results obtained via a one-dimensional map technique, thereby confirming the experimental results, are also included.

In all cases considered in this paper the quantization has been neglected. In reality, upon taking the fine structure of the nonlinear characteristic into account the filters considered become *finite-state* systems and no chaotic oscillations can occur. This problem has been addressed in[357] and[268].

There are still many open questions regarding dynamics of digital signal processing systems. The two cases presented in this paper and some others found in the literature show that complex dynamics belong to the class of phenomena encountered typically digital systems. Existence of such phenomena opens new areas for research work and possibly new unconventional applications.

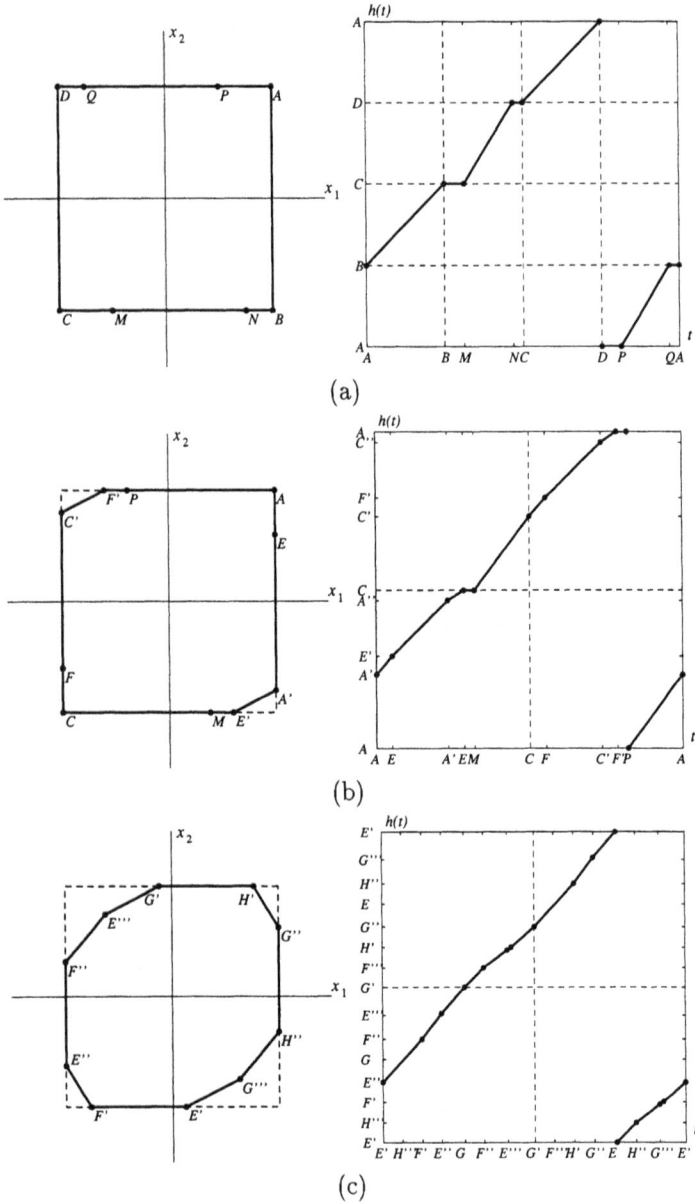

(a)

(b)

(c)

Figure 9.19: Examples of invariant polygons and graphs of associated one-dimensional maps defined on these polygons. Mapping of a square (a), a hexagon (b) and a decagon (c) - into itself.

CHAPTER 10

SYNCHRONIZATION OF CHAOTIC CIRCUITS AND APPLICATIONS

Looking at a variety of systems exhibiting chaotic behavior one could wonder what happens when such systems interact, whether it is possible at all to find mechanisms of their interactions and possibly explain how in many cases interconnected chaotic systems perform useful tasks like signal processing (in many cases with irreproducible excellence), show organized collective behavior out of cooperation of a large number of chaotic subsystems (eg. as in the case of the brain[305,314]).

It seems that extremely complex systems operating in chaotic mode "organize themselves" performing useful tasks using the mechanisms of mutual synchronization. The underlying mechanisms enabling the interactions between subsystems, each operating in a chaotic mode, produce such useful behavior. These principles could possibly be used in building other systems of practical importance.

This chapter provides an overview of existing synchronization concepts and gives some answers to this kind of questions - the basic answers for the simplest possible systems and their interconnections. We describe the concept of synchronization of chaotic systems which can be viewed as the simplest kind of "useful" cooperation of chaotic systems. We will describe also possible applications of such synchronized chaotic systems. As we will see synchronized chaotic systems could find some interesting applications in signal processing and communication. It is also believed that synchronization plays a crucial role in information processing in living organisms and learning synchronization mechanism in large interconnections (arrays) of chaotic systems could lead to even more exciting applications eg. in image or speech processing.

10.1 Synchronization in chaotic systems

The possibility of two or more chaotic systems oscillating in a coherent, synchronized way is not an obvious one. Considering one of the main features often mentioned in the definitions of chaotic behavior, namely the property of sensitive dependence on initial conditions (instability in the Lyapunov sense) one could conclude that synchronization is not possible because it is not possible in real systems to reproduce exactly neither the same starting conditions nor to match exactly the parameters of two systems. We are able to build "nearly" identical systems, but there is an inevitable technological mismatch and noise impeding exact reproduction of all parameters. Thus any, even infinitesimal change of any parameter will eventually result in divergence of nearby starting orbits.

However the above reasoning is not correct - the Lyapunov stability concept of trajectories in a single system is not the proper one to analyze synchronization of two or more systems. In such a case one should ask what are the conditions which imply the convergence of trajectories of the two systems rather then consider the stability of each one alone. In other words, having two (or more) nonlinear systems ($N \geq 2$):

$$\dot{x}_i = f_i(x_i) \qquad , x \in R^n, \ 1 \leq i \leq N \qquad (10.1)$$

we would like to find conditions under which their solutions will converge to each other ie.:

$$\lim_{t \to \infty} (x_i - x_j) = 0 \qquad i \neq j \qquad (10.2)$$

in a more general setting one could allow some scaling or transformation of the the variables e.g.

$$\lim_{t \to \infty} (x_i - K x_j) = 0 \qquad i \neq j \qquad (10.3)$$

or

$$\lim_{t \to \infty} (x_i - \mathcal{F}(x_j)) = 0 \qquad i \neq j \qquad (10.4)$$

In the latter case we consider so-called *generalized synchronization* .

There is no general answer to this problem. Below we will describe some concepts for obtaining coherent (synchronous) operation of chaotic systems.

10.2 Linear coupling

10.2.1 *Uni-directional coupling*

The simplest possibility considered in several papers[86,250] is the linear coupling of the two systems we would like to synchronize - injection of a scaled error signal into the system to be synchronized:

$$\begin{aligned} \dot{x} &= f_1(x) \\ \dot{y} &= f_2(y) + \Delta(x - y) \end{aligned} \qquad (10.5)$$

where: $x, y \in R^n$, $\Delta = diag[\delta_1, ..., \delta_n]^T$.

The synchronization problem is formulated as follows: Find Δ such that $y(t) \longrightarrow x(t)$ for $t \longrightarrow \infty$ (i.e. the solution $y(t)$ will synchronize with the signal $x(t)$).

This kind of linear coupling has been used for some particular types of systems (the example of Chua's circuit will be described below). Kocarev *at al.*[250] have given some theorems concerning the convergence of solutions x and y. The following are the most interesting results:

Case $f_1 = f_2$

Theorem 10.1 [250] *If $f_1 = f_2$ and $|x(t = 0) - y(t = 0)|$ is sufficiently small then there exist finite values $\tilde{\delta}_i$ with $i = 1, 2, ..., n$ such that for $\delta_i > \tilde{\delta}_i$, $y(t)$ approaches the goal $x(t)$.*

Case $f_1 \neq f_2$

For simplicity, we will assume that $\delta_i = k$, for all $i = 1, 2, ..., n$. Equation (10.5) can be rewritten as:

$$\begin{aligned} \dot{x} &= f_1(x) \\ \dot{y} &= f_2(y) + k(x - y) \end{aligned} \qquad (10.6)$$

where k is a real nonnegative parameter.

Theorem 10.2 *For $\epsilon = \delta^{-1}$ and sufficiently small $|x(t = 0)| - |y(t = 0)|$ there exists t_0 such that $y(t)$ converges uniformly to $x(t)$ as $\epsilon \to 0^+$ on all closed subsets of $t_0 < t < \infty$.*

These two theorems give some very general conditions for synchronization. Theorem 2 not only enables us to treat the case when there is a parameter mismatch but the systems are nearly identical but could also be applied for obtaining synchronization of systems with quite different dynamics !!! One should note however that the problem of choosing the initial conditions of the two systems is of particular importance – the theorems do not tell us much about the regions of convergence.

The linear coupling concept can be generalized to the case where the two chaotic systems influence each-other continuously i.e. the linear coupling is mutual.

10.2.2   Mutual (bi-directional) coupling

Let us consider the following system:

$$\left.\begin{array}{l} \dot{x} = f(x) + \lambda(y - x) \\ \dot{y} = g(y) + \mu(x - y) \end{array}\right\} \tag{10.7}$$

where $x, y \in R^n$ and λ, μ are $n \times n$ diagonal matrices with nonnegative elements, $\lambda = diag[\lambda_i]$, $\mu = diag[\mu_i]$, $i = 1, \ldots, n$. We will assume without loss of generality, that $f(0) = g(0) = 0$.

Equation (10.7) describes the dynamics of two mutually coupled (sub) systems x and y. This type of so-called *diffusively coupled* dynamical systems are commonly proposed for modeling various phenomena in Nature and describing many engineering applications. Examples come from physics, chemistry, biology and medicine, as well as from mechanical, chemical and electrical engineering. Let us give two examples only. The first one comes from developmental and control problems in biology; (1) is in fact the simplest equation[421] in Turing's reaction-diffusion theory of morphogenesis[438]. In this case x is the vector of morphogen concentrations of the first subsystem (cell), y is the vector of morphogen concentrations of the second subsystem (cell), and f represents the reaction kinetics. The second one comes from electrical engineering: (1) describes the time -evolution of two resistively coupled electrical circuits[250]. Recently it has been also demonstrated that this type of coupling (which we call *mutual error feedback*) could be useful in synchronization and control problems for chaotic systems[250].

Does the model (1) give any kind of cooperative or synchronized behavior when each of the subsystems operates in a chaotic regime? In other words – is it possible to choose λ and μ in such a way that $lim_{t \to \infty}(x(t) - y(t)) = 0$.

Let us consider first a special case of equation (10.7), namely $f = g$, $\lambda = \mu$:

$$\left.\begin{array}{l} \dot{x} = f(x) + \mu(y - x) \\ \dot{y} = f(y) + \mu(x - y) \end{array}\right\} \tag{10.8}$$

where $x, y \in R^n$ and μ is an $n \times n$ diagonal matrix with positive constant diffusion coefficients. For this system we have the following result which is the extension of Theorem 1 of[250].

Theorem 10.3 *There exist finite values of μ_i , say μ_i^* , such that for $\mu_i > \mu_i^*$, $x(t)$ approaches $y(t)$ as $t \to \infty$, that is the two subsystems are synchronized.*
Proof : Denote $u = x - y$, so that from (10.8) we have

$$u = [-f + Df|_0]u + O(x, y) = Au + O(x, y) \tag{10.9}$$

where Df is the Jacobian matrix of f, and $O(x, y)$ represents the higher-order terms. It is obvious that one can find μ_i such that $|a_{jj}| > \sum_{\substack{i=1 \\ i \neq j}}^{n} |a_{ij}|\ i = 1 \ldots n$, as a consequence A is a stable matrix. Therefore, $u = 0$ is asymptotically stable, and $x(t)$ approaches $y(t)$ as $t \to \infty$.

Let us consider next the coupling of two systems with different dynamics as described by equation (10.7). In this case we have the following theorem:

Theorem 10.4 *For sufficiently small $|x(t = 0)| + |y(t = 0)|$ and ε, there exists t_0 such, that $x(t)$ converges uniformly to $y(t)$ as $\varepsilon \to 0^+$ on all subsets of $t_0 < t < \infty$.*

10.2.3 Synchronization via linear coupling - examples

Example 1. - Chua's circuits

As a first example let us consider linear coupling of two Chua's circuits. Several experimental studies have been carried out to show chaotic synchronization[86]. Extensive experimental studies were done using Kennedy's[239] implementations of Chua's circuits. It seems that experiments using Chua's circuits are the easiest to perform and we give some of the details below.

Let us consider a linear coupling of two Chua's circuits:

$$
\begin{aligned}
\dot{x}_1 &= \alpha(X_2 - x_1 - f(x_1)) + \delta_1(y_1 - x_1) \\
\dot{x}_2 &= x_1 - x_2 - x_3 + \delta_2(y_2 - x_2) \\
\dot{x}_3 &= -\beta x_2 + \delta_3(y_3 - x_3) \\
\dot{y}_1 &= \alpha(y_2 - y_1 - f(y_1)) + \delta_1(x_1 - y_1) \\
\dot{y}_2 &= y_1 - y_2 - y_3 + \delta_2(x_2 - y_2) \\
\dot{y}_3 &= -\beta y_2 + \delta_3(x_3 - y_3)
\end{aligned}
\tag{10.10}
$$

Chua *et al.*[86] have shown that two coupled Chua's circuits characterized by $\alpha = 10.0$, $\beta = 14.87$, $a = -1.27$, $b = -0.68$ will synchronize ie. the solutions of the two systems will approach each other asymptotically for the following sets of parameters δ:

1. $\delta_1 > 0.5$, $\delta_2 = \delta_3 = 0$,
2. $\delta_2 > 5.5$, $\delta_1 = \delta_3 = 0$,
3. $2 > \delta_3 > 0.7$, $\delta_1 = \delta_2 = 0$,

In the laboratory experiments the x-coupling could be realized by inserting a R_x $(\delta_x = \frac{C_2 R}{C_1 R_x})$ resistor between the $+$ terminals of the nonlinear resistors. The y-coupling by connecting an R_y $(\delta_y = \frac{R}{R_y})$ resistor between the $+$ terminals of C_2 capacitors.

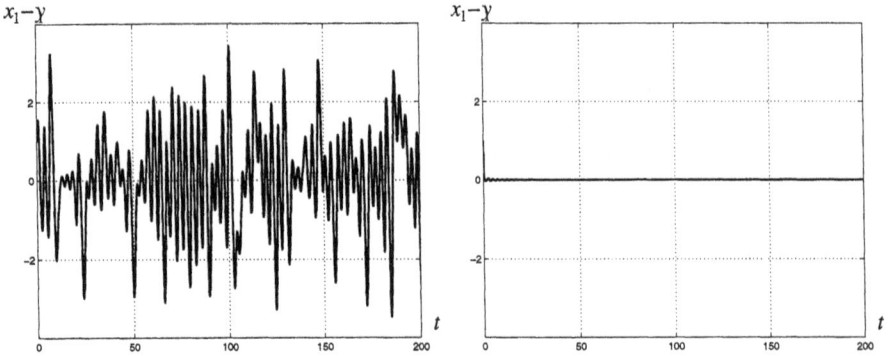

Figure 10.1: Tow plots showing change of synchronization error $x_1 - y_1$ in time in the case of two Chua's circuits coupled via the first variables for $\delta_1 = 0.2$ (a) and $\delta_1 = 5.0$ (b). In the first case synchronization does not occur. When the coupling strength is sufficient one can see rapid convergence of the synchronization error towards 0.

Fig.10.1 and 10.2 show typical plots confirming synchronization of two chaotic Chua's circuits in the case of linear coupling. In the Fig.10.1 the circuits were coupled via the first variables while in Fig.10.2 the case of coupling via the second variables is shown. In the latter case and small coupling coefficient we observe chaotic bursts between some periods of synchronization. Transient behavior before reaching synchronization is very short as shown in Fig.10.1b and Fig.10.2b..

10.2.4   Partial synchronization

Finally, we consider the following special case of coupling of the two systems when some of the elements of matrices $\mu = 0$ or λ are 0. In a particular case we obtain "one-way" coupling ($\mu = 0$):

$$\left.\begin{array}{l} \dot{x} = f(x) + \lambda(y - x) \\ \dot{y} = g(y) + \end{array}\right\} \tag{10.11}$$

Let us assume that two subsystems for $\lambda = 0$, have chaotic attractors. The corresponding undriven system of (10.11) is

$$\dot{x} = f(x) - \lambda x \tag{10.12}$$

We assume that (10.12) is asymptotically stable at the origin for given matrix λ. Let us fix all elements of the matrix λ, except one, say λ_n . A typical

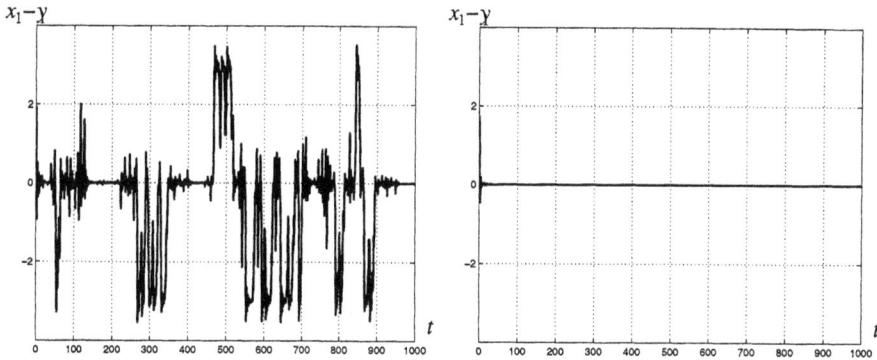

Figure 10.2: Tow plots showing change of synchronization error $x_1 - y_1$ in time in the case of two Chua's circuits coupled via the first variables for $\delta_1 = 0.5$ (a) and $\delta_1 = 1.0$ (b). In the first case synchronization does not occur – chaotic burst are visible between periods of synchronization. When the coupling strength is sufficient one can see rapid convergence of the synchronization error towards 0.

vector field (10.12) undergoes various bifurcations from chaos to a fixed point as λ_n increases from 0 to ∞. On the other hand the driven system (10.11) also bifurcates but for typical vector fields, the dynamical regime remains chaotic. In fact, we have found that a typical system (10.11) bifurcates from chaos to hyper-chaos and vice versa. As λ_n increases, x_n and y_n become more and more synchronized. The two subsystems will be not synchronized (i.e. $x_i \neq y_i$, for $i = 1, ..., n - 1$), but only partial synchronization between two components of x and y will occur (i.e. $x_n = y_n$).

10.2.5 Examples of partial synchronization

To show partial synchronization we will use two commonly recognized paradigmatic chaotic systems : f - Chua's system (exhibiting the Double scroll attractor) and g - Lorenz (Lorenz attractor) (the values of the parameters are standard). In this case the equations read:

$$\left.\begin{aligned}
\dot{x}_1 &= 6.3(x_2 - x_1) - 9g(x_1) &&+ \lambda_1(y_1 - x_1) \\
\dot{x}_2 &= 0.7(x_1 - x_2) + x_3 &&+ \lambda_2(y_2 - x_2) \\
\dot{x}_3 &= -7x_2 &&+ \lambda_3(y_3 - x_3) \\
\dot{y}_1 &= 10(-y_1 + y_2) &&+ \mu_1(x_1 - y_1) \\
\dot{y}_2 &= 28y_1 - y_2 - y_1y_3 &&+ \mu_2(x_2 - y_2) \\
\dot{y}_3 &= y_1y_2 - 2.666y_3 &&+ \mu_3(x_3 - y_3)
\end{aligned}\right\} \quad (10.13)$$

We obtained the following results:

- First case : $\lambda_1 = 10$, $\lambda_2 = 0$, $\lambda_3 = 0$ partial synchronization between x_1 and y_1 - the quality of synchronization is poor (Fig.10.3); $\lambda_1 = 1000$, $\lambda_2 = 0$, $\lambda_3 = 0$ partial synchronization between x_1 and y_1 - the quality of synchronization is good (Fig.10.4).

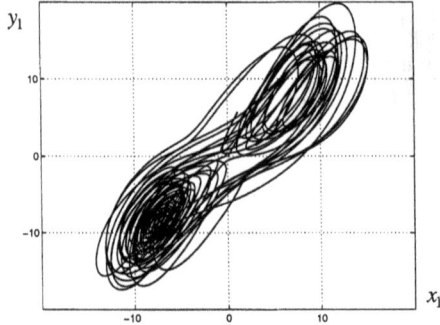

Figure 10.3: Partial synchronization. (x_1, y_1) - projection of the attractor A, with $\lambda_1 = 10$, $\lambda_2 = \lambda_3 = 0$, and $\mu_1 = \mu_2 = \mu_3 = 0$.

- Second case : $\lambda_1 = 10$, $\lambda_2 = 10$, $\lambda_3 = 0$ partial synchronization between x_1 and y_1, and x_2 and y_2 - the quality of synchronization is poor; The results are shown in Fig.10.5
  $\lambda_1 = 1000$, $\lambda_2 = 100$, $\lambda_3 = 0$ partial synchronization between x_1 and y_1, and x_2 and y_2 - the quality of synchronization is good; The results are shown in Fig.10.6.

In both cases x_3 and y_3 are not synchronized.

There is also a striking similarity of the x_1-x_2 projections in the cases of poor and good synchronization. We cannot give any explanation for this phenomenon.

Linear coupling between two chaotic systems, each operating in a chaotic mode,it is possible to synchronize their oscillations. Apart from full synchronization of all corresponding (coupled) variables the investigations carried out so far enabled us to discover a very interesting phenomenon of partial synchronization of the systems. Namely, two systems operating in different chaotic modes when coupled in a suitable way can influence each-other in such a way that only some variables synchronize (one, two or more) while the remaining variables oscillate chaoticly without showing any convergence properties (co-

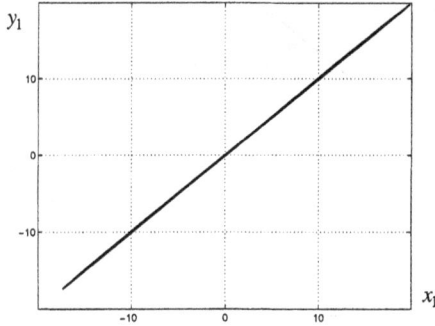

Figure 10.4: Partial synchronization. (x_1, y_1) - projection of the attractor A, with $\lambda_1 = 1000$ $\lambda_2 = \lambda_3 = 0$, and $\mu_1 = \mu_2 = \mu_3 = 0$.

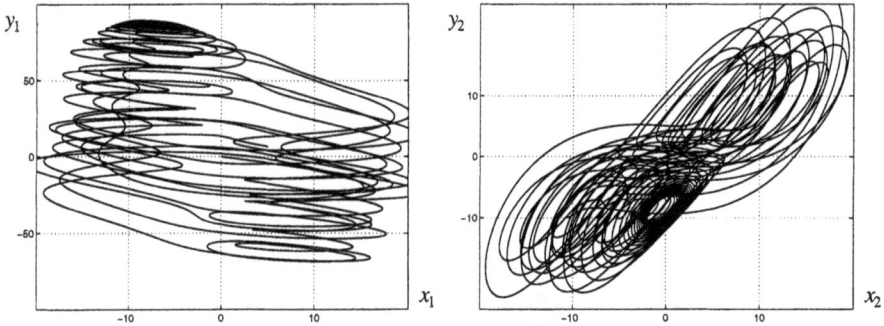

Figure 10.5: Partial synchronization. (x_1, y_1) and (x_2, y_2) - projections of the attractor A, with $\lambda_1 = \lambda_2 = 10$, $\lambda_3 = 0$, and $\mu_1 = \mu_2 = \mu_3 = 0$.

herence). Extensive simulation and laboratory experiments are being carried out to find further confirmation of the described phenomena.

10.2.6 Pecora-Carroll drive-response concept

So far the most effective and widely studied approach is due to Pecora and Carroll[58,59,60,365,366] who proposed a solution to a class of synchronization problems.

They considered an n-dimensional autonomous system governed by a state equation of the form:

$$\frac{dx}{dt} = f(x(t)) \tag{10.14}$$

Pecora and Carroll proposed a decomposition of the original system into two

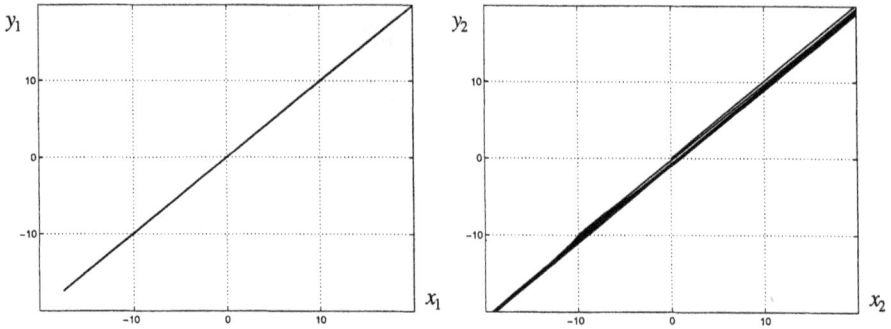

Figure 10.6: Partial synchronization. (x_1, y_1) and (x_2, y_2)- projections of the attractor A, with $\lambda_1 = \lambda_2 = 1000$, $\lambda_3 = 0$, and $\mu_1 = \mu_2 = \mu_3 = 0$.

the parts (subsystems) in an arbitrary way thus dividing the state vector into $x = \begin{bmatrix} x_D \\ x_R \end{bmatrix}$. The "D"-part will be referred to as the driving subsystem and the "R"-part as the response subsystem respectively. Then:

$$\begin{aligned} \dot{x}_D &= g(x_D, x_R) \\ \dot{x}_R &= h(x_D, x_R) \end{aligned} \qquad (10.15)$$

where: $x_D = [x_1, \ldots, x_m]^T$, $x_R = [x_{m+1}, \ldots, x_n]^T$, $g = [f_1(x), \ldots, f_m(x)]^T$, $h = [f_{m+1}(x), \ldots, f_n(x)]^T$

Pecora and Carroll suggested building an identical copy of the response subsystem and drive it with the x_D variables coming from the original system. In such a situation we obtain the following equations:

$$\begin{aligned} \dot{x}_D &= g(x_D, x_R) \\ \dot{x}_R &= h(x_D, x_R) \\ \dot{x}'_R &= h(x_D, x'_R) \end{aligned} \qquad (10.16)$$

Let us next examine the difference $\Delta x_R = x'_R - x_R$. The subsystem components x_R and x'_R will asymptotically approach each-other (synchronize) if $\Delta x_R \longrightarrow 0$ for $t \longrightarrow \infty$. In the limit this leads to the variational equations for the response subsystem:

$$\frac{d\Delta x_R}{dt} = D_{x_R} h(x_D(t), x_R(t)) \Delta x_R + o((\Delta x_R)^2) \qquad (10.17)$$

where: D_{x_R} denotes the Jacobian of the response subsystem with respect to x_R only. The behavior of the solutions of this system depends on the so-called conditional (depending on x_D) Lyapunov exponents measuring the average convergence/divergence rate of nearby points in the state space.

Pecora and Carroll proposed the following necessary condition for chaotic synchronization.

Theorem. The subsystems x_R and x'_R will synchronize only if the conditional Lyapunov exponents are all negative.

This methodology has been successfully applied to obtain chaos synchronization in coupled Lorenz systems[365], Rössler systems[365], hysteretic circuit[58]. Finally Pecora and Carroll proposed[366] a specific laboratory circuit for studying synchronization phenomena[59]. Interesting results have been also obtained for coupled Chua's circuits. Some of these are presented in the following section.

Example - synchronization of Chua's circuits using the drive-response concept

Considering the Pecora–Carroll approach it has been confirmed[86] that in two configurations:

x-drive configuration for which the state equations read:

$$\dot{x}_1 = \alpha(x_2 - x_1 - f(x_1))$$
$$\dot{x}_2 = x_1 - x_2 - x_3$$
$$\dot{x}_3 = -\beta x_2 \qquad\qquad (10.18)$$
$$\dot{y}_2 = x_1 - y_2 - y_3$$
$$\dot{y}_3 = -\beta y_2$$

and the conditional Lyapunov exponents are $(-0.05, -0.05)$. Fig.10.7 shows a typical phase plot confirming excellent synchronization of chaotic trajectories in this case.

y-drive configuration for which the state equations become:

$$\dot{x}_1 = \alpha(x_2 - x_1 - f(x_1))$$
$$\dot{x}_2 = x_1 - x_2 - x_3$$
$$\dot{x}_3 = -\beta x_2 \qquad\qquad (10.19)$$
$$\dot{y}_1 = \alpha(x_2 - y_1 - f(y_1))$$
$$\dot{y}_3 = -\beta x_2$$

and the conditional Lyapunov exponents were found $(-2.5, 0)$

In the **z-drive configuration** the subsystems do not synchronize - one of the conditional Lyapunov exponents was found to be positive.

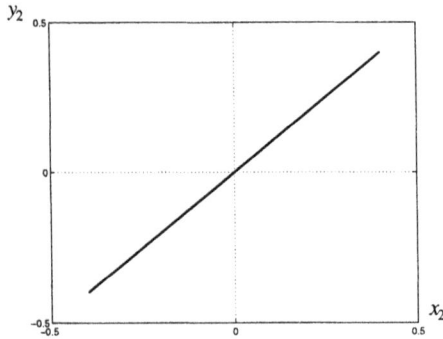

Figure 10.7: Phase plot showing synchronization in the x-drive configuration of Chua's circuits (horizontal axis: x_2, vertical axis: y_2).

10.3 Possible applications in communication

First attempts to use random signals in secure communications date back to 1926 when Vernam published his paper[453]. (For the topic of cryptography we refer the reader to the Special Issue of The Proceedings of IEEE[470]). [a]

These ideas were explored again in the context of chaotic signals and used nearly simultaneously by independent research groups at the ONR[60,367], University of California, Berkeley[177,249,364], MIT[353], a joint team from Swiss Federal Institute of Technology and University College Dublin[100,240].

It is the group of Prof. Leon O. Chua who published the first real circuit implementation and test results proving that the ideas of using chaotic signals and synchronized chaotic circuits in communication problems is not only useful but technically feasible, offering possibly competitive solutions to secure communication problems.

There is also an interesting alternative approach based on the information theoretic formalism of chaos reported recently in[192] – however no implementations have been reported so far.

We will see on the simple examples that a chaos producing system (Chua's circuit) could be used as the enciphering 'key'. This 'key' is fully identified by the actual circuit parameters.

---

[a]Vernam proposed to use binary alphabet and the key only one time, i.e. to encipher each bit of the text with a new randomly chosen bit of the key. The enciphering principles were not changed since Caesar times. Each of the letters of the text to be enciphered - X, has to be replaced by a symbol Y obtained via chosen "modulo" summation with a secret key Z i.e. $Y = X \oplus Z$.

10.3.1 Chaotic switching

The simplest idea of how chaotic systems could be used in data transmission is the parameter modulation or chaotic switching[353]. The basic idea is to encode the binary signal in terms of different attractors existing for different system parameter values in the system (eg. the 1- corresponds to parameter value μ_1 and further – chaotic attractor $\mathcal{A}_1$, 0- corresponds to parameter value μ_2 and a chaotic attractor $\mathcal{A}_2$). The chaotic system behavior is switched between $\mathcal{A}_1$ and $\mathcal{A}_2$ thus the time responses system are modulated by parameter changes.

The usefulness of this simple idea has been demonstrated by Parlitz *et al.*[364]. The block diagram of the proposed transmission system is shown in Fig.10.8.

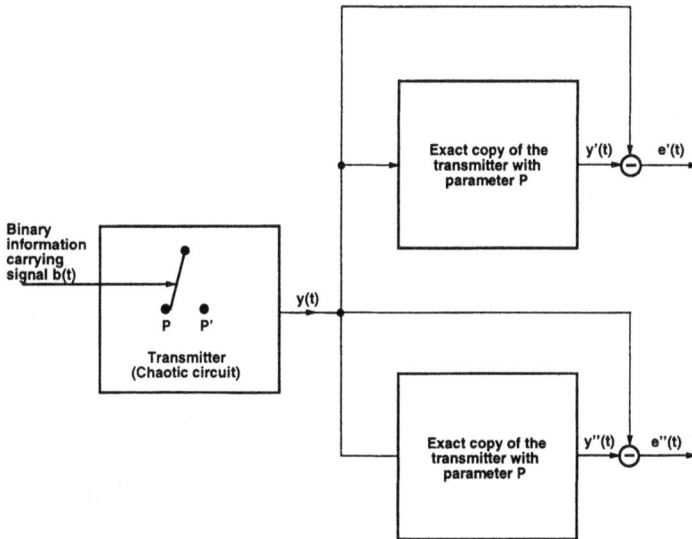

Figure 10.8: Block diagram of the transmission system utilizing chaos switching technique.

Chua's circuit has been used as a source of chaotic signals (Fig.10.9). In the simulation experiments the parameters $R = 1001\Omega$, $R_0 = 20\Omega$, $G_a = -1.139mS$, $G_b = -0.711mS$, $B_p = 1V$ were fixed while the other parameters were switched between $L = 12mH$, $C_1 = 17nF$ and $C_2 = 178nF$ for $b_{in} = $ "1" (first parameter set) and $L = 13.3mH$, $C_1 = 18.8nF$ and $C_2 = 197nF$ for $b_{in} = $ "0" (second parameter set) depending on the binary input signal $b_{in}(t)$. In both cases the system possesses qualitatively similar attractors. The voltage

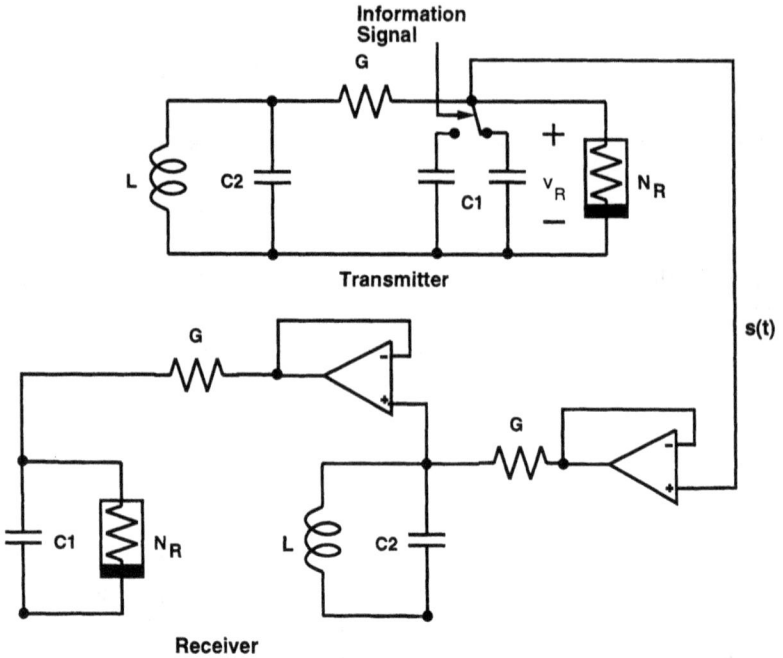

Figure 10.9: Transmitter circuit — Chua's circuit with switchable capacitor and receiver circuit built using two partial Chua's circuits following Pecora-Carroll concept.

across the capacitor C_1 has been chosen as the signal to be transmitted $s(t)$. The transmitted signal in both cases is chaotic and thus broad–band.

Following the Pecora–Carroll principles the receiver (Fig.10.9) is built as a copy of a part of the transmitter Chua's circuit (y_A, z_A - subsystem # 1) with the first set of parameters. To determine whether there is indeed synchronization, reference signals have to be generated using the known variables $x = v_{C_1}$, y_A, z_A – for this purpose the subsystem # 2 has been added (reproducing the variable x_{B1}). This system synchronizes only for one of the transmitted states i.e. when the quantity $\Delta_0 = x - x_{B1} = 0$. Second Chua's circuit (built with the second set of parameters) can also be added to reproduce the variable x_{B2} and synchronize with the chaotic signal in the second state only ($\Delta_1 = x - x_{B2} = 0$). The use of two chaotic signals with mutually exclusive synchronization properties improves the reliability of the system.

Fig.10.10 presents the waveforms obtained in simulations. The waveforms

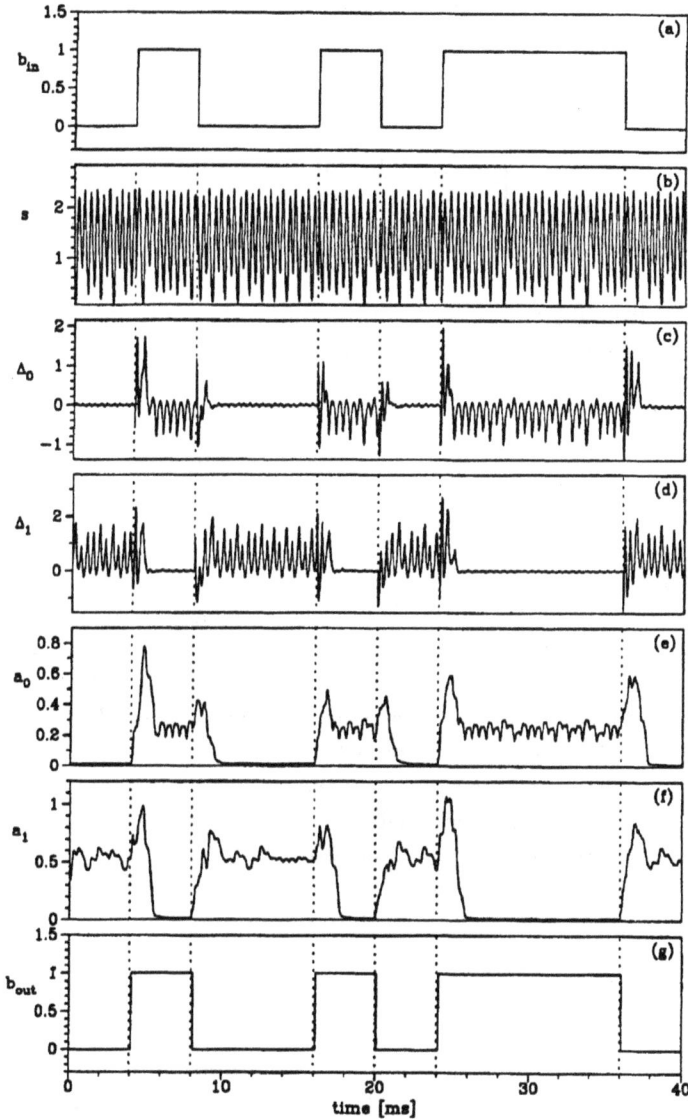

Figure 10.10: Transmission of digital signals via parameter modulation (chaos switching) technique. 0 and 1 states are coded by two different chaotic attractors. (a) – binary input signal b_{in}, (b) – transmitted signal $s(t)$, (c) – response Δ_0, (d) – response Δ_1, (e) – 40-points moving average of Δ_0, (f) – 40-points moving average of Δ_1, (g) – output binary signal b_{out}.

represent respectively: (a) – binary input signal b_{in}, (b) – transmitted signal $s(t)$, (c) – response Δ_0, (d) – response Δ_1, (e) – 40-points moving average of Δ_0, (f) – 40-points moving average of Δ_1, (g) – output binary signal b_{out}, $\varepsilon = 0.1$. b_{out} was derived using the rule:

$$b_{out} = \begin{cases} 0, b_{old} = 0 & \text{for} \quad a_0 < \varepsilon, a_1 > \varepsilon \\ 1, b_{old} = 1 & \text{for} \quad a_0 > \varepsilon, a_1 < \varepsilon \\ b_{old} & \text{for} \quad a_0 < \varepsilon, a_1 < \varepsilon \\ 1 - b_{old} & \text{for} \quad a_0 > \varepsilon, a_1 > \varepsilon \end{cases} \qquad (10.20)$$

The resulting digital signal b_{out} agrees up to a small time delay with the original input signal n_{in}. Parlitz *et al.* in their study[364] for the first time presented also results of laboratory experiments demonstrating the applicability of the proposed method and showing that secure communication using chaotic switching is possible and might lead to new developments in communication techniques.

10.3.2   Chaotic masking – secure communication

Another possibility of using chaotic signals for secure communication is using them for masking the information–carrying signal. This idea has been described by Oppenheim *et al.*[353] and Kocarev *et al.*[250]. The information–carrying signal is simply added to the masking chaotic signal. In[250] the authors report on using Pecora–Carroll approach and building an experimental set up for secure signal transmission based on the masking principle. Again Chua's circuit has been used as a universal chaotic building block. The diagram of the proposed system is shown in Fig.10.11. It contains a Chua's circuit in

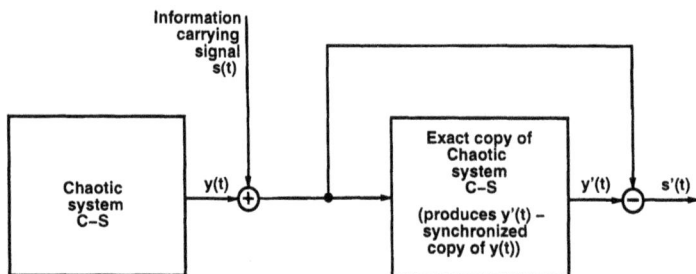

Figure 10.11: Block diagram of secure communication system employing the masking principle.

the transmitter part and two partial Chua's circuits in the receiver part. The receiver has exactly the same structure as in the case of the previous example (Fig.5). The first sub-circuit serves as a decoding key and synchronizes only when exactly matched with the transmitter circuit thus reproducing the v_{C_2} signal. The second sub-circuit is used for restitution of the missing variable v_{C_1} needed for recovering the information signal by simple subtraction as shown in the block diagram. In all laboratory tests in a real circuit implementation it has been confirmed that chaotic signal can be used as a masking signal and that it is possible to decode such a signal successfully using Pecora–Carroll synchronization concept.

10.3.3 Chaotic modulation - spread–spectrum transmission

The most complex issue offered for secure communication has been described in a recent paper by Halle *et al.*[177]. The proposed idea is to modulate the information signal by a broad–spectrum, noise–like chaotic signal. The principle of operation of such a communication scheme relays on so-called **inverse system approach**.

Inverse system approach

In this approach proposed by Boehme *et al.*[34] and elaborated by Feldmann *et al.*[120] the basic idea is to construct an inverse nonlinear system which using the output signal from the original system (transmitter) will reproduce the input signal as shown in Fig.10.12. In terms of the input-output relation the receiving system realizes an inverse operation on the signals.

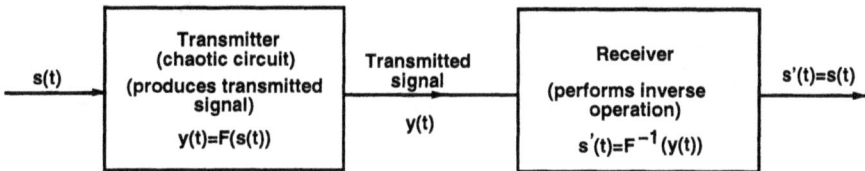

Figure 10.12: Schematic explanation of the inverse system approach.

In general finding of an inverse of a nonlinear system (inverse of a nonlinear operator realized by the circuitry) is a very difficult task. Several approaches to solving this problem are know from the automatic control literature (see references in[120]) however, there is no general solution to this problem. In the next section we will show an example of inverse system when discussing transmission of signals by chaos modulation.

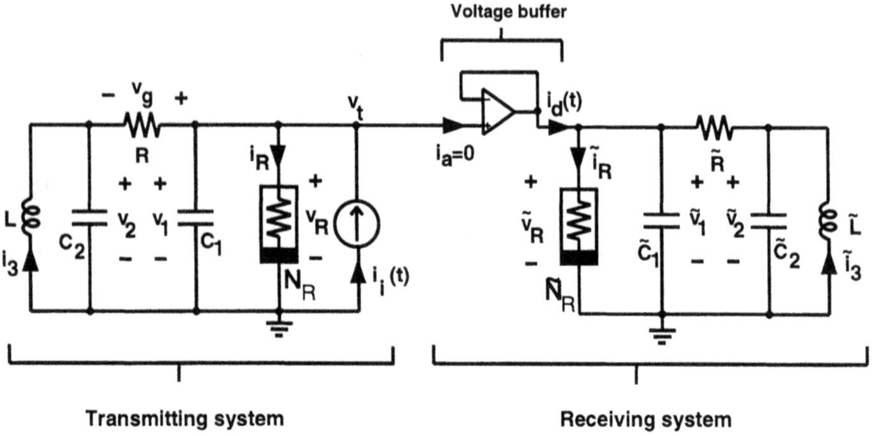

Figure 10.13: Schematic diagram of the chaos modulation communication system employing two self-synchronizing Chua's circuits.

Example

The transmission system shown in Fig.10.13 utilizes two Chua's circuits. In the transmitting system a current signal $i_i(t)$ is injected into the circuit and modifies the voltage across the capacitor C_1. This current signal depends on the input information $v_s(t)$ to be transmitted $i_i(t) = c(v_s(t))$ (where c is an invertible coding function). The detected current signal i_d is then decoded through $v_r(t) = c^{-1}(i_d(t))$. For proper operation it is necessary that $v_r(t) \approx v_s(t)$. Coding function c should be chosen in such a way that during transmitter operation for all $v_s(t)$ the transmitted signal remains chaotic and looks the same. The voltage across the capacitor C_1 is further used as the signal transmitted through the channel to the receiver circuit and used as a forcing voltage on the second Chua's circuit capacitor $\tilde{C}_1$.

Assuming that all circuit components of the transmitter and receiver are matched exactly and inserting a voltage buffer to separate the two subsystems we have $\tilde{v}_1 = v_1$. Using this condition and subtracting the circuit equations describing dynamics of each of the Chua's circuits one obtains:

$$\begin{aligned} C_2 \frac{d(v_2 - \tilde{v}_2)}{dt} &= -\frac{1}{R}(v_2 - \tilde{v}_2)) + (i_3 - \tilde{i}_3)) \\ L \frac{d(i_3 - \tilde{i}_3)}{dt} &= -(v_2 - \tilde{v}_2) \end{aligned} \tag{10.21}$$

For $R, L, C_2 > 0$ we have[177] $(v_2(t) - \tilde{v}_2(t)) \longrightarrow 0$ for $t \longrightarrow \infty$. This implies also $i_d(t) \longrightarrow i_i(t)$ and $v_r(t) \longrightarrow v_s(t)$ for $t \longrightarrow \infty$.

This means that the current flowing into the second Chua's circuit must be equal (after possibly some transient) to the current injected into the first Chua's circuit.

Halle *et al.*[177] describe the laboratory implementation of such a transmission system based on two synchronized Chua's circuits with the division operation $c(v_s(t)) = \frac{v_s(t)}{v_1(t)}$ chosen as the coding function and multiplication operation $v_r = c(v_s(t))v_1$ as the decoding one. The block diagram of the implemented system is shown in Fig.10.14. This diagram could serve as a general

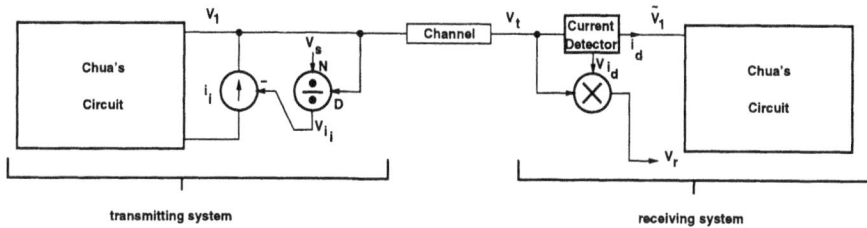

Figure 10.14: Block diagram of the laboratory chaos modulation system. In the divider, N stands for the numerator and D for the denominator.

principle of a transmission system utilizing chaotic modulation.

The results presented in[177] demonstrate the feasibility of using self–synchronizing circuits (and in particular Chua's circuits) to implement spread–spectrum communication systems. For the full account of the experimental results we refer the reader to the original paper[177]. We would like to stress here that this kind of chaotic signal modulation offers several advantages over the parameter modulation or simple masking techniques. Firstly, the whole range of chaotic signal spectrum is used for "hiding" the information. Secondly, the sensitivity to parameter variation is increased thus offering increased security.

10.4 Conclusions

Synchronization principles as described in the previous sections enable us to build chaotic systems operating coherently and use them to solve real communication problems. Understanding synchronization phenomena of simple interconnections of chaotic oscillators enabled several interesting developments as switching, masking and modulation using chaotic signals and could serve also as a basis for further studies.

There exists already a widespread interest in studies of higher dimensional coupled chaotic systems in particular arrays[5,10,26,91,113,226,227,310,372,373,426] of chaotic oscillators.

This kind of systems are important as models of biological and physical systems and also from the information processing point of view offering possible engineering applications (eg.[373]). One can expect a rapid development of research in this area.

CHAPTER 11

CHAOS CONTROL

Research efforts during the past decade have led to a general understanding of a variety of nonlinear phenomena in physical, biological, chemical and many other systems. In particular in electronic and electrical engineering many unwanted phenomena as excess noise, false frequency lockings, squegging, phase slipping have been found to be associated with bifurcations and chaotic behavior. Also many nonlinear phenomena in other science and engineering disciplines have a strong link with "electronic chaos" - let us mention here the most spectacular ones: heart fibrillation (Electro-Cardio-Gram waveforms) and epileptic foci in Electro-Encephalo-Graphic patterns. Chaos, so commonly encountered in physical systems, represents rather a peculiar type of behavior commonly considered as causing malfunctioning, disastrous and thus unwanted in most applications. It is obvious that an amplifier, a filter, an A/D converter, a phase-locked loop or a digital filter generating chaotic responses is of no use - at least for its original purpose. Similarly we would like to avoid situations where the heart does not pump blood properly (fibrillation or arrythmias) or epileptic attacks. Considering even more spectacular potential applications one could think about influencing rainfall, avoiding hurricanes and other atmospheric disasters believed to be associated with large-scale chaotic behavior.

The most common goal of control for chaotic systems is suppression of oscillations of the "bad" kind and influencing it in such a way that it will produce a prescribed, desired motion. The goals vary depending on a particular application. The most common goal is to convert chaotic motion into a stable periodic or constant one. It is not at all obvious how such a goal could be achieved as one of the fundamental features of chaotic systems - the sensitive dependence on initial conditions seems to contradict any stable system operation. Recently several applications have been mentioned in the literature where the desired state of system operation is chaotic — the control problems in such cases are defined as: convert unwanted chaotic behavior into another kind of chaotic motion with prescribed properties (this is as a matter of fact the goal of chaos synchronization) or change periodic behavior into chaotic motion (which might be the goal in the case of removal of epileptic seizures).

The last-mentioned type of control is often referred to as *anti-control* of chaos.

Many chaotic systems display so-called multiple basins of attraction and fractal basin boundaries. This means that, depending on the initial conditions, trajectories can converge to different steady states [a](chaotic or otherwise).

In many cases the sets of initial states leading to a particular type of behavior are inter-wound in a complicated way forming fractal structures. Thus one could consider elimination of multiple basins of attraction as another kind of control goal.

In some cases chaos is the dynamic state in which we would like the system to operate. We can imagine that mixing of components in a chemical reactor would be much quicker in a chaotic state than in any other one, or chaotic signals could be useful for hiding information. In such cases however we need a "wanted kind" of chaotic behavior with precisely prescribed features and/or we need techniques to switch between different kinds of behavior (chaos-order or chaos-chaos).

Considering the possibilities of influencing the dynamics of a chaotic circuit one can distinguish four basic approaches:

- variation of an existing accessible system parameter,

- change in the system design - modification of its internal structure,

- injection of an external signal(s),

- introduction of a controller (classical PI, PID, linear or nonlinear, neural, stochastic etc.).

Due to very rich dynamic phenomena encountered in typical chaotic systems, there exist a large variety of approaches to controlling such systems. This paper presents selected methods developed for controlling chaos in various aspects—starting from the most primitive concepts like parameter variation, through classical controller applications (open- and closed-loop control), to quite sophisticated ones like stabilization of unstable periodic orbits embedded within a chaotic attractor.

---

[a]Mathematically speaking the steady state behavior in a real electronic circuit corresponds to motion on an attracting limit set. Trajectories in nonlinear systems may possess several different limit sets and thus exhibit a variety of steady-state behaviors depending on the initial condition.

11.1 Fundamental properties of chaotic systems and goals of the control

As already mentioned in previous chapters systems displaying chaotic behavior posses specific properties. Now we will exploit these properties when attacking the control problem. In what way a chaotic system differs from any other object of control—what are its specific properties which could be useful?

The first observation concerns the origins of chaos in the system. If one could vary one or more of the parameters of the system, qualitative changes in the underlying dynamics could be observed. We could see examples of bifurcation diagrams constructed by numerical integration of system equations in previous chapters. Chaotic state usually results from a series of bifurcations. A route to chaos via a sequence of bifurcations has important implications for chaos control: firstly, it provides us with the insight into other accessible behaviors that can be obtained by changing parameters (this may be used for redesigning the system); secondly, a common feature of bifurcations is that stable and unstable orbits are created or annihilated—some of them may still exist in the chaotic range and constitute potential goals for control.

There are two fundamental properties of chaotic systems which are of potential use for control purposes.

The first property, already mentioned in this book many times is the *sensitive dependence on initial conditions*. For a long time this fundamental property has been considered as the main hindrance to control — How could one visualize successful control if the dynamics may change drastically with small changes of initial conditions or parameters? How could one produce a prescribed kind of behavior if errors in initial conditions will be exponentially amplified ?

This instability property does not however necessarily mean that control is impossible! It has been shown that despite the fact that nearby starting trajectories diverge they can be convergent to another prescribed kind of trajectory— one simply has to employ a different notion of stability. In fact, we do not need that the nearby trajectories converge—the requirement is quite different—the trajectories should merely converge to some goal trajectory $g(t)$, i.e.:

$$lim_{t \to \infty} |x(t) - g(t)| = 0 \qquad (11.1)$$

Depending on a particular application $g(t)$ could be one of the solutions existing in the system or any external waveform we would like to impose. Extreme sensitivity may even be of prime importance as control signals are in such cases very small.

The second important property of chaotic systems is the *existence of a countable infinity of unstable periodic orbits within the attractor* already considered earlier. These orbits, although invisible during experiments, constitute a skeleton of the attractor. Indeed, the trajectory passes at some time arbitrarily close to every such orbit. This invisible structure of unstable periodic orbits plays a crucial role in many methods of chaos control—using specific methods the chaotic trajectory can be perturbed in such a way that it will stay in the vicinity of a chosen unstable orbit from the skeleton.

These two fundamental properties of chaotic signals and systems offer some very interesting issues for control not available in other classes of systems. Namely:

- due to sensitive dependence on initial conditions it is possible to influence the dynamics of the systems using very small perturbations; moreover, the response of the system is very fast, and

- the existence of an countable infinity of unstable periodic orbits within the attractor offers extreme flexibility and a wide choice of possible goal behaviors for the same set of parameter values.

11.2   Simple techniques for suppressing chaotic oscillations — change in the system design

11.2.1   *Effects of large parameter changes*

The simplest way of suppressing chaotic oscillations is to change the system parameters (redesign) in such a way as to produce the desired kind of behavior. The influence of parameter variations on the asymptotic behavior of the system can be studied using a standard tool used in analysis of chaotic systems—the bifurcation diagram. Typical bifurcation diagram reveals a variety of dynamic behaviors for appropriate choices of system parameters and tells us what parameter values should be chosen to obtain the desired goal behavior. In electronic circuits, changes in the dynamic behavior are obtained by changing the value of one of its passive elements values (which means replacing one of the resistors, capacitors, or inductors). In Fig.11.1 we show a sample bifurcation diagram revealing a variety of dynamic behaviors observed in the RC chaos generator (studied by Ogorzałek ,1989a,b) when changing the value of capacitor C_2 . For example, choosing $C_2 = 2.05F$, one can expect a period one orbit in the system while for $C_2 = 1.45F$ (the units here are scaled in such a way that the fundamental frequency of the circuit is 1Hz), a stable

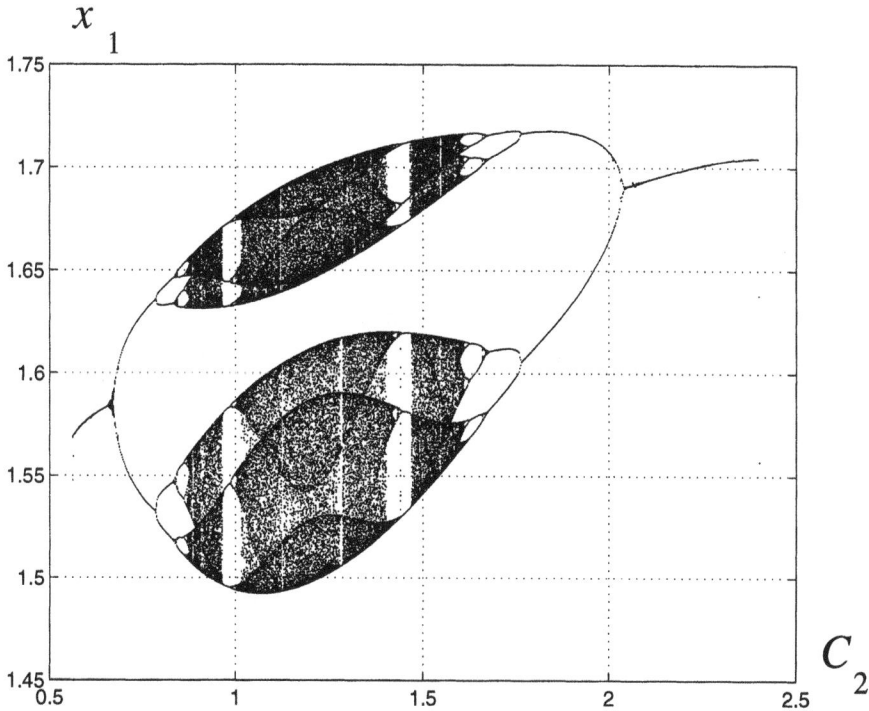

Figure 11.1: Sample bifurcation diagram obtained for the RC-ladder chaos generator showing attainable states for different values of the tuning parameter. From a chaotic state one can change the dynamic behavior to any kind of periodic state by appropriately adjusting the variable parameter.

period-3 orbit could be observed. Thus having the generator operating in a chaotic range one can tune (control) it using a variable capacitor to obtain any periodic state existing and displayed in the bifurcation diagram.

This method, although intuitively simple has a major drawback - it requires large parameter variations ("large energy control"). This requirement cannot be met in many physical systems where the construction parameters are either fixed or can be changed over very small ranges. This method is also difficult to apply on the design stage as there are no simulation tools for electronic circuits allowing bifurcation analysis (eg. SPICE has no such capability). On the other hand programs offering such types of analysis require a description of the problem in closed mathematical form such as differential or difference equations. Changes of parameters are even more difficult to introduce once the circuitry is fabricated or breadboarded and if possible at all can be done on trial-and-error basis.

11.2.2 *"Shock absorber" concept—change in system structure*

This simple technique is being used in a variety of applications. The motivation comes from mechanical engineering where devices absorbing unwanted vibrations are commonly used (eg. beds of machine-tools, shock-absorbers in vehicle suspensions etc). The idea is to modify the original chaotic system design (add the "absorber" without major changes in the design or construction) in order to change its dynamics in such a way that a new stable orbit appears in a neighborhood of the original chaotic attractor. In an electronic system, the absorber can be as simple as an additional shunt capacitor or an LC tank circuit. Kapitaniak *et al.*[229], proposed such a "chaotic oscillation absorber" for Chua's circuit - it is a parallel RLC circuit coupled with original Chua's circuit via a resistor (Fig.11.2)- depending on its value the original chaotic behavior can be converted to a chosen stable oscillation. The equations describing dynamics of this modified system can be given in a dimensionless form:

$$
\begin{aligned}
\dot{x} &= \alpha[y - x - g(x)] \\
\dot{y} &= x - y + z + \varepsilon(y^1 - y) \\
\dot{z} &= -\beta y \\
\dot{y'} &= \alpha'[-\gamma'y' + z' + \varepsilon(y - y')] \\
\dot{z'} &= -\beta'y'
\end{aligned}
\tag{11.2}
$$

In terms of circuit equations we have an additional set of two equations for the "absorber" (y^1, z^1) and a small term $(\varepsilon(y^1 - y))$ via which the original equations of Chua's circuit are modified.

11.3 External perturbation techniques

Several authors have demonstrated that a chaotic system can be forced to perform in a desired way by injecting external signals that are independent on the internal variables or structure of the system. Three types of such signal have been considered: (a) aperiodic signals ("resonant stimulation"); (b) periodic signals of small amplitude; (c) external noise.

11.3.1 *"Entrainment"—Open loop control*

Aperiodic external driving was one of the first methods introduced by Hübler 1989 (termed "resonant stimulation") and developed in the works of Hübler

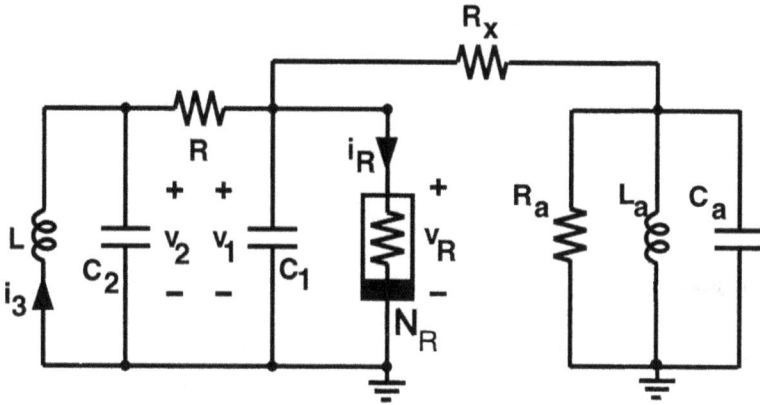

Figure 11.2: A parallel RLC circuit connected to the chaotic Chua's circuit acting as a chaotic oscillation absorber.

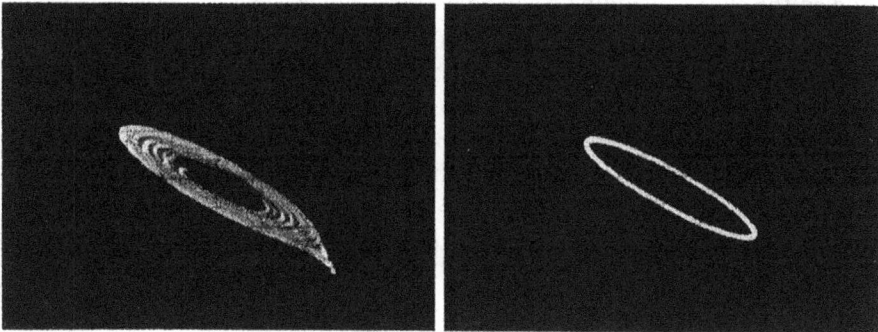

Figure 11.3: Spiral-type Chua's attractor and a period-one orbit obtained using parallel RLC oscillation absorber.

and Jackson. A mathematical model of the considered experimental system is needed (eg. in the form of differential equation: $\frac{dx}{dt} = F(x)$, $x \in R^n$, where $F(x)$ is differentiable, and a unique solution exists for every $t > 0$).

The goal of the control is to entrain the solution $x(t)$ to an arbitrarily chosen behavior g(t) i.e.

$$\lim_{t \to \infty} |x(t) - g(t)| = 0 \qquad (11.3)$$

Entrainment can be obtained by injecting the control signal:

$$\frac{dx}{dt} = F(x) + (\dot{g} - F(g))\mathbf{1}(t) \qquad (11.4)$$

The Entrainment method has the great advantage that no feedback is required and no parameters are being changed—thus the control signal can be computed in advance and no equipment for measuring the state of the system is needed. The goal does not depend on the considered system and in fact it could be any signal at all (except solutions of the autonomous system since $\dot{g} - F(g) \equiv 0$ in this case and there is no control signal). It should be noticed however that this method has limited applicability since a good model of the system dynamics is necessary, and the set of initial states for which the system trajectories will be entrained is not known.

11.3.2   Weak periodic perturbation

Interesting results have been reported by Breiman and Goldhirsch 1991, who studied the effects of adding a small periodic driving signal to a system behaving in a chaotic way. They discovered that external sinusoidal perturbation of small amplitude and appropriately-chosen frequency can eliminate chaotic oscillations in a model of the dynamics of a Josephson junction and cause the system to operate in some stable periodic mode. Unfortunately, there is little theory behind this approach and the possible goal behaviors can only be learned by trial-and-error. Some hope for further understanding and applications can be based on using theoretical results known from the theory of synchronization.

11.3.3   Noise injection

A noise signal of small amplitude injected in a suitable way into the circuit (system) offers potentially new possibilities for stabilization of chaos. First observations date back to the work of Herzel[193] and effects of noise injection were also studied in an RC-ladder chaotic oscillator[337,338]. In particular we observed that injection of noise of sufficiently high level can eliminate multiple domains of attraction. In our experiments with the RC-ladder chaos generator we found that the two main branches, representing two distinct, coexisting solutions as shown in Fig.3, will join together if white noise of high level is added.

This approach although promising needs further investigations as little theory is available to support experimental observations.

11.4 Control engineering approaches

Several attempts have been made to use known methods belonging to the "control engineer's toolkit". For example the use of PI and PID controllers for chaotic circuits, applications of stochastic control techniques, Lyapunov-type methods, robust controllers and many other methodologies including intelligent control and neural controllers have been described in literature (the paper by Chen and Dong, 1993a and Chapter 5 in Madan's book , 1993 give an excellent review of applications of such methods). In electronic circuits two schemes—linear feedback and time-delay feedback—seem to find most successful applications.

11.4.1 Error - feedback control

Several methods of chaos control have been developed which rely on a common principle that the control signal is some function ϕ of the difference between the actual system output $x(t)$ and the desired goal dynamics $g(t)$. This control signal could be an actual system parameter:

$$p(t) = \phi(x(t) - g(t)) \qquad (11.5)$$

as proposed in the "adaptive control scheme" described by Huberman and Lumer, or an additive signal produced by a linear controller:

$$u(t) = K(x(t) - g(t)) \qquad (11.6)$$

as in the methodology developed by Pyragas and Chen and Dong. The control term is simply added to the system equations. One can readily see that although mathematically simple such an "addition" operation might pose serious problems in real applications. The block diagram of the control scheme is shown in Fig.11.4.

Using error feedback chaotic motion has been successfully converted into periodic one both in discrete- and continuous-time systems. In particular

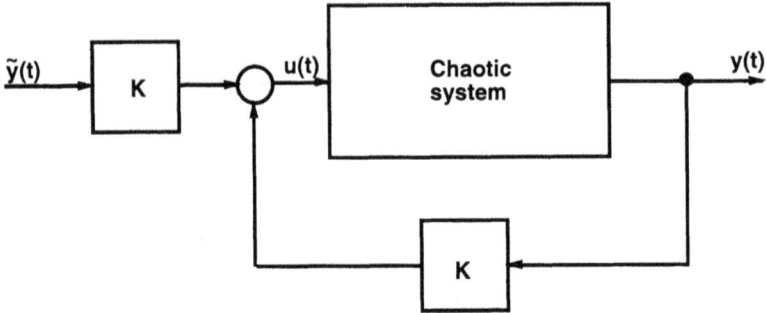

Figure 11.4: Block diagram of the linear feedback control scheme proposed by Chen and Dong.

chaotic motions in Duffing's oscillator and Chua's circuit have been controlled (directed) towards fixed points or periodic orbits (Chen and Dong 1993). The equations of the controlled circuit read:

$$
\begin{aligned}
\dot{x} &= \alpha[y - x - g(x)] \\
\dot{y} &= x - y + z - K_{22}(y - \tilde{y}) \\
\dot{z} &= -\beta y
\end{aligned}
\tag{11.7}
$$

Thus we have a single term added to the original equations.

Fig.7 shows a double scroll Chua's attractor and large saddle-type unstable periodic orbit towards which the system has been controlled.

The important properties of linear feedback chaos control method are that the controller has a very simple structure and access to the system parameters is not required. The method is immune to small parameter variations but might be difficult to apply in real systems (interactions of many system variables are needed). The choice of the goal orbit poses the most important problem— usually the goal is chosen in multiple experiments or can be specified on the basis of model calculations.

11.4.2 *Time-delay feedback control (Pyragas method)*

Pyragas[379] proposed an interesting feedback structure employing a delayed copy of the output signal. He obtained very promising results in the control of many different chaotic systems. Despite the lack of mathematical rigor, this method is being successfully used in several applications. The control signal applied to the system is proportional to the difference between the output and

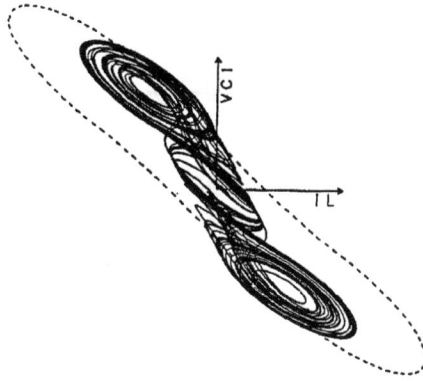

Figure 11.5: Double scroll Chua's attractor and large saddle type unstable periodic orbit stabilized using linear feedback.

a delayed copy of the same output.

$$\frac{dx}{dt} = F[x(t)] + K[y(t) - y(t - \tau)] \tag{11.8}$$

A block diagram of the control scheme is shown in Fig.11.6. Depending on the delay constant τ and the linear factor K, various kinds of periodic behaviors can be observed in the chaotic system.

An interesting application of this technique is described by Mayer-Kress et al.[289] —Pyragas' control scheme has been used for tuning chaotic Chua's circuits to generate musical tones and signals. More recently Celka[64] used Pyragas' method to control a real electro-optical system.

The positive features of the delay feedback control method are: self-control (no external signals are injected) and no access to system parameters is required. The control action is immune to small parameter variations. In real electronic systems the required variable delay element is readily available (for example, analog delay lines are available as of-the-shelf components). The primary drawbacks of the method is that there is no a priori knowledge of the goal (the goal is arrived at by trial-and-error).

11.5 Control in terms of stabilizing unstable periodic orbits

11.5.1 Ott-Grebogi-Yorke approach (OGY)

Ott, Grebogi and Yorke[354,355] proposed in 1990 a feedback method exploiting one of the key properties of chaotic attractors, namely the existence of a dense

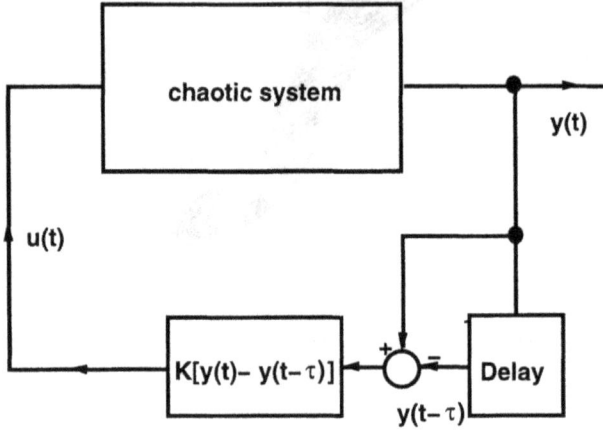

Figure 11.6: Block diagram of the delay feedback control scheme proposed by Pyragas.

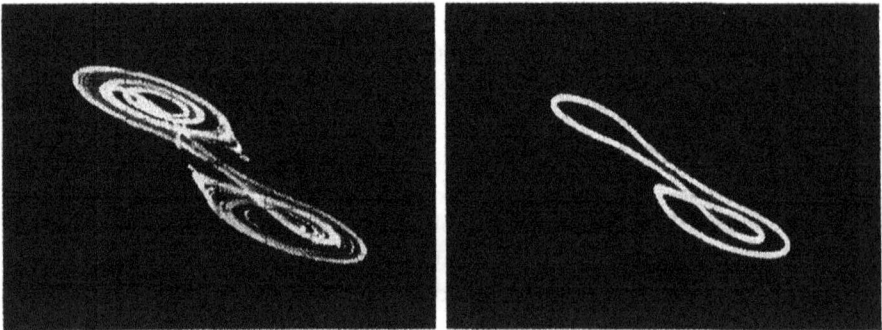

Figure 11.7: Double scroll Chua's attractor and one of the periodic orbits stabilized during experiments in the delayed feedback control.

set of unstable periodic orbits. To visualize best how the method works let us assume that the dynamics of the system are described by a k-dimensional map: $x_{n+1} = F(x_n, p)$, $x_i \in R^k$. This map in the case of continuous-time systems can be constructed eg. by introducing a transversal surface of section for system trajectories, p is some accessible system parameter which can be changed in some small neighborhood of its nominal value p^*. To explain the method we will concentrate now on stabilization of period-one orbit. Let $x_F = F(x_F, p^*)$ be the chosen fixed point (period-1) of the map around which we would like to stabilize the system. Assume further that the position of this orbit changes smoothly with p parameter changes (i.e. p^* is not a bifurcation value), and there are little changes in the local system behavior for small variations of p. In a close vicinity of this fixed point with good accuracy we can assume that the dynamics is linear and can be expressed approximately by:

$$x_{n+1} - x_0 = A(x_n - x_0) + g(p_n - p^*) \tag{11.9}$$

The elements of the matrix $A = \frac{\partial F}{\partial x}(x_F, p^*)$ and vector $g = \frac{\partial F}{\partial p}(x_F, p^*)$ can be calculated using the measured chaotic time series and analyzing its behavior in the neighborhood of the fixed point. Further the eigenvalues λ_s, λ_u and eigenvectors e_s, e_u of this matrix can be found.

$$Ae_u = \lambda_u e_u \text{ and } Ae_s = \lambda_s e_s \tag{11.10}$$

where the subscripts "u" and "s" correspond to unstable and stable directions respectively. These eigenvectors determine the stable and unstable directions in the small neighborhood of the fixed point (Fig.11.8).

$$A = \begin{bmatrix} e_u & e_s \end{bmatrix} \begin{bmatrix} \lambda_u & 0 \\ 0 & \lambda_s \end{bmatrix} \begin{bmatrix} e_u & e_s \end{bmatrix}^{-1} \tag{11.11}$$

Let us denote by f_s, f_u the contravariant eigenvectors ($f_s^T e_s = f_u^T e_u = 1$, $f_s^T e_u = f_u^T e_s = 0$ — comp. Fig.11.9). Thus

$$A = \begin{bmatrix} e_u & e_s \end{bmatrix} \begin{bmatrix} \lambda_u & 0 \\ 0 & \lambda_s \end{bmatrix} \begin{bmatrix} f_u^T \\ f_s^T \end{bmatrix} = \lambda_u e_u f_u^T + \lambda_s e_s f_s^T \tag{11.12}$$

This implies that f_u^T is a left eigenvector of A with the same eigenvalue e_u:

$$f_u^T A = f_u^T (\lambda_u e_u f_u^T + \lambda_s e_s f_s^T) = \lambda_u f_u^T \tag{11.13}$$

The control idea now is to monitor the system behavior until it comes close to the desired fixed point (we assume that the system is ergodic and the

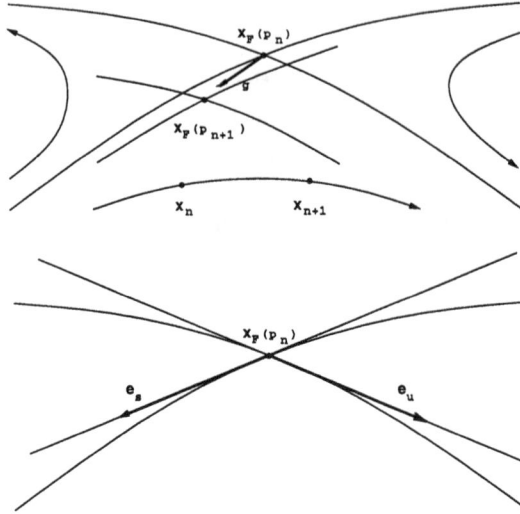

Figure 11.8: Explanation of the linearization technique used by the OGY method. (a) – displacement of the fixed point due to parameter change; (b) – Stable and unstable eigenvectors of the matrix A.

trajectory fills densely the attractor - thus eventually it will pass arbitrarily close to any chosen point) and then change p by a small amount so the next state x_{n+1} should fall on the stable manifold of x_0 i.e. choose p_n such that $f_u^T(x_{n+1} - x_F) = 0$:

$$p_n = -(\frac{\lambda_u}{f_u^T g})f_u^T(x_n - x_F) + p^* \tag{11.14}$$

which can be expressed as a local linear feedback action:

$$p_{n+1} = p_n + Cf_u^T(x_n - x_F(p_n)) \tag{11.15}$$

The actuation of the value of the control signal to be applied at the next iterate is proportional to the distance of the system state from the desired fixed point — $(x_n - x_F(p_n))$, projected onto the perpendicular unstable direction f_u. The constant C depends on the magnitude of the unstable eigenvalue λ_u and the shift g of the attractor position with respect to the change of the system parameter projected onto the unstable direction f_u. Fig.11.10 schematically explains the action of the OGY algorithm in the case k=2. The OGY technique has the notable advantage of not requiring analytical models of the system dynamics and is well-suited for experimental systems. One can use either full

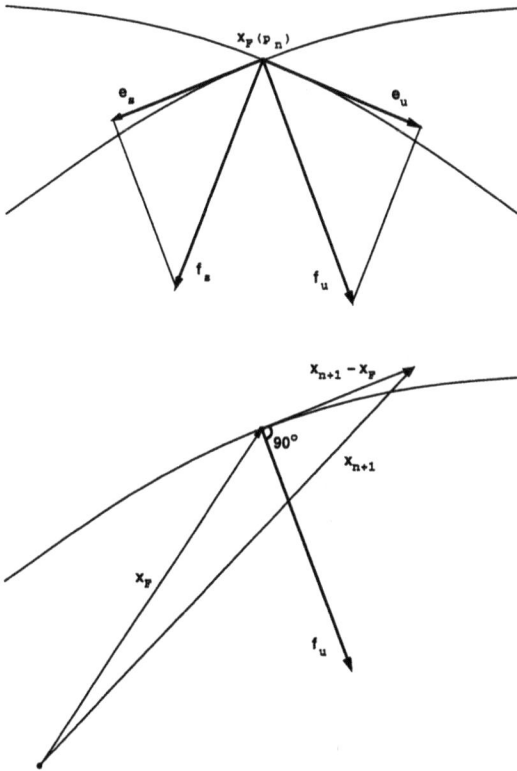

Figure 11.9: Explanation of the linearization technique used by the OGY method. (c) – New contravariant basis vectors; (d) – Action of the control.

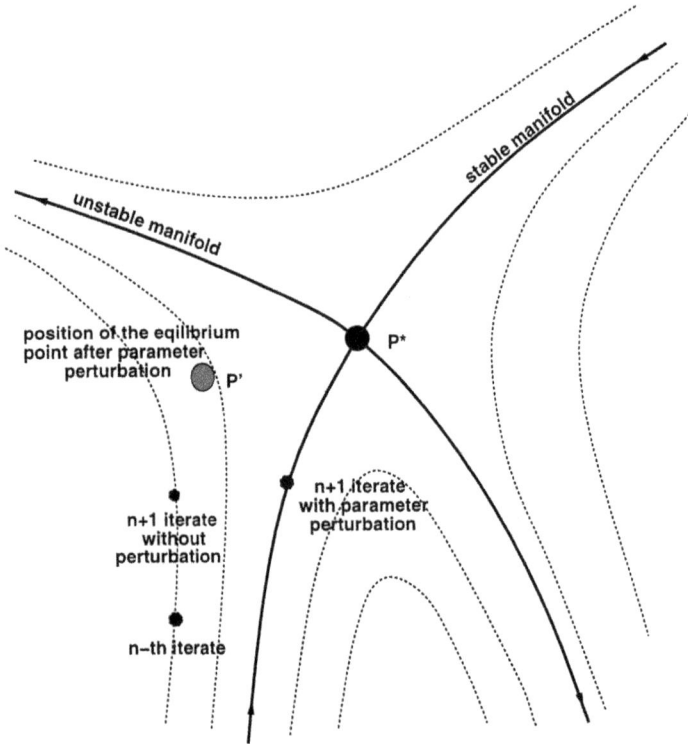

Figure 11.10: Schematic explanation of the OGY algorithm - a linear perturbation is applied in such a way that the successive iterate falls onto the stable manifold of the fixed point.

information from the process or use delay co-ordinate embedding technique using single variable experimental time series (see Dressler and Nitsche 1992). The procedure can also be extended to higher-period orbits. Any accessible variable (controllable) system parameter can be used for applying perturbation and the control signals are very small.

We have carried out an extensive study of application of the OGY technique to controlling chaos in Chua's circuit. Using an application-specific software package (Dabrowski *et al.*[96,97]) we were able to find some of the unstable periodic orbits embedded in the Double Scroll Chua's chaotic attractor and use them as control goals. Plates 15 and 16 show time evolution of the voltages when attempting to stabilize unstable period-one and period-two orbits. Before control is achieved, the trajectories exhibit chaotic transients — red trajectory starting from the point indicated by a blue circle represents transient behavior before entering close neighborhood of the chosen orbit. Yellow orbits indicate achieved stabilized orbits.

Implementation problems for OGY technique

When applying the OGY method to control chaos in a real physical circuit, the main problem encountered was the error introduced by: inevitable noise of the circuit elements, A/D and D/A conversion of signals (quantification), rounding operations in the computer calculations etc. The method was found to be very sensitive to the noise level—very small control signals sometimes are hidden within the noise and control was impossible. We will look into details of electronic implementations of chaos controllers in one of the next sections.

11.5.2 Sampled input waveform method

A very simple, robust and effective method of chaos control in terms of stabilization of an unstable periodic orbit has been proposed by Dedieu and Ogorzałek[98]. A sampled version of the output signal, corresponding to a chosen unstable periodic trajectory uncovered from a measured time series, is applied in the chaotic system causing the system to follow this desired orbit. In real systems, this sampled version of the unstable periodic orbit can be programmed into a programmable waveform generator and used as the forcing signal.

The block diagram of this control scheme is shown in Fig.11.11. For controlling chaos in Chua's circuit we try to force the system with a sampled version of a signal $\hat{V}_{C1}(t)$ ($\hat{V}_{C1}(t) = C^T \hat{\mathbf{x}}(t)$). Forcing the system with a continuous signal $\hat{V}_{C1}(t)$ will force the system to exhibit a solution $\mathbf{x}(t)$, which

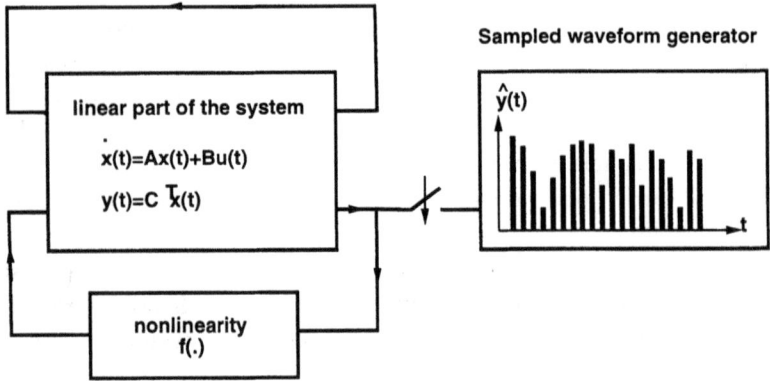

Figure 11.11: Block diagram of the sampled input chaos control system.

tends asymptotically towards $\hat{\mathbf{x}}(t)$. This is obvious since forcing $V_{C1}(t)$ will instantaneously force the current through the piecewise-linear resistance to a 'desired' value $\hat{i}_R(t)$. The remaining sub-circuit (R, L, C_2) which is a RLC stable circuit will then exhibit a voltage $V_{C2}(t)$ and a current $i_L(t)$ which will asymptotically converge towards $\hat{V}_{C2}(t)$ and $\hat{i}_L(t)$.

The sampled input control method is very attractive as the goal of the control can be specified using analysis of output time-series of the system; access to system parameters is not required. The control technique is immune to parameter variations, noise, scaling and quantization. Instead of a controller, we need a generator to synthesize the goal signal. Signal sampling reduces the memory requirements for the generator. Fig.11.12 shows the chaotic attractor and two sample orbits controlled within the chaos range.

11.5.3   OPF (occasional proportional feedback) - analog chaos controller

In real applications, a "one-dimensional" version of the OGY method—the so-called Occasional Proportional Feedback (OPF) method — has proven to be most efficient.

To explain the action of the OPF method let us consider a return map eg. as shown in Fig.11.13. For our considerations we take an approximate one-dimensional map obtained for the RC-ladder chaos generator as described in Chapter 7. For nominal parameter values the position of the graph of the map is as shown by the rightmost curve—all periodic points are unstable. In particular, the point P is an unstable equilibrium. Looking at the system operation starting from point v_n, at the next iterate (the next passage of the

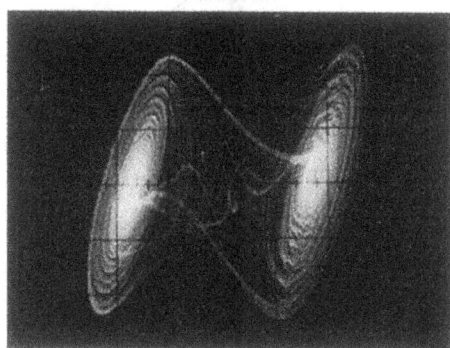

(a)

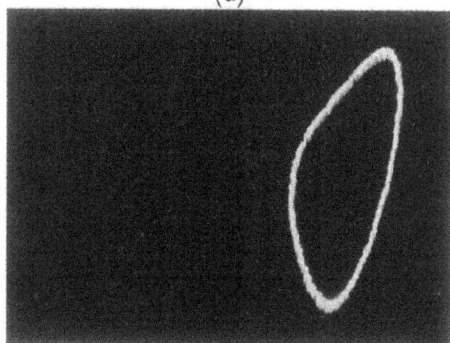

(b)

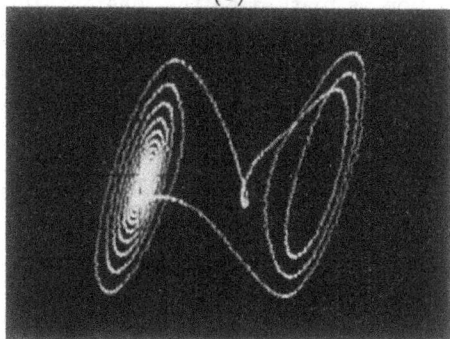

(c)

Figure 11.12: Double scroll Chua's attractor (a) observed in the experimental system and an two examples of orbits controlled during laboratory experiments (b and c).

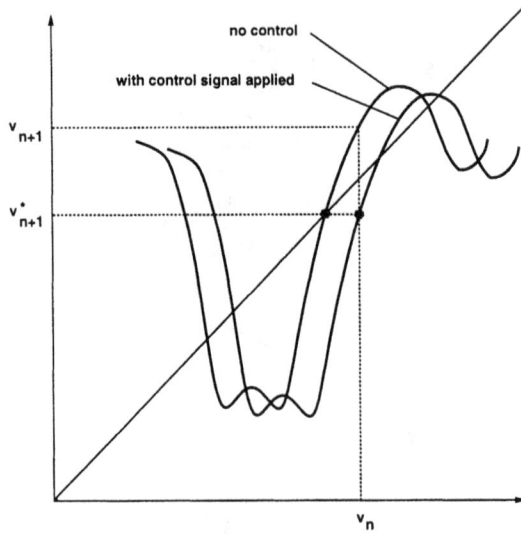

Figure 11.13: Schematic representation of the Occasional Proportional Feedback method.

trajectory through the Poincaré plane), one would obtain v_{n+1}. We would like to direct the trajectories towards the fixed point P. This can be achieved by changing a chosen system parameter such that the graph of the return map moves to a new position as marked on the diagram, thus forcing the next iterate to fall at v_{n+1}^*—after this is done the perturbation can be removed and activated again if necessary.

In mathematical terms we can compute the control signal using only one variable, for example ξ_1:

$$p(\xi) = p_0 + c(\xi_1 - \xi_{F1}). \qquad (11.16)$$

This method has been successfully implemented in a continuous-time analog electronic circuit and used in a variety of applications ranging from stabilization of chaos in laboratory circuits[203,204,215] to stabilization of chaotic behavior in lasers[89,374,386].

The OPF method (see Fig. 11.14) may be applied to any real chaotic system (also higher-dimensional ones) where the output can be measured in an electronic way and the control signal can be applied via a single electrical variable. The signal processing is analog and therefore is fast and efficient. Processing here means detecting position of a one-dimensional projection of a

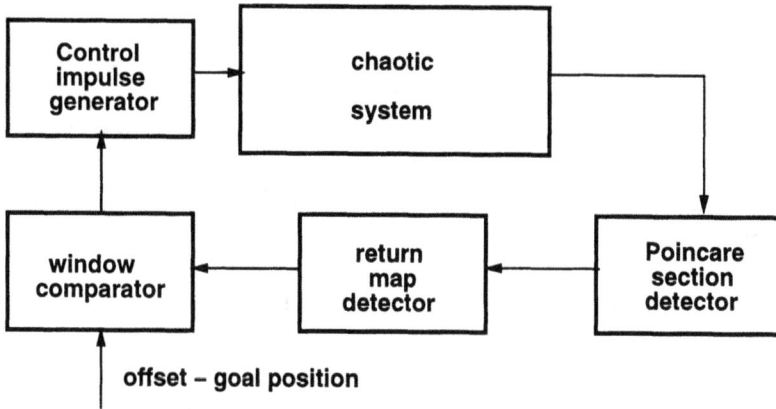

Figure 11.14: Block diagram of the OPF chaos controller.

Poincaré section (map) which can be accomplished by the window comparator, taking the input waveform. The comparator gives a logical high when the input waveform is inside the window. This is then ANDed with the delayed output from the external frequency generator. This logical signal drives the timing block which triggers the sample-and-hold and then the analog gate. The output from the gate, which represents the error signal at the sampling instant, is then amplified and applied to the interface circuit which transforms the control pulse into a perturbation of the system. The frequency, delay, control pulse width, window position, width and gain are all adjustable. The interface circuit used depends on the chaotic system under control.

One of the major advantages of Hunt's controller over OGY is that the control law depends on only one variable and does not require any complicated calculations in order to generate the required control signal. The disadvantage of the OPF method is that there is no systematic method for finding the embedded unstable orbits (unlike OGY). The accessible goal trajectories have to be determined by trial-and-error. The applicability of the control strategy is limited to systems in which the goal is suppression of chaos without more strict requirements.

11.6 Improved electronic chaos controller

Recently in collaboration with colleagues from University College Dublin we have proposed an improved electronic chaos controller which uses Hunt's method without the need for an external synchronizing oscillator.

Hunt OPF controller used the peaks of one of the system variables to generate the 1^D map. He then uses a window around a fixed level to set the region where control is applied. In order to find the peaks, Hunt's scheme uses a synchronizing generator.

In our modified controller [306,136,137], we simply take the derivative of the input signal and generate a pulse when it passes through zero. We use this pulse instead of Hunt's external driving oscillator as the "synch" pulse for our Poincaré map. This obviates the need for the external generator and so makes the controller simpler and cheaper to build.

A circuit diagram for our modified controller is given in Fig. 11.15. The variable level window comparator is implemented using a window comparator around zero and a variable level shift. Two comparators and three logic gates form the window around zero. The synchronizing generator used in Hunt's controller is replaced by an inverting differentiator and a comparator. A rising edge in the comparator's output corresponds to a peak in the input waveform. We use the rising edge of the comparator's output to trigger a monostable flip-flop. The falling edge of this monostable's pulse triggers another monostable, giving a delay. We use the monostable's output pulse to indicate that the input waveform peaked a fixed time earlier. If this pulse arrives when the output from the window comparator is high then a monostable is triggered. The output of this monostable triggers a sample-and-hold on its rising edge which samples the error voltage; on its falling edge, it triggers another monostable. This final monostable generates a pulse which opens the analog gate for a specific time (the control pulse width). The control pulse is then applied to the interface circuit, which amplifies the control signal and converts it into a perturbation of one of the system parameters, as required.

We tested our controller using a chaotic Colpitts oscillator[241,242] and laboratory implementation of Chua's circuit. The actual implementation of the control for the case of Colpitts[136,137] oscillator is shown in Fig.11.16. A typical chaotic attractor which can be observed in the Colpitts oscillator is shown in Fig. 11.6a.

By choosing the emitter resistor R_{EE} as the control parameter, we succeeded in stabilizing several periodic orbits of this system. The orbits shown in Fig. 11.6b,c were stabilized using a window of $0.4V$ placed at the bottom left hand side of the attractor. Implementation of a laboratory Chua's circuit together with interface circuit to connect the controller is shown in Fig.11.18. Fig.11.19 shows an example of a period-4 orbit stabilized using the improved chaos controller. Below the period-4 orbit in Fig.11.19 we show oscilloscope

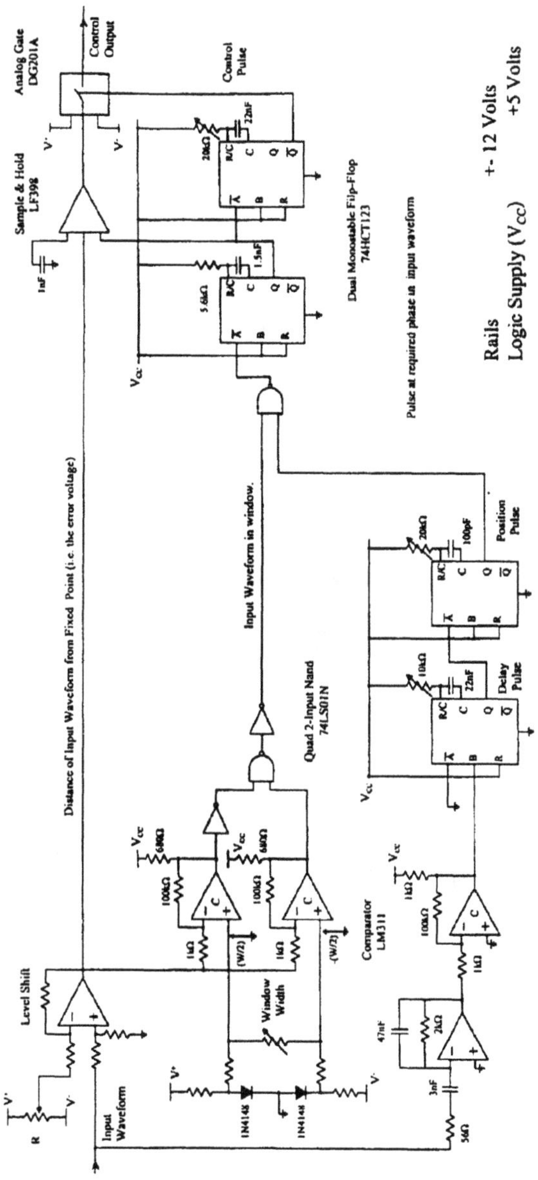

Figure 11.15: Improved analog chaos OPF controller without external synchronization.

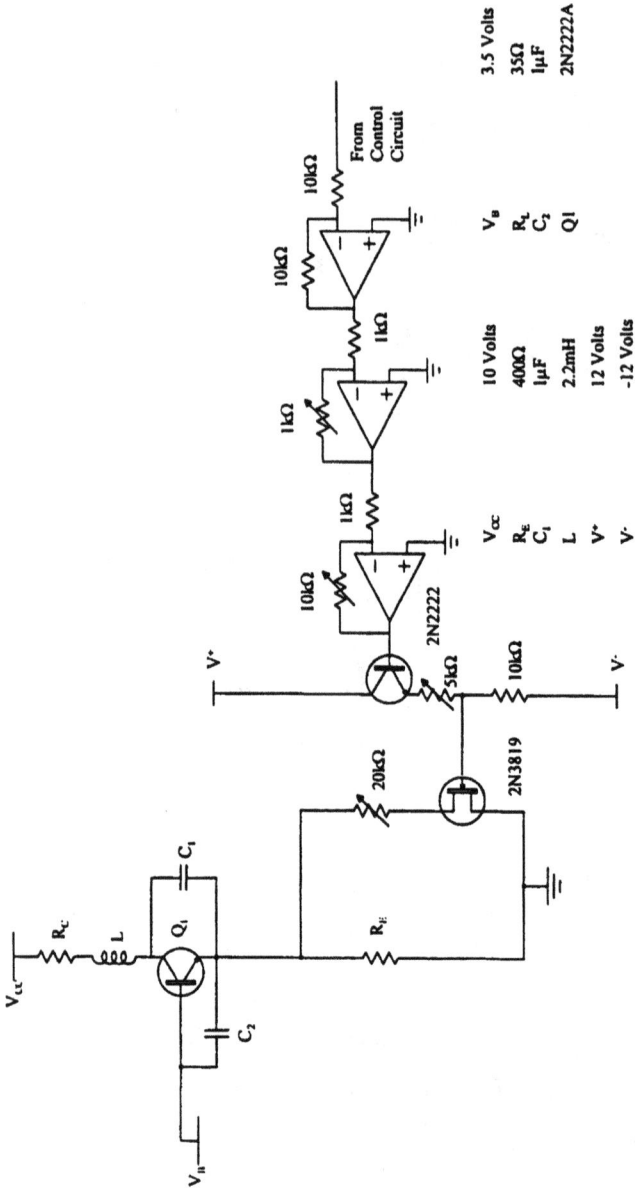

Figure 11.16: Circuit diagram for the chaotic Colpitts oscillator and the interface circuit.

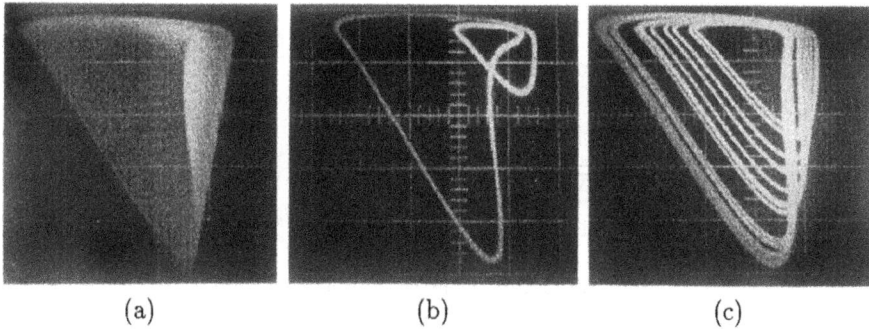

(a) (b) (c)

Figure 11.17: (a) Chaotic attractor observed in the Colpitts oscillator, (b) stabilized period-1 orbit, (c) stabilized long periodic orbit

traces for the goal trajectory and the control signal (bottom trace). It is interesting to notice the impulsive action of the controller. As with Hunt's implementation of OPF control, the orbits were found by trial and error.

11.7 "Chaos-to-chaos" control—Synchronization as a control problem

It should be pointed out that synchronization and control problems of chaotic systems have common features. In particular, synchronization can be considered as a particular type of control problem in which the goal of the control scheme is to track (follow) the desired (input) chaotic trajectory. It is only very recently that such a control problem has been recognized in control engineering. The linear coupling technique and the linear feedback approach to controlling chaos can be applied for obtaining any chosen goal—regardless of whether it is chaotic, periodic or constant in time. Using the approach described in Chapter 10, Section 10.2 we can synchronize/control chaotic systems to chaotic trajectories being solutions of a qualitatively different chaotic system. An impressive example of this kind of control/influence could be in generating Lorenz-like behavior in Chua's circuit, as described in Section 10.2. We believe that this kind of chaotic synchronization—control to a chaotic goal—could lead to new developments and possibly new applications of chaotic systems.

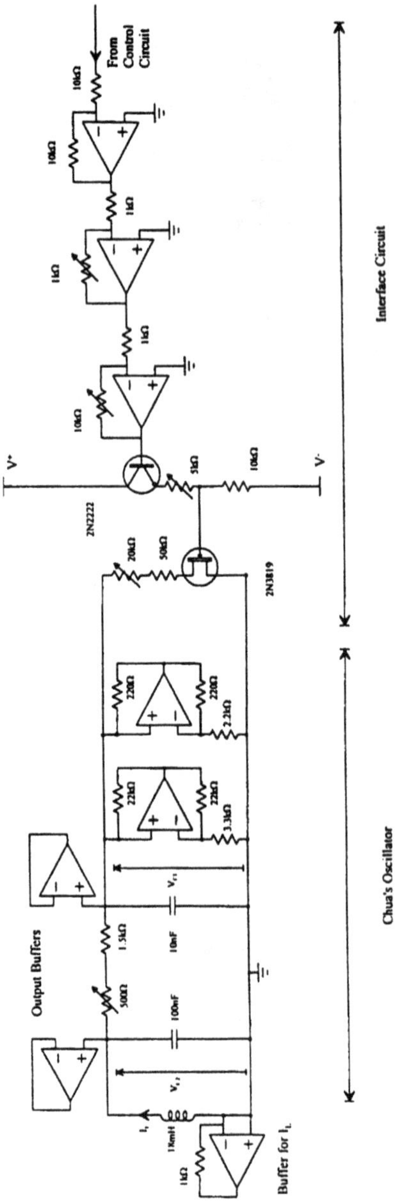

Figure 11.18: Circuit diagram for the implementation of Chua's circuit and the interface circuit.

(a)

(b)

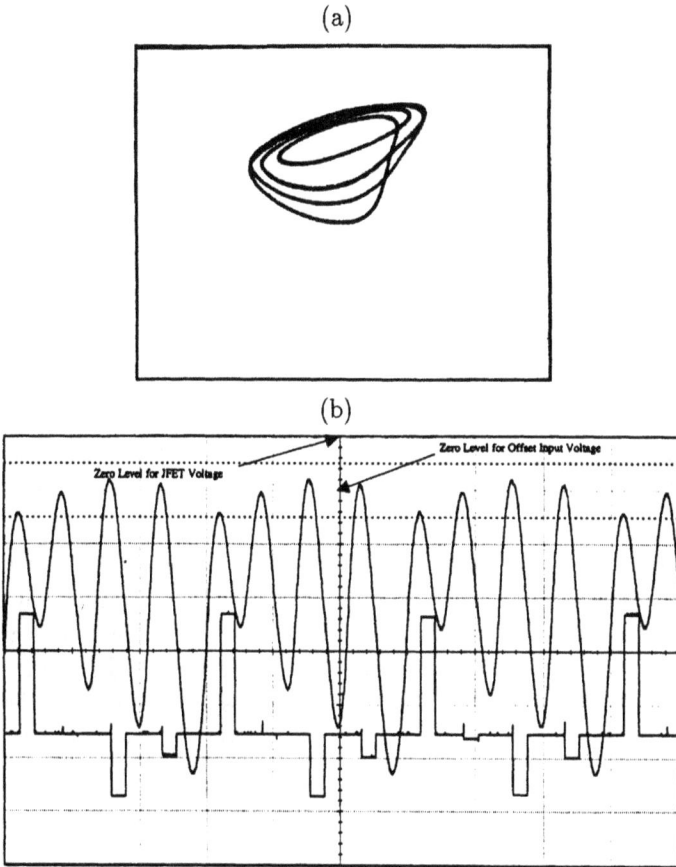

Figure 11.19: Period-4 orbit stabilized in Chua's circuit using improved chaos controller

11.8   Electronic implementations – are these possible?

The widespread interest in chaos control is basically due to extremely interesting and important possible applications. These applications range from
bio-medical ones (eg. defibrillation or removal of epileptic seizures), through
solid-state physics, lasers, aircraft wing vibrations and even weather control,
just to name a few of the daring attempts. Looking at the possible applications alone it becomes obvious that chaos control techniques and their possible
implementations will greatly depend on the nature of the process under consideration. Looking from the control implementation perspective, real systems
exhibiting chaotic behavior show many differences. The main ones are[346]:

- speed of the phenomenon (frequency spectrum of the signals),

- amplitudes of the signals,

- existence of corrupting noises, their spectrum and amplitudes,

- accessibility of the signals to measurement,

- accessibility of the control (tuning) parameters,

- acceptable levels of control signals

In most cases electronic equipment will play a crucial role. Looking for an
implementation of a particular chaos controller we must first look at the above
system-induced limitations: How can we measure and process signals from
the system? Are there any accessible system variables and parameters which
could be used for the control task? How to choose the ones that offer the best
performance for achieving control? At what speed do we need to compute
and apply the control signals? What is the lowest acceptable precision of
computation? Can we achieve control in real-time?

A slow system like a bouncing magneto-elastic ribbon (with eigen-frequencies
below 1Hz) is certainly not as demanding as a telecommunication channel
(running possibly at GHz) or a laser when it comes to control!

Considering electronic implementations, one must look at several closely
linked areas: sensors (for measurements of signals from a chaotic process), electronic implementation of the controllers, computer algorithms (if computers
are involved in the control process) and actuators (introducing control signals
into the system). External to the implementation (but directly involved in the
control process and usually fixed based on the measured signals) is finding of
the goal of the control.

However, many methods have been developed and described in the literature [66,344,345], most of them are still only of academic interest because of lack of success in implementation. A control method cannot be accepted as successful if computer simulation experiments are not followed by further laboratory tests and physical implementations. Only very few results of such tests are known - among the exceptions are: control of a green light laser [386], control of a magneto-elastic ribbon [103] and a few other examples.

11.8.1 Example - Implementation problems for the OGY method

When implementing the OGY method for a real world application one has to do the following series of elementary operations[346]:

1. data acquisition — measurement of a (usually scalar) signal from the chaotic system under consideration. This operation should be performed in such a way as not to disturb the existing dynamics. For further computerized processing, measured signals must be sampled and digitized (A/D conversion);

2. selection of appropriate control parameter;

3. finding unstable periodic orbits using experimental data (measured time series) and fixing the goal of control;

4. finding parameters and variables necessary for control;

5. application of the control signal to the system — this step requires continuous measurement of system dynamics in order to determine the moment at which to apply the control signal ie. the moment when the actual trajectory passes in a small vicinity of the chosen periodic orbit, and immediate reaction of the controller (application of the control pulse) in such an event.

In computer experiments, it has been confirmed that all the above-mentioned steps of OGY can be carried out successfully in a great variety of systems, achieving stabilization of even long-period orbits.

There are several problems which arise when attempting to build an experimental setup. Despite the fact that the variables and parameters can be calculated off-line, one has to consider that the signals measured from the system are usually corrupted due to noise, several nonlinear operations associated with the A/D conversion (possibly rounding, truncation, finite word-length, overflow correction etc.). Using corrupted signal values and the introduction of additional errors by the computer algorithms and linearization used for the control calculation may result in a general failure of the method. Additionally there are time delays in the feedback loop (eg. waiting for reaction of the computer, interrupts generated when sending and receiving data etc.)

Effects of calculation precision

In a simple example below, we consider the case of calculating control parameters to stabilize a fixed point in the Lozi map and show how the A/D conversion accuracy and the resulting calculations of limited precision affect the possibilities for control. In the tests described below we consider the quality of computations alone, without looking at other problems like time delays in the control loop[346].

To be able to compare the results of digital manipulations, we first computed the interesting parameters using analytical formulas — we will assume that these are the required true values of the parameters:

A. **Conditions of calculations: analytical (theoretical results)**
   Coordinates of the fixed point: $(0.8879418373, 0.8879418373)$
   Control vector g: $[0.4038961828, 0.4038961828]$
   Jacobian eigenvalues: $-1.913225419, -0.1553417742$
   Stable direction: $[0.1535007507, 09881485105]$
   Unstable direction: $[0.8880129457, -0.459818393]$
   Possibilities of control: successful

Further we calculated the same parameters using different word-length and different arithmetics (overflow rules, rounding or truncation etc.). The results are summarized in B through F below. Note, that all these numbers due to limited word-length have only two or three significant digits — all the remaining digits represent "calculation noise" — accumulation errors added in the computer calculations.

B. **Conditions of calculations: floating point representation, 32bit word-length**
   Coordinates of the fixed point: $(0.883074402809143, 0.881085395812988)$
   Control vector g: $[0.420236974954605, 0.411248594522476]$
   Jacobian eigenvalues: $-1.913303494453430, -0.0375206992030014$
   Stable direction: $[0.111510016024113, 0.993763267993927]$
   Unstable direction: $[0.882781267166138, -0.4698422999382]$
   Possibilities of control: successful

C. **Conditions of calculations: fixed point representation, 12bit precision, rounding**
Coordinates of the fixed point: (0.883055806159973, 0.881017923355103)
Control vector g: [0.435197770595551, 0.46562001191232]
Jacobian eigenvalues: −1.922089338302612, −0.03144733607769
Stable direction: [0.115565232932568, 0.993299901485443]
Unstable direction: [0.88289874792099, −0.469563484191895]
Possibilities of control: successful

D. **Conditions of calculations: fixed point representation, 12bit precision, truncation**
Coordinates of the fixed point: (0.882387578487396, 0.880320370197296)
Control vector g: [0.456262081961496, 0.477374047040939]
Jacobian eigenvalues: −1.933647394180298, 0.01997028850019
Stable direction: [0.115881502628326, 0.993263065814972]
Unstable direction: [0.896894037723541, −0.442245543003082]
Possibilities of control: successful

E. **Conditions of calculations: fixed point representation, 10bit precision, rounding**
Coordinates of the fixed point: (0.883261978626251, 0.881381988525391)
Control vector g: [0.349784225225449, 0.3618243932724]
Jacobian eigenvalues: −1.89931046962738, −0.020768757909536
Stable direction: [0.136759996414185, 0.990604162216187]
Unstable direction: [0.890513718128204, −0.454956352710724]
Possibilities of control: often fails

F. **Conditions of calculations: fixed point representation, 8bit precision, rounding**
Coordinates of the fixed point: (0.886274695396423, 0.886274695396423)
Control vector g: [0.000011920940779, 0.000011920940779]
Jacobian eigenvalues: 0, 0
Stable direction: Impossible to determine
Unstable direction: Impossible to determine
Possibilities of control: impossible

Comparing the results of computations summarized above we can easily see that if we are able to achieve an accuracy of two to three decimal digits the calculations are precise enough to ensure proper functioning of the OGY algorithm in the case of the Lozi system. To have some safety margin and robustness in the algorithm the acceptable A/D accuracy cannot be lower than 12 bit and probably it would be best to apply 16bit conversion. This kind of accuracy is nowadays easily available using general purpose A/D converters even at speeds in the MHz range. Implementing the algorithms, one must consider the cost of implementation — with growing precision and speed requirements, the cost grows exponentially. This issue might be a great limitation when it comes to IC implementations.

Approximate procedures for finding periodic orbits

Another possible source of problems in the control procedure are errors introduced by algorithms for finding periodic orbits (goals of the control). Using experimental data we can only find approximations to unstable periodic orbits [17, 261, 403].

Simple technique proposed by Lathrop and Kostelich[261] for recovering unstable periodic orbits from an experimental time series described in Chapter 3 has several drawbacks. Firstly, the results strongly depend on the choice of ε and the length of the measured time series. Further, they depend on the choice of norm and the number of state variables analyzed; Secondly, the stopping criterion ($||x_{m+k} - x_m|| < \varepsilon$) in the case of discretely sampled continuous-time systems is not precise enough. This means that one can never be sure of how many orbits have been found or whether all orbits of a given period have been recovered.

In our experiments we varied ε between 0.000001 and 0.001 and fixed the threshold for distinguishing between orbits at 0.001. Although with greater ε more orbits with given period were detected, most of them were later recognized as identical — there was no significant difference in the number and shape of different unstable periodic orbits found.

As this step is typically carried out off-line it does not badly affect the whole control procedure. It has been found in experiments that when the tolerances for detection of unstable orbits were chosen too large, the actual trajectory stabilized during control showed greater variations and the control signal had to be applied every iteration to compensate for inaccuracies. Clearly, making the tolerance large could cause failure of control.

Some new methods have been proposed recently [423] which could possibly improve localization procedures for unstable periodic orbits.

Effects of time delays

Several elements in the control loop may introduce time delays that can be detrimental to the functioning of the OGY method. Although all calculations may be done off-line, two steps are of paramount importance:

- detection of the moment when the trajectory passes the chosen Poincaré section

- determination of the moment at which the control signal should be applied (close neighborhood of chosen orbit).

When these two steps are carried out by a computer with a data acquisition card, at least a few interrupts (and therefore a time delay) must be generated in order to detect the Poincaré section, take the decision of being in the right neighborhood, and to send the correct control signal.

Most experiments with OGY control of electronic circuits have been able to achieve control when the systems were running in the 10-100 Hz range. We found out that for higher frequency systems time delays become a crucial point in the whole procedure. The failure of control was mainly due to the late arrival of the control pulse –the system was being controlled at a wrong point in state space where the formulas used for calculations were probably no longer valid; trajectory was already far away from the section plane when the control pulse arrived.

To compensate for some of the delays, we have proposed a hardware solution for a detector of the Poincaré section and vicinity detector. Block diagrams of these two pieces of equipment are shown in Figures 1 and 2. The Poincaré section system here uses all (three in our application) state variables to simplify detection. To implement this function using just one variable, delay coordinates must be introduced — realization in hardware would become much more complicated in this case if possible at all (one could think of calculating suitable time delay by a computer algorithm and memorizing the necessary time-delayed samples in special-purpose registers).

11.9   Conclusions

The control problems existing in the domain of chaotic systems are far from being fully identified, to say nothing about their solutions. Due to extreme richness of the phenomena, one can treat every single such problem as a new challenge for scientists and engineers. Among many problems to be solved, we mention here the basic ones: How can the methods already developed be used in real applications? What are the limitations of these techniques in terms of convergence, initial conditions etc.? What are the limitations in terms of system complexity and possibilities of implementation? Are these methods useful in biology or medicine?

New application areas have opened up thanks to these new developments in various aspects of controlling chaos—these include neural signal processing (Mpitsos and Burton 1992), biology and medicine (Nicolis[314], Garfinkel *et al.*[141], Schiff *et al.*[402] and many others. We expect in near future a breakthrough in cardiac disfunctioning treatment thanks to new generation of defibrillators and pace-makers functioning on the chaos control principles. Sensitive dependence on initial conditions—another one of the key features of chaotic systems—offers yet another fantastic control possibility called "targetting" (Kostelich *et al.*[254], Shinbrot *et al.*[417])—a desired point in the phase space is reached by piecing together in a controlled way fragments of chaotic trajectories. This method has already been applied successfully for directing satellites to desired positions using infinitesimal amounts of fuel (see Farquhar *et al.*[117] and papers *ibidem*).

Finally, we stress that almost all chaotic systems known to date have strong links with electronic circuits—variables are sensed in an electric or electronic way, identification, modeling and control is carried out using electric analogs, electronic equipment and electronic computers as sensors and transducers. This presents an infinite wealth of opportunities for researchers in the domain of electronics, and not only for theoreticians.

REFERENCES

1. N.B. Abraham, J.P. Gollub, H.L. Swinney : "Testing Nonlinear Dynamics". *Physica D*, **11D**, pp.252-264, 1984. ,
2. R.H. Abraham, C.D. Shaw : *The Visual Mathematics Library - VISMATH*, Aerial Press Inc. Santa Cruz, 1983. • Volume 0 : Manifolds and Mappings. • Volume 1 : Dynamics - The Geometry of Behavior. Part I. Periodic Behavior. • Volume 2 : Dynamics - The Geometry of Behavior. Part II. Chaotic Behavior. • Volume 3 : Dynamics - The Geometry of Behavior. Part III. Global Behavior. • Volume 4 : Dynamics - The Geometry of Behavior. Part IV. Bifurcation Behavior.
3. V.S. Afraimovich, L.P. Shil'nikov : "On singular trajectories of dynamical systems", *Uspiechy Matiematiczeskich Nauk*, (in Russian) **27**, No.3(165), pp.189-190, 1972.
4. V.S. Afraimovich, L.P. Shil'nikov : "On singular manifolds of Morse-Smale systems", (in Russian) *Trudy Moskowskogo Matiematiczeskogo Obshchestva*, **28**, pp.181-214, 1973.
5. V.S. Afraimovich, V.I. Nekorkin, G.V. Osipov, V.D. Shalfeev *Stability, Structure and Chaos in Nonlinear Synchronisation Arrays*, A.V. Gaponov–Grekhov and M.I. Rabinovich Eds., IAP, USSR Academy of Sciences, 1989 (in Russian)
6. K.T. Alligood, E.D. Yorke, J.A. Yorke : "Why Period-Doubling Cascades Occur : Periodic Orbit Creation Followed by Stability Shedding", Report Dept. of Mathematics, University of Maryland, College Park, 1985.
7. D. Amrani : "Limit Cycles and Chaotic Motion in Relay Systems", MPh Thesis. The University of Sussex, Brighton 1988.
8. D. Anastassiou, M. Vetterli : "Television by the bit", *IEEE Circuits and Devices Magazine*, **7**, pp.16-21, 1991.
9. A. Antoniou : *Digital Filters: Analysis and Design* (McGraw-Hill, New York), 1979.
10. V.S. Anishchenko, T.E. Vadivasova, D.E. Postnov and M.A. Safonova, "Synchronization of Chaos", *Int. J. Bifurcation and Chaos*, **2**, No.3, pp.633-644, 1992.
11. A. Arneodo, P. Coullet, C. Tresser : "Possible New Strange Attractors with Spiral Structure", *Comm. Mathematical Physics*, **79**, pp.573-579, 1981.
12. A. Arneodo, P. Coullet, C. Tresser : "Oscillators with Chaotic Behavior : An Illustration of a Theorem by L.P.Šilnikov", *J.Stat. Physics*, **27**, No.1, pp.171-182, 1982.

13. A. Arneodo, P.H. Coullet, E.A. Spiegel, C. Tresser : "Asymptotic Chaos", *Physica D*, **14D**, pp.327-347, 1985.
14. W.I. Arnold : *Ordinary differential equations*, (Polish edition) PWN, Warszawa 1975.
15. W.I. Arnold : *Theory of differential equations*, (in Polish PWN, Warszawa 1983.
16. R. Artuso, E. Aurell, P. Cvitanović : "Recycling of Strange Sets: , I. Cycle Expansions, II.Applications", Preprint of The Niels Bohr Institute NBI - 89 - 41 oraz NBI - 89 - 42, 1989.
17. D. Auerbach, P. Cvitanović, J.-P. Eckmann, G. Gunaratne, I. Procaccia : "Exploring Chaotic Motion Through Periodic Orbits", *Phys. Rev. Lett.*, **58**, No.23, pp.2387-2389, 1987.
18. A. Azzouz, M. Hasler : "Chaos in the RL- Diode Circuit", in *Proc. ECCTD'87*, **1**, pp.295-300, Paris 1987.
19. A. Azzouz, R. Duhr, M. Hasler : "Transition to Chaos in a Simple Nonlinear Circuit Driven by a Sinusoidal Voltage Source", *IEEE Trans. Circuits Systems*, **CAS-30**, No.12, pp.913-914, 1983.
20. A. Azzouz, R. Duhr, M. Hasler : " Bifurcation Diagram for a Piecewise - Linear Circuit", *IEEE Trans. Circuits Systems*, **CAS-31**, No.6, pp.587-588, 1984.
21. J. Baillieul : "Chaotic Dynamics and Nonlinear Feedback Control", in *Proc.XVI Banach Center Semester on Optimal Control*, Polish Academy of Sciences 1984.
22. J. Baillieul, R.W. Brockett, R.B. Washburn : "Chaotic Motion in Nonlinear Feedback Systems", *IEEE Trans. Circuits Systems*, **CAS-27**, No.11, pp.990-997, 1980.
23. P. Bak : "The devil's staircase", *Physics Today*, **39**, pp.38-45, 1986.
24. M.F. Barnsley : *Fractals Everywhere*, Academic Press, Orlando 1988.
25. P. Bartissol, L.O. Chua : "The Double Hook", *IEEE Trans. Circuits and Systems*, **CAS-35**, No.12, pp.1512-1522, 1988.
26. V.N. Belykh, N.N. Verichev, L. Kocarev, L.O.Chua "On chaotic synchronisation in a linear array of Chua's circuits". Memorandum No. UCB/ERL M93/11, 29 January 1993.
27. P. Bergé, Y. Pomeau, C. Vidal : *Order within Chaos. Towards a Deterministic Approach to Turbulence*, John Wiley & Sons 1986.
28. M.-Y. Bernard, J.-C. Neau : "Les phénoménes chaotiques non-linéaires? c'est très simple... au début", *Annales des Telecommunications*, **42**, No.5-6, pp.197-209, 1987.
29. M. Bier, T.C. Bountis : "Remerging Feigenbaum Trees in Dynamical Systems", *Physics Letters*, **103A**, No.5, pp.239-244, 1984.
30. G.D. Birkhoff : "Dynamical systems", in *Amer. Math. Soc. Coll. Publ. USA.*, **9**, 1927.
31. J.S. Birman, R.F. Williams : "Knotted periodic orbits in dynamical systems. I: Lorenz's equations", *Topology*, **22**, pp.47-82, 1983.
32. L.S. Block, W.A. Coppel : *Dynamics in One Dimension*, Springer-Verlag, Berlin Heidelberg, 1992.

33. R. Boite, H. Leich : *Les filtres numériques*, (Masson, Paris), 1980.
34. F. Boehme, U. Feldmann, W. Schwarz, A. Bauer : "Information transmission by chaotizing", in *Proc. NDES'94*, pp.163-168, Krakow 1994.
35. R. Boite, X.-Q. Yang : "On the free overflow oscillations in digital filters" 1989 (preprint).
36. N.K. Bose : *Digital Filters. Theory and Applications* (North-Holland/Elsavier, New York), 1985.
37. R. Bowen : *Equilibrium States and Ergodic Theory of Anosov Diffeomorphisms*, Springer Lecture Notes in Mathematics. **470**, 1975.
38. Y. Breiman, I. Goldhirsch : "Taming Chaotic Dynamics with Weak Periodic Perturbation", *Phys. Rev. Letters*, **66**, pp.2545-2548, 1991.
39. R.W. Brockett : "On the Asymptotic Properties of Solutions of Differential Equations with Multiple Equilibria", *J.Diff. Equations*, **44**, No.2, pp.249-262, 1982.
40. R.W. Brockett : "On Conditions Leading to Chaos in Feedback Systems", in *Proc. 21 IEEE Conf. Decision and Control*, Florida. **2**, pp.932-936, 1982.
41. R.W. Brockett , W. Cebuhar : "A Prototype Chaotic Differential Equation", 1984 (preprint).
42. R.W. Brockett , J. Loncaric : "Chaos and Randomness in Dynamical Systems", in *Proc. 22nd IEEE Conference on Decision and Control*, **1**, pp.1-4, 1983.
43. R.W. Brockett, J.R. Wood : "Understanding Power Converter Chaotic Behavior Mechanisms in Protective and Abnormal Modes", MIT 1984 (preprint).
44. D.S. Broomhead, G. Rowlands : "On The Use of Perturbation Theory in the Calculation of the Fractal Dimension of Strange Attractors", *Physica D*, **10D**, pp.340-352, 1984.
45. M.E. Broucke : "One Parameter Bifurcation Diagram for Chua's Circuit", *IEEE Trans. Circuits Systems*, **CAS-34**, No.2, pp.208-209, 1987.
46. R. Brown, L.O. Chua : "Clarifying chaos: Examples and Counterexamples", *Int. J. Bif. Chaos*, **6**, No.2, pp.219-249, 1996
47. P. Bryant, C. Jeffries : "The Dynamics of Phase Locking and Points of Resonance in a Forced Magnetic Oscillator", *Physica D*, **25D**, pp.196-232, 1987.
48. I.M. Burkin, G.A. Leonow : "On existence of non-trivial periodic solutions in auto-oscillatory systems", *Sibirskij Matiematiczeskij Zurnal*, (in Russian), **XVIII**, No.2, pp.251-262, 1977.
49. I.M. Burkin, G.A. Leonow : "On existence of periodic solutions in a nonlinear system of third order", *Differencyjalnyje Urownienja*, (in Russian), **XX**, No.12, pp.2036-2042, 1982.
50. N.W. Butenin, Ju.I. Nejmark, N.A. Fufajew : *Introduction to the theory of nonlinear oscillations*, (in Russian) Nauka, Moskwa 1976.
51. H.J. Butterweck : "On the quantization noise contributions in digital filters which are uncorrelated with the output signal". *IEEE Trans.*

Circuits Systems, **CAS-26**, pp.901-910, 1989.

52. H.J. Butterweck, F.H.R. Lucassen, G. Verkroost : "Subharmonics andd other quantization effects in periodically excited recursive digital filters". *IEEE Trans. Circuits Systems*, **CAS-33**, pp.958-964, 1986.

53. H.J. Butterweck, J. Ritzerfeld, M. Werter : "Finite wordlength effects in digital filters". *Archiv für Elektronik und Übertragungstechnik*, **43**, pp.76-89, 1989.

54. V. Capellini, A.G. Constantinides, P. Emiliani : *Digital Filters and Their Applications* (Academic Press, London-New York-San Francisco), 1978.

55. G.-C. Cardarilli, M. Hasler : "Nonlinear effects in digital transmission systems". *Mitteilung AGEN*, **46**, pp.37-40, 1987.

56. T.L. Carroll, I. Triandaf, I.B. Schwartz, L. Pecora : Tracking unstable orbits in an experiment. *Physical Review A*, **46**, pp.6189-6192, 1992.

57. T.L. Carroll "Synchronising chaotic systems using filtered signals" *Physical Review E*, **50**, No.4, 1994.

58. T.L. Carroll and L.M. Pecora, "Synchronizing Chaotic Circuits", *IEEE Trans. Circuits Syst.*, **CAS-38**, No.4, pp.453-456, 1991.

59. T.L. Carroll and L.M. Pecora, "A Circuit for Studying the Synchronisation of Chaotic Systems", *Int. J. Bifurcation and Chaos*, **2**, No.3, pp.659-667, 1992.

60. T.L. Carroll and L.M. Pecora, "Cascading Synchronized Chaotic Systems", *Physica D*, **D67**, pp.126-140, 1993.

61. T.L. Carroll and L.M. Pecora, "Synchronising nonautonomous chaotic circuits", *IEEE Trans. Circuits Syst. Part II*, **CAS-40**, No.10, pp.646-650, 1993.

62. T.L. Carroll, L. Pecora Eds. : *Nonlinear Dynamics in Circuits*, World Scientific 1995.

63. J. Cascais, N. Dilao, A. Noronha da Costa : "Chaos and Reverse bifurcation in a RCL Circuit", *Physics Letters*, **93A**, No.5, pp.213-216, 1983.

64. P. Celka : "Control of Time-Delayed Feedback Systems with Application to Optics", In *Proc. NDES'94* , Krakow 1994, pp.141-146.

65. G. Chen : "Control and synchronization of chaotic systems", (bibliography), EE Dept, Univ of Houston, TX – available from ftp: "uhoop.egr.uh.edu/pub/TeX/chaos.tex" (login name and password: both "anonymous").

66. G. Chen, X. Dong : "From chaos to order — Perspectives and methodologies in controlling chaotic nonlinear dynamical systems", *Int. J. Bifurcation and Chaos*, **3**, pp.1363-1409, 1993.

67. G. Chen, X. Dong : "Controlling Chua's circuit", em J. Circuits, Systems and Computers, **3**, pp.139-148, 1993.

68. S.-N. Chow , J.K. Hale, J. Mallet-Paret : "An Example of Bifurcation to Homoclinic Orbits", *J.Diff.Equations*, **37**, No.3, pp.351-373, 1980.

69. L.O. Chua : "Dynamic Nonlinear Networks : State of The Art", *IEEE*

Trans. Circuits Systems, **CAS-27**, No.11, pp.1059-1087, 1980.

70. L.O. Chua : "Nonlinear Circuits", *IEEE Trans. Circuits Systems*, **CAS-31**, pp.69-81, 1984.

71. L.O. Chua, F. Ayrom : "Designing Nonlinear Single Op-Amp Circuits: A Cook-Book Approach", *Int.J.Circuit Theory and Applications*, **13**, pp.235-268, 1985.

72. L.O. Chua, M. Hasler, J. Neirynck, P. Verburgh : "Dynamics of a Piecewise-Linear Resonant Circuit", *IEEE Trans. Circuits Systems*, **CAS-29**, No.8, pp.535-547, 1982.

73. L.O. Chua, M. Komuro, T. Matsumoto : "The Double Scroll Family: • Part I : Rigorous Proof of Chaos. • Part II : Rigorous Analysis of Bifurcation Phenomena", *IEEE Trans. Circuits Systems*, **CAS-33**, No.11, pp.1073-1118, 1986.

74. L.O. Chua, M. Komuro, T. Matsumoto : "The Double Scroll Family • Part I and •Part II", Memorandum No.UCB/ERL M85/102, University of California Berkeley, 4 December 1985.

75. L.O. Chua, T. Lin : "Chaos in Digital Filters", *IEEE Trans. Circuits Syst.*, **CAS-35**, No.6, pp.648-658, 1988.

76. L.O. Chua, T. Lin : "Fractal Pattern of Second - Order Nonlinear Digital Filters : A New Symbolic Analysis", *Int. J. Circuit Theory and Applications*, **18**, pp.541-550, 1990.

77. L.O. Chua, T. Lin : "Chaos and Fractals from Third - Order Digital Filters", *Int. J. Circuit Theory and Appl.*, **18**, No.3, pp.241-256, 1990.

78. L.O. Chua, T. Lin : "Fractal patterns of second order non-linear digital filters: A new symbolic analysis". *Int.J.Circ.Theory Appl.*, **18**, pp.541-550, 1990.

79. L.O. Chua, R,N. Madan : "Sights and Sounds of Chaos", *IEEE Circuits and Devices Magasine*, **4**, No.1, pp.3-13, 1988.

80. L.O. Chua, T. Sugawara : "3-D Rotation Instrument for Displaying Strange Attractors", Memo. UCB/ERL M86/61, University of California Berkeley, 1986.

81. L.O. Chua, T. Sugawara : "Panoramic Views of Strange Attractors", *Proc. IEEE*, **75**, No.8, pp.1107-1120, 1987.

82. L.O. Chua, Y. Yao, Q. Yang : "Devil's Staicase Route to Chaos in a Nonlinear Circuit", *Int. J. Circuit Theory and Appl.*, **14**, pp.315-329, 1986.

83. L.O. Chua, Y. Yao, Q. Yang : "Generating Randomness from Chaos and Constructing Chaos with Desired Randomness", *Int. J. Circuit Theory and Appl.*, **18**, No.3, pp.215-240, 1990.

84. L.O. Chua : "Global unfolding of Chua's circuit", *IEICE Trans. Fund. Elect. Comm. Comp. Sci.*, **E76-A**, pp.704-734, 1993.

85. L.O.Chua : "The genesis of Chus'a circuit", *Archiv für Elektronik und Übertragungstechnik*, **46**, No.4, pp.250-257, 1992.

86. L.O.Chua, L. Kocarev, K.Eckert and M.Itoh, "Experimental Chaos Synchronisation in Chua's Circuit", *Int. J. Bifurcation and Chaos*, **2**,

No.3, pp.705-708, 1992.
87. T. Claasen, F.G. Mecklenbräuker, J.B.H. Peek : "Frequency domain criteria for the absence of zero-input limit cycles in nonlinear discrete-time systems, with applications to digital filters", *IEEE Trans. Circuits Systems*, **CAS-22**, pp.232-239, 1975.
88. P. Collet, J.-P. Eckmann : *Iterated Maps of the Interval as Dynamical Systems*, Birkhäuser, Basel 1980.
89. Corcoran, E. 1992 Kicking Chaos out of Lasers, *Scientific American*.
90. J.D. Crawford, S. Omohundro : "On the Global Structure of Period Doubling Flows", *Physica D*, **13D**, pp.161-180, 1984.
91. J.P. Crutchfield "Spatio–temporal complexity in nonlinear image processing". IEEE Trans. Circuits Systems, vol.CAS.35, No.7, pp.770–780, 1988.
92. P. Cvitanović : "Invariant Measurement of Strange Sets in Terms of Cycles", *Physical Review Letters*, **61**, No.24, pp.2729-2732, 1988.
93. W.J. Dallas : "A digital prescription for X-ray overload". *IEEE Spectrum*, **4**, pp.33-36, 1990.
94. A.C. Davies : "Limit Cycles and Chaos from Overflow non-linearities of Digital Filters". Paper presented at Dresden, Germany, Feb. 12-14, 1990.
95. A.C. Davies, R. Sriranjan : "Chaotic Signals generated by Overflow non-linearity in Digital-filters". IEE Saraga Colloquium "Electronic Filters", Digest No. 1989/97, 9 June 1989.
96. A. Dąbrowski, Z. Galias, M.J. Ogorzałek : "A Study of Identification and Control in a Real Implementation of Chua's Circuit", In *System Structure and Control - 2nd IFAC Workshop*, pp.278-281, 1992.
97. A. Dąbrowski, Z. Galias, M.J. Ogorzałek : "On-line identification and control of chaos in a real Chua's circuit", *Kybernetika - Czech Academy of Sciences*, **30**, pp.425-432, 1994.
98. H. Dedieu, M.J. Ogorzałek : "Controlling chaos in Chua's circuit via sampled inputs", *Int. J. Bifurcation and Chaos*, **4**, pp.447-455 1994.
99. H. Dedieu, M. Ogorzalek : "Identification and Control of a Particular Class of Chaotic Systems", in em SPIE Proceedings, Chaotic Circuits for Communication, Vol. 2612, pp. 148-156, 1995.
100. H. Dedieu, M.P.Kennedy, M. Hasler : "Chaos Shift Keying: Modulation and Demodulation of a Chaotic Carrier using Self-Synchronising Chua's Circuits", *IEEE Trans. Circuits Systems Part II*, **CAS-40**, No.10, pp.634-642, 1993
101. D.D. Delchamps : "Some Chaotic Consequences of Quantization in Digital Filters and Digital Control Systems", in *Proc 1989 IEEE International Symposium on Circuits and Systems*, **1**, pp.602-605, Portland, Oregon 1989.
102. R.L. Devaney : *An Introduction to Chaotic Dynamical Systems*, The Benjamin/Cummings Publishing Co. Inc. 1986.
103. W.L. Ditto, M.L. Pecora : "Mastering chaos", *Scientific American*, pp.62-68, 1993.

104. W.L. Ditto, S.N. Rauseo, M.L. Spano : "Experimental Control of Chaos", *Phys. Rev. Letters*, **65**, pp.3211-3214, 1990.
105. W.L. Ditto, M.L. Spano, J.F. Lindner : "Techniques for the control of chaos", *Physica*, **D86**, pp.198-211, 1995.
106. U. Dressler, G. Nitsche : "Controlling Chaos Using Time Delay Coordinates", *Phys. Rev. Letters*, **68**, pp.1-4, 1992.
107. P.M. Ebert, J.E. Mazo, M.G. Taylor : "Overflow oscillations in digital filters". *Bell System Technical Journal*, **48**, pp.2999-3020, 1969.
108. J.P. Eckmann, D. Ruelle : "Ergodic theory of chaos and strange attractors", *Rev. Modern Phys.*, **57**, No.1, pp.617-656, 1985.
109. S. Espejo-Meana, A. Rodriguez - Vasquez, J.L. Huertas, J.M. Quintana : "Application of Chaotic Switched - Capacitor Circuits for Random Number Generation", in *Proc 1989 European Conference on Circuit Theory and Design*, IEE Publication 308, pp.440-444, Brighton 1989.
110. T. Endo, L.O. Chua : "Chaos from Phase - Locked Loops", *IEEE Trans. Circuits Syst.*, **CAS-34**, No.8, pp.987-1003, 1988.
111. T. Endo, L.O. Chua : "Chaos from Phase - Locked Loops - Part.II : High - dissipation case", *IEEE Trans. Circuits and Systems*, **CAS-35**, No.2, pp.255-263, 1989.
112. T. Endo, L.O. Chua : "Bifurcation and Fractal Basin Boundaries of Phase - Locked Loop Circuits", in *Proc. 1989 IEEE International Symposium on Circuits and Systems*, **2**, pp.820-823, Portland - Oregon 1989.
113. L. Fabiny, K. Wiesenfeld : "Clustering behavior of oscillator arrays", *Physical Review A*, **43**, No.6, pp.2640-2648, 1991.
114. J.D. Farmer, E. Ott, J.A. Yorke : "The Dimension of Chaotic Attractors", *Physica D*, **7D**, pp.153-180, 1983.
115. J.D. Farmer, J.J. Sidorowich : "Exploiting Chaos to Predict the Future and Reduce Noise", Los Alamos 1988 (preprint).
116. J.D. Farmer, J.J. Sidorowich : "Predicting Chaotic Time Series", *Phys. Rev.Letters*, **59**, No.8, pp.845-848, 1987.
117. R. Farquhar, D. Muhonen, S. Davies, D. Dunham : "Trajectories and orbital meneuvers for the ISEE-3/ICE comet mission", *J. Astronautical Sci.*, **33**, pp.235-254, 1985.
118. S. Fauve, F. Heslot : "Stochastic Resonance in a Bistable System. Physics Letters", **97A**, No.1,2, pp.5-7, 1983.
119. M.J. Feigenbaum : "Universal Behavior in Nonlinear Systems", *Physica D*, **7D**, pp.16-39, 1983.
120. U. Feldmann, M. Hasler, W. Schwarz : "Communication by chaotic signals: The inverse system approach", in *Proc. IEEE ISCAS'95*, pp.680-683, Seattle 1995.
121. P. Fisher, W.R. Smith : *Chaos, Fractals and Dynamics*, M.Dekker, New York 1985.
122. S.W. Fomin, I.P. Kornfeld, J.G. Sinaj : *Edgodic Theory*, (in Polish) PWN, Warszawa 1987.
123. D. Fournier, H. Kawakami, C. Mira : "Sur les bifurcations d'un

difféomorphisme quadratique bi-dimensionnel. Situations homoclines et heteroclines. Zone Morse-Smale", *C.R.Acad.Sci.Paris*, **298**, Serie I, No.11, pp.253-256, 1984.

124. D. Fournier, H. Kawakami, C. Mira : "Feuilletage du plan des bifurcations d'un difféomorphisme bi-dimensionnel. Doublement de l'ordre (periode) des zones sources et des zones echangeurs", *C.R.Acad.Sci.Paris*, **301**, Serie I, No.5, pp.223-228, 1985.

125. D. Fournier, H. Kawakami, C. Mira : "Sequences de Myrberg et communications entre feuillets du plan des bifurcations d'un difféomorphisme bi-dimensionnel", *C. R. Acad. Sci. Paris*, **301**, Serie I, No.6, pp.325-328, 1985.

126. D. Fournier, H. Kawakami, C. Mira : "La chaine de liaisons : image symbolique d'un ensemble complet de communications entre feuillets du plan des bifurcations d'un difféomorphisme bi-dimensionnel", *C.R.Acad.Sci. Paris*, **303**, Serie I, No.7, pp.315-320, 1986.

127. T.B. Fowler : "Application of Stochastic Control Techniques to Chaotic Nonlinear Systems", *IEEE Trans. Automatic Control*, **AC-34**, pp.201-205, 1989.

128. S. Fraser, R. Kapral : "Universal Vector Scaling in One- Dimensional Maps", *Physical Review A*, **30**, No.2, pp.1017-1025, 1984.

129. A. Fraser, H. Swinney : "Independent coordinates for strange attractors from mutual information", *Phys. Rev. A*, **33**, pp.1134, 1986.

130. P. Frederickson, J.L. Kaplan, E.D. Yorke, J.A. Yorke : "The Lyapunov Dimension of Strange Attractors", *J.Diff. Equations*, **49**, pp.185-209, 1983.

131. W.J. Freeman, "Srange attractors that govern mammalian brain dynamics shown by trajectories of electroencephalographic (EEG) potential", *IEEE Trans. Circuits Systems*, **CAS-35**, No.7, pp.781-783, 1988.

132. E. Freire, L.G. Franquelo, J. Aracil : "Periodicity and Chaos in an Autonomous Electronic System", *IEEE Trans. Circuits Systems*, **CAS-31**, No.3, pp.237-247, 1984.

133. Z. Galias, M.J. Ogorzałek : "Bifurcation Phenomena in Second Order Digital Filter with Saturation-Type Adder Overflow Characteristic", *IEEE Trans. Circuits Syst.*, **CAS-37**, No.8, 1990.

134. Z. Galias, M.J. Ogorzałek : "On symbolic dynamics of a chaotic second-order digital filter". *Int. J. Circuit Theory and Applications*, 1991.

135. Z. Galias, S. Mitkowski, M.J. Ogorzałek : "KRAKFIL - A Software Toolkit for Unconventional Digital Filter Analysis", in *Proceedings of The 1991 European Conference on Circuit Theory and Design*, Copenhagen, 1991.

136. Z. Galias, C.A. Murphy, M.P.Kennedy and M.J. Ogorzałek : "Electronic Chaos Controller", *Chaos Solitons and Fractals* (in press) 1997.

137. Z. Galias, C.A. Murphy, M.P. Kennedy, M.J. Ogorzałek : "A feedback Chaos Controller: Theory and Implementation". in *Proc. 1996 IEEE*

ISCAS, vol.3, str.120-123, Atlanta 1996.

138. Z. Galias : "On a discrete time nonlinear system associated with the second-order digital filter", *SIAM Journal on Applied Mathematics*, Nov. 1995.

139. J.M. Gambaudo : "Ordre, désordre et frontière des systèmes Morse - Smale", Thèse d'Etat, Université de Nice 1987.

140. J.M. Gambaudo, P.A. Glendinning, C. Tresser : "Stable Cycles with Complicated Structure", *Le Journal de Physique Lettres*, **46**, No.15, pp.L653-L657, 1985.

141. A. Garfinkel, M.L. Spano, W.L. Ditto, J.N. Weiss : "Controlling Cardiac Chaos", *Science*, **257**, pp.1230-1235, 1992.

142. P. Gaspard : "Generation of a Countable Set of Homoclinic Flows Through Bifurcation", *Physics Letters*, **97A**, No.1, 2, pp.1-4, 1983.

143. P. Gaspard : "Generation of a Countable Set of Homoclinic Flows Through Bifurcation in Multidimensional Systems", *Physique Theorique - Mathematiques Appliquees, Bulletin de la Classe des Sciences*, Academie Royale de Belgique, pp.61-83, 1985.

144. P. Gaspard, G. Nicolis : "What Can We Learn from Homoclinic Orbits in Chaotic Dynamics?", *J.Stat. Physics*, **31**, No.3, pp.499-518, 1983.

145. P. Gaspard, R. Kapral, G. Nicolis : "Bifurcation Phenomena near Homoclinic Systems : A Two Parameter Analysis", *J.Stat. Physics*, **35**, No.5/6, pp.697-727, 1984.

146. P. Gaspard, X.-J. Wang : "Sporadicity : Between Periodic and Chaotic Dynamical Behaviours", *Proc. National Academy of Sciences USA*, **85**, pp.4591-4595, 1988.

147. N.K. Gavrilov, L.P. Shil'nikov : 'On three-dimensional systems close to systems with a homoclinic orbit I", *Matiematiczeskij Sbornik*, (in Russian), **88(130)** No.4(8), pp.475-492, 1973.

148. N.K. Gavrilov, Shil'nikov L.P. : 'On three-dimensional systems close to systems with a homoclinic orbit II", *Matiematiczeskij Sbornik*, (in Russian), **90(132)** No.1, pp.139-156, 1973.

149. J.Y. Gay : "Un dispositif d'obtention de sections de Poincaré et d'applications de premier retour sur calculateur analogique", in *Proc. IMACS-IFAC Symposium, Modelling and Simulation for Control of Distributed Parameter Systems*, Villeneuve d'Ascq., pp.681-683, 1986.

150. J.-Y. Gay : "A Hybrid Device for Obtaining Poincaré Sections and First Return Maps", in *Applied Modelling and Simulation*, ed.: P.Borne, S.G.Tzafestas, North -Holland, Amsterdam 1987.

151. J.-Y. Gay : "A High-Speed Device for Obtaining First Return Maps", in *Proc. 12 IMACS World Congress*, pp.283-284, Paris 1988.

152. R. Genesio, A. Tesi : "Distortion Control of Chaotic Systems: The Chua's circuit, *J. Circuits Syst. Comput.*, **3**, pp.151-171, 1993.

153. D.P. George : "Bifurcations in a Piecewise-Linear System", *Physics Letters A*, **118**, No.1, pp.17-21, 1986.

154. N.N. Georgijew : "Investigations of a nonlinear differential equation of third order", (in Bulgarian), *Comptes Rendus de l'Academie Bulgare*

des Sciences, **29**, No.1, pp.17-19, 1976.

155. N. Gershenfeld : "An Experimentalist's Introduction to the Observation of Dynamical Systems", in *Directions in Chaos*, Hao-Bai Lin Ed. World Scientific,

156. P. Glendinning : "Bifurcations near Homoclinic Orbits with Symmetry",, *Physics Letters*, **103A**, No.4, pp.163-166, 1984.

157. P. Glendinning, C. Sparrow : "Local and Global Behavior near Homoclinic Orbits", *J.Stat. Physics*, **35**, No.5/6, pp.645-696, 1984.

158. I. Goldhirsch, P.-L. Sulem, S.A Orszag : "Stability and Lyapunov Stability of Dynamical Systems : A Differential Approach and a Numerical Method", *Physica D*, **27D**, pp.311-337, 1987.

159. P. Grabowski : "Dissipative systems of the Lur'e type", (in Polish), Proc. Nat. Conference Network Theory and Electronic Systems, Poznań 1985.

160. P. Grabowski : "An Application of Shilnikov's Theorem to Linear Systems with Piecewise-Linear Feedback", in *Structure, Coherence and Chaos in Dynamical Systems*, Ed.: P.L.Christiansen, R.D.Parmentier, pp.473-479 . Manchester University Press , Manchester-New York 1989

161. P. Grabowski, M.J. Ogorzałek : "Multilevel Oscillations and Chaos", in *Proc. 1989 European Conference on Circuit Theory and Design*, Brighton, IEE Publication 308, pp.376-380, Brighton 1989.

162. P. Grassberger, I. Procaccia : "Measuring the Strangeness of Strange Attractors", *Physica D*, **9D**, pp.189-208, 1983.

163. P. Grassberger, I. Procaccia : "Characterisation of Strange Attractors", *Physical Rev. Letters*, **50**, No.5, pp.346-349, 1983.

164. P. Grassberger, I. Procaccia : "Dimensions and Entropies of Strange Attractors from a Fluctuating Dynamics Approach", *Physica D*, **13D**, pp.34-54, 1984.

165. C. Grebogi, E. Ott, J.A. Yorke : "Basin Boundary Metamorphoses : Changes in Accessible Boundary Orbits", *Physica*, **24D**, pp.243-262, 1987.

166. C. Grebogi, E. Ott, J.A. Yorke : "Crises, Sudden Changes in Chaotic Attractors, and Transient Chaos", *Physica D*, **7D**, pp.181-200, 1983.

167. C. Grebogi, E. Ott, S. Pelikan, J.A. Yorke : "Strange Attractors that Are Not Chaotic", *Physica D*, **13D**, pp.261-268, 1984.

168. J.M. Greene, J.-S. Kim : "The Calculation of Lyapunov Spectra", *Physica D*, **24D**, pp.213-225, 1987.

169. J. Grudniewicz, J. Kudrewicz, T. Barczyk : "Synchronization and Chaos in a Real Discrete Phase - Locked Loop", in *Proc. 1988 IEEE International Symposium on Circuits and Systems*, **1**, pp.269-272, Helsinki 1988.

170. J. Gruendler : "The Existence of Homoclinic Orbits and the Method of Melnikov for Systems in R^n", *SIAM J.Math.Anal.*, **16**, No.5, pp.907-931, 1985.

171. J. Guckenheimer, P. Holmes : *Nonlinear Oscillations, Dynamical Sys-*

tems and Bifurcations of Vector Fields, Ed.4. Springer Verlag, New York-Berlin-Heidelberg 1993.

172. J. Guckenheimer, R.F. Williams : "Structural Stability of Lorenz Attractors", *Publications IHES*, **50**, pp.307-320, 1980.

173. J. Guckenheimer : "Toolkit for Nonlinear Dynamics", *IEEE Trans. Circuits Syst.*, **CAS-30**, No.8, pp.586-590, 1983.

174. I. Gumowski, C. Mira : *Dynamique Chaotique. Transformations ponctuelles. Transition Ordre - Desordre*, Collection Nabla, Cepadues Editions 1980.

175. F. Guo, L. Pei, S. Wu : "Chaotic Behavior in a Nonlinear Circuit", *Acta Electronica Sinica*, **14**, No.1, pp.30-35, 1986.

176. H. Haken : "At Least One Lyapunov Exponent Vanishes if the Trajectory of an Attractor Does not Contain a Fixed Point", *Physics Letters*, **94A**, No.2, pp.71-72, 1983.

177. K.S. Halle, C.-W. Wu, M.Itoh and L.O.Chua, "Spread Spectrum Communication through Modulation of Chaos", *Int. J. Bifurcation and Chaos*, **3**, No.2, 1993.

178. D.C. Hamill, D.J. Jeffries : "Subharmonics and Chaos in a Controlled Switch-Mode Power Converter", *IEEE Trans. Circuits Systems*, **CAS-35**, No.8, pp.1059-1061, 1988.

179. T.T. Hartley, F. Mossayebi : "Control of Chua's circuit", *J. Circuits Syst. Comput.*, **3**, pp.173-194, 1993.

180. M. Hasler : "Electrical Circuits with Chaotic Behavior", *Proc. IEEE*, **75**, No.8, pp.1009-1021, 1987.

181. M. Hasler : "Avoiding complex behavior". *Helvetica Physica Acta*, **62**, pp.552-572, 1989.989.

182. M. Hasler, J. Neirynck : *Circuits non linéaires*, Presse Polytechniques Romandes, Lausanne 1985.

183. M. Hasler "Synchronisation principles and applications" Chapter 6.2 in Circuits Syst., C. Toumazou Ed., Tutorials of ISCAS'94.

184. M. Hasler : "Bruit de Quantification et Cycles Limites", Chapter 6 in *Techniques modernes de traitement nomérique des signaux*, ed. M. Kunt (Presses Polytechniques et Universitaires Romandes, Lausanne) pp.279-348, 1991.

185. C. Hayashi : *Oscillations in physical systems*, WNT, Warszawa 1973.

186. C. Hayashi, M. Abe, K. Oshima, H. Kawakami : "The Method of Mapping as Applied to the Solution for Certain Types of Nonlinear Differential Equations", in *Proc. 9th Conf. Nonlin. Osc.*, pp.40-44, Kiev Naukowa Dumka 1984.

187. C. Hayashi, H. Kawakami : "Behavior of Solutions for Duffing-van der Pol Equation", in *Proc. 10th International Conf. Nonlinear Oscillations*, pp.164-167, Varna 1984.

188. C. Hayashi, H. Kawakami : "Bifurcations and the Generation of Chaotic States in the Solutions of Nonlinear Differential Equations", *Proc. 4th Nat. Congress Theor. and Apl. Mech.*, pp.1-6, Varna 1981.

189. C. Hayashi, Y. Ueda, H. Kawakami : "Solution of Duffing's Equation

Using Mapping Concept", in *Proc.4th Conf. Nonlin. Osc.*, pp.25-40, Prague 1967.

190. C. Hayashi, Y. Ueda, H. Kawakami : "Transformation Theory as Applied to the Solutions of Nonlinear Differential Equations of the Second Order", *Int.J.Nonlinear Mech.*, 4, pp.235-255, 1969.

191. C. Hayashi, Y. Ueda, H. Kawakami : "Periodic Solutions of Duffing's Equation with Reference to Doubly Asymptotic Solutions", in *Proc. 5th Conf. Nonlin. Osc.*, 2, pp.507-521.-521, Kiev 1969.

192. S. Hayes, C. Grebogi, E. Ott : "Communicating with Chaos", *Physical Review Letters*, 70, No.20, pp.3031-3034, 1993.

193. H. Herzel : "Stabilization of chaotic orbits by random noise", *ZAMM*, 68, pp.1-3, 1988.

194. K. Hockett, P.J. Holmes : "Nonlinear Oscillations, Iterated Maps, Symbolic Dynamics, and Knotted Orbits", *Proc. IEEE*, 75, No.8, pp.1071-1080, 1987.

195. K. Hockett, P.J. Holmes : "Josephson's Junction, Annulus Maps, Birkhoff Attractors, Horseshoes and Rotation Sets", *J.Ergod. Theory Dynam. Syst.*, 6, pp.205-239, 1986.

196. P.J. Holmes : "Knotted Periodic Orbits in Suspensions of Smale's Horseshoe : Period Multiplying and Cabled Knots", *Physica D*, 21D, pp.7-41, 1986.

197. P.J. Holmes, D. Whitley : "Bifurcations of One- and Two- Dimensional Maps", *Phil. Trans. Royal Society London*, 311A, pp.43-102, 1984.

198. P.J. Holmes, R.F. Williams : "Knotted Periodic Orbits in Suspensions of Smale's Horseshoe : Torus Knots and Bifurcation Sequences", *Archive for Rational Mechanics and Analysis*, 90, No.2, pp.115-194, 1985.

199. C.S. Hsu : *Cell-to-Cell Mapping*, Applied Mathematical Sciences 64, Springer Verlag 1988.

200. B.A. Huberman, J. Rudnick : "Scaling Behavior of Chaotic Flows", *Phys. Rev.Letters*, 45, pp.154-156, 1980.

201. B.A. Huberman, A.B. Zisook : "Power Spectra of Strange Attractors", *Phys. Rev.Letters*, 46, pp.626-628, 1981.

202. B.A. Huberman, E. Lumer : "Dynamics of Adaptive Systems", *IEEE Trans. Circuits Systems*, CAS-37, pp.547-550, 1990.

203. E.R. Hunt : "Stabilizing High-Period Orbits in a Chaotic System: The Diode Resonator", *Phys. Rev. Letters*, 67, pp.1953-1955, 1991.

204. E.R. Hunt : "Keeping chaos at bay", *IEEE Spectrum*, 30, pp.32-36, 1993.

205. A. Hübler : "Adaptive control of chaotic systems", *Helvetica Physica Acta*, 62, pp.343-346, 1989.

206. A. Hübler, E. Lüscher : "Resonant stimulation and control of nonlinear oscillators", *Naturwissenschaft*, 76, pp.76, 1989.

207. N. Inaba, S. Mori : "Chaotic Phenomena in a Circuit with a Diode Due to Change of the Oscillation Frequency", *Trans. IEICE Japan*, E71, No.9, pp.842-849, 1988.

208. N. Inaba, S. Mori : "Chaos Via Torus Breakdown in a Forced Oscillator Including a Diode", in *Proc 1989 European Conference on Circuit Theory and Design*, Brighton , IEE Publication 308, pp.152-156, 1989.
209. N. Inaba, T. Saito, S. Mori : "Chaotic Phenomena in Circuits with a Linear Negative Resistance and an Ideal Diode", Sagami Institute of Technology, Fujisawa 1987 (preprint).
210. N. Inaba, S. Mori : "Chaotic Phenomena in Four Circuits with an Ideal Diode Due to Change of the Oscillation Frequency", in *Proc. 1989 International Symposium on Circuits and Systems*, **3**, pp.2147-2150, Portland-Oregon 1989.
211. E.A. Jackson : "Control of dynamic flows with attractors", *Physical Review A*, **44**, pp.4839-4853, 1991.
212. E.A. Jackson : "On the control of complex dynamic systems", *Physica D*, **D 50**, pp.341-366, 1991.
213. W.A. Jakubowich : "Frequency domain conditions of self-osciallations in systems with one stationary nonlinearity", *Sibirskij Matiematiczeskij Zurnal*, **XIV**, No.5, pp.1110-1129, 1973.
214. S. Jangi, Y. Jain : "Embedding spectral analysis in equipment", *IEEE Spectrum*, **2**, pp.40-43, 1991.
215. G.E. Johnson, T.E. Tigner, E.R. Hunt : "Controlling chaos in Chua's Circuit", *J. Circuits Syst. Comput.*, **3**, 109-117, 1993.
216. R.K. Jurgen : "The challenges of digital HDTV", *IEEE Spectrum*, pp.28, 1991.
217. C. Kahlert, O.E. Rössler : "Analytical Properties of Poincaré Halfmaps in a Class of Piecewise - Linear Dynamical Systems", *Z. Naturforsch.*, **40a**, pp.1011-1025, 1985.
218. C. Kahlert : "Existence and Uniqueness of Solutions of Piecewise- Defined Continuous Dynamical Systems", *Z. Naturforsch.*, **41a**, pp.567-568, 1986.
219. C. Kahlert, O.E. Rössler : "The Saparating Mechanism for Poincaré Halfmaps", *Z. Naturforsch.*, **41a**, pp.1369-1380, 1986.
220. C. Kahlert : "Inverse Poincaré Halfmaps", *Z. Naturforsch.*, **42a**, pp.143-152, 1987.
221. C. Kahlert, L.O. Chua : "Transfer Maps and Return Maps for Piecewise- Linear Three- Region Dynamical Systems", *Int. J. Circuit Theory Appl.*, **15**, pp.23-49, 1987.
222. C. Kahlert : "The Chaos Producing Mechanism in Chua's Circuit", *Int. J. Circuit Theory Appl.*, **16**, pp.227-232, 1988.
223. C. Kahlert : "The Limiting Behavior of the Curves Ω_k in Poincare Halfmaps", University of Tübingen 1987 (preprint).
224. C. Kahlert : "The Ranges of Transfer and Return Maps in Three- Region Piecewise- Linear Dynamical Systems", University of Tübingen, 1987 (preprint).
225. C. Kahlert : "Dynamics of the Inclusions Appearing in the Return Maps of Chua's Circuit. I. The Creation Mechanism", University of Tübingen , 1987 (preprint).

226. K. Kaneko, "Clustering, coding, switching, hierarchical ordering, and control in a network of chaotic elements", *Physica D*, **D41**, pp.137-172, 1990.

227. K. Kaneko, "Pattern dynamics in spatio–temporal chaos", *Physica D*, **D34**, pp.1-41, 1989.

228. H. Kao, J.C. Huang, Y.S. Gou : "Routes to Chaos in Duffing Oscillator with One Potential Well", in *Proc. ISCAS 1988*, Helsinki 1988.

229. T. Kapitaniak, L. Kocarev, L.O. Chua : "Controlling chaos without feedback and control signals", *Int. J. Bifurcation Chaos*, **3**, pp.459-468, 1993.

230. R. Kapral, S. Fraser : "Bistable Oscillating States in Disspative Dynamical Systems : Scaling Properties and One- dimensional Maps", *Journal of Physical Chemistry.* **88**, No.21, pp.4845-4852, 1984.

231. H. Kawakami : "Strange Attractors in Duffing's Equation : 50 Phase Portraits of Chaos. Dept. of Electronics", Tokushima University, October 1986.

232. H. Kawakami : "Recurrence : construction numérique dans le cas d'un système continu", in *Actes du seminaire de Grenoble , Traitement Numérique des Attracteurs Etranges*, Editions CNRS, ed.: M.Cosnard, 1985.

233. H. Kawakami, G. Shichino : "An Invariant of Chaotic Response in a Josephson Junction Circuit", in *Proc. IEEE 1985 ISCAS*, **2**, pp.843-846, Kyoto 1985.

234. H. Kawakami, G. Shichino : "Average Voltage of Chaotic Response in a Josephson Junction Circuit", Proc. Meeting The Theory of Dynamical Systems and its Applications to Nonlinear Problems RIMS - Kyoto University 1984.

235. H. Kawakami : "Bifurcation of Periodic Responses in Forced Dynamic Nonlinear Circuits : Computation of Bifurcation Values of the System Parameters", *IEEE Trans. Circuits Systems*, **CAS-31**, No.3, pp.248-260, 1984.

236. H. Kawakami : "Fuzzy Boundary of Domain of Attraction in Forced Oscillatory Circuit", in *Proc. ECCTD'87*, **1**, pp.393-398, Paris 1987.

237. M.P. Kennedy : "Chaos in the Colpitts Oscillator", *IEEE Trans. Circuits Syst.–Part I*, **CAS-41**, 1994.

238. M.P. Kennedy, L.O. Chua : "Van der Pol and Chaos", Memorandum No.UCB/ERL M85/97, University of California, Berkeley, 15 November, 1985.

239. M.P. Kennedy "Robust Op-amp realization of Chua's circuit", *Frequenz*, **46**, No.3–4, pp.66-80, 1992.

240. M.P. Kennedy and H. Dedieu "Experimental Demonstration of Binary Chaos—Shift–Keying using Self-Synchronising Chua's Circuits"

241. M.P. Kennedy, "Chaos in the Colpitts oscillator", *IEEE Trans. Circ. Syst.*, bf CAS-41, no. 11, pp. 771–774, 1994.

242. M.P. Kennedy, "On the relationship between the chaotic Colpitts oscillator and Chua's oscillator", *IEEE Trans. Circ. Syst.*, **CAS-42**,

no. 6, pp. 376–379, 1995.

243. M.P. Kennedy : "Experimental chaos from autonomous electronic circuits", *Phil. Trans. Royal Soc. London*, **353A**, No.1701, pp.13-32, 1995.

244. J.H. Kim, J. Stringer : *Applied chaos (lectures and discussions from the Int. Workshop on Applications of Chaos)*, John Wiley & Sons.

245. P.E. Kloeden : "Chaotic Difference Equations in R^n ", *J.Australian Math. Society. Series A*, **31**, pp.217-225, 1981.

246. H. Koçak : *PHASER - Differential and Difference Equations through Computer Experiments*, Springer Verlag, 1988.

247. L. Kocarev, A. Shang, L.O. Chua : "Transition in Dynamical Regime by Driving: A Method of Control and Sysnchronisation of Chaos", *Int. J. Bifurcation and Chaos*, **3**, pp.479-483, 1993.

248. Kocarev, L., Ogorzalek, M.J. 1993 Mutual synchronisation between different chaotic systems. In *Proc. 1993 International Symposium on Nonlinear Theory and Applications*, **3**, 835-840.

249. L. Kocarev, K.S.Halle, K.Eckert, L.O.Chua and U.Parlitz, "Experimental Demonstration of Secure Communications via Chaotic Synchronisation", Int. J. Bifurcation and Chaos, vol.2, No.3, pp.709-713, 1992.

250. L. Kocarev, A. Shang, L.O.Chua, "Transitions in dynamical regimes by driving: a unified method of control and synchronization of chaos", *Int.J. of Bifurcation and Chaos*, **3**, No.2, pp.479-483, 1993.

251. L. Kocarev, L.O. Chua, "*On chaos in digital filters: Case $b = -1$*", to appear in the IEEE Trans. Circuits Syst., Part II, vol.40, 1993.

252. M. Komuro, T. Matsumoto, L.O. Chua : "Poincaré Maps of The Double Scroll", in *Dynamical Systems and Nonlinear Oscillations* (ed. G.Ikegami). **1**, pp.238-246, World Scientific Advanced Series in Dynamical Systems, 1986.

253. M. Komuro : "Normal Forms of Piecewise Linear Vector Fields and Chaotic Attractors. ● Part I -Linear Vector Fields with a Section. ● Part II -Chaotic Attractors", *Japan Journal Appl.Math.*, **5**, No.2, pp.257-304, 1988 and No.3, pp.503-549, 1988,

254. E.J. Kostelich, C. Grebogi, E. Ott, J.A. Yorke : "Higher-dimensional targeting", *Physical Review E*, **47**, pp.305-310 1993.

255. G. Kubin : "What is a chaotic signal?", in *Proc. 1995 IEEE Workshop on Nonlinear Signal and Image Processing*, Vol.I, pp.141-144, Halkidiki 1995.

256. J. Kudrewicz : *Frequency domain methods in the theory of nonlinear dynamical systems*, (in Polish), WNT, Warszawa 1070.

257. J. Kudrewicz : *Dynamics of Phase-Locked loops*, WNT, Warszawa 1992.

258. J. Kudrewicz, J. Grudniewicz : "Investigations of periodic and chaotic oscillations in an impulsive pahe-locked loop", Technical Report, Institute of Electonics Fundamentals, Warsaw Technical University, Warszawa 1983.

259. J. Kudrewicz, J. Grudniewicz, B. Świdzińska B. : "Chaos, Phase Slipping and Cantor -like Sets in a Discrete Phase-Locked Loop", in *Proc. ECCTD'87*, **2**, pp.507-512, Paris 1987.

260. J. Kudrewicz, Z. Michalski : "Dynamics of generators described by third-order differential equations", in *Materia/ly XII Krajowej Konferencji Teoria Obwod/ow i Uk/lady Elektroniczne*, (in Polish), pp.67-72, Myczkowce-Rzesz/ow 1989.

261. D.P. Lathrop, E. Kostelich : "Characterization of an experimental strange attractor by periodic orbits", *Phys. Rev A*, **40**, No.7, pp.4028-4031, 1989.

262. F. Ledrappier : "Some Relations Between Dimension and Lyapunov Exponents", *Commun. Math.Phys.*, **81**, pp.229-238, 1981.

263. G.A. Leonow : "Frequency domain conditions for existence of periodic solutions in autonomous systems", *Sibirskij Matiematiczeskij Zurnal*, **XIV**, No.6, pp.1259-1265, 1973.

264. G.A. Leonow : "On an analogue of Bendixon's criterion for third-order equations", *Differencyjalnyje Urawnienija*, (in Russian) **XIII**, No.2, pp.367-368, 1977.

265. T.-Y. Li, J.A. Yorke : "Period Three Implies Chaos", *American Math. Monthly*, **82**, pp.985-992, 1975.

266. T.-Y. Li, J.A. Yorke : "The Simplest Dynamical System", in *Dynamical Systems an Int. Symposium*, **2**, eds.: L.Cesari, J.K.Hale, J.P.LaSalle, Academic Press, pp.203-206, New York 1976.

267. T.-Y. Li, M. Misiurewicz, G. Pianigiani, J.A. Yorke : "No Division Implies Chaos", *Trans. Amer. Math. Soc.*, July 1981.

268. T. Lin, L.O. Chua : "On Chaos of Digital Filters in the Real World", *IEEE Trans. Circuits Systems*, **CAS-38**, pp.557-558, 1991.

269. E.N. Lorenz : "Lyapunov Numbers and the Local Structure of Strange Attractors", *Physica D*, **17D**, pp.279-294, 1985.

270. J. Luprano, M. Hasler : "More Details on the Devil's Staircase Route to Chaos", *IEEE Trans. Circuits and Systems*, **CAS-36**, No.1, pp.146-148, 1989.

271. E. Lüscher, A. Hübler :"Resonant stimulation of complex systems", *Helvetica Physica Acta*, **62**, p.543, 1989.

272. O. Macchi, M. Jaidane-Saidane : "Bifurcations in Adaptive ARMA Predictors", in *Proc. ECCTD'87*, **1**, pp.287-294, Paris 1987.

273. O. Macchi, M. Jaidane-Saidane : "Adaptive IIR Filtering and Chaotic Dynamics : Application to Audiofrequency Coding", *IEEE Trans. Circuits and Systems*, **CAS-36**, No.4, pp.591-599, 1989.

274. R.S. MacKay, C. Tresser : "Transition to Topological Chaos for Circle Maps". *Physica*,**19D**, pp.206-237, 1986.

275. Madan, R.N. (ed.) : *Chua's Circuit; A Paradigm for Chaos*. World Scientific 1993.

276. J. Mallet-Paret, J.A. Yorke : "Snakes : Oriented Families of Periodic Orbits, Their Sources, Sinks and Continuation", *J.Differential Equations*, **43**, pp.419-450, 1982.

277. F.R. Marotto : "Snap-Back Repellers Imply Chaos in R^n", *J. Math. Analysis and Appl.*, **63**, pp.199-223, 1978.

278. J. L. Massey : "An Introduction to Contemporary Cryptology", *Proc. of IEEE*, **76**, No. 5, pp. 533-549, 1988.

279. T. Matsumoto : "A Chaotic Attractor from Chua's Circuit", *IEEE Trans. Circuits Syst.*, **CAS-31**, No.12, pp.1055-1058, 1984.

280. T. Matsumoto : "Chaos in Electronic circuits", *Proc. IEEE*, **75**, No.8, pp.1033-1057, 1987.

281. T. Matsumoto, L.O. Chua, K. Kobayashi : "Hyperchaos : Laboratory Experiment and Numerical Confirmation", *IEEE Trans. Circuits Systems*, **CAS-33**, No.11, pp.1143-1147, 1986.

282. T. Matsumoto, L.O. Chua, M. Komuro : "The Double Scroll", *IEEE Trans. Circuits Systems*, **CAS-32**, No.8, pp.798-818, 1985.

283. T. Matsumoto, L.O. Chua, M. Komuro : " The Double Scroll Bifurcations", *Int.J.Circuit Theory Appl.*, **14**, pp.117-146,1986.

284. T. Matsumoto, L.O. Chua, M. Komuro : "Birth and Death of the Double Scroll", *Physica D*, **24D**, pp.97-124, 1987.

285. T. Matsumoto, L.O. Chua, S. Tanaka : "Simplest Chaotic Nonautonomous Circuit", *Physical Review A*, **30**, No.2, pp.1155-1157, 1984.

286. T. Matsumoto, L.O. Chua, R. Tokunaga : "Chaos Via Torus Breakdown", *IEEE Trans. Circuits Syst.*, **CAS-34**, No.3, pp.240-253, 1987.

287. T. Matsumoto, L.O. Chua, K. Tokumasu : "Double Scroll Via a Two-Transistor Circuit", *IEEE Trans. Circuits Systems*, **CAS-33**, No.8, pp.828-835, 1986.

288. T. Matsumoto, M. Komuro, H. Kokubu, R. Tokunaga : *Bifurcations. Sights, Sounds and Mathematics*, Springer Verlag, New York 1993.

289. G. Mayer-Kress, I. Choi, N. Weber, R. Bargar, A. Hübler : "Musical signals from Chua's circuit", *IEEE Trans. Circ. Systems Part II*, **40**, pp.688-695, 1993.

290. A.I. Mees, P.B. Chapman : "Homoclinic and Heteroclinic Orbits in the Double Scroll Attractor", *IEEE Trans. Circuits Systems*, **CAS-34**, No.9, pp.1115-1120, 1987

291. A.I. Mees, C.T. Sparrow : "Some Tools for Analysing Chaos", *IEEE Proc.*, **75**, No.8, 1987

292. G.B. Mindlin, X.-J. Hou, H. Solari, R. Gilmore, N.B. Tufillaro, "Classification of strange attractors by integers", *Phys. Rev. Lett.*, **64**, pp.2350-2353, 1990.

293. C. Mira : "La structure de bifurcations boites-emboitées. Ses conséquences", Academie Royale de Belgique, Bulletin de la Classe des Sciences. 5 Série, T. LXXI, No.1-2, pp.92-126, 1985,

294. C. Mira : "Quelques situations fondamentales dans les systèmes dynamiques non linéaires et chaotiques. Exemples", *Annales des Telecommunications*, **42**, No.5-6, pp.217-238, 1987.

295. C. Mira : *Chaotic Dynamics. From the One-Dimensional Endomorphism to the Two-Dimensional Diffeomorphism*, World Scientific, 1987.

296. W. Mitkowski, M.J. Ogorzałek : "Chaotic Motion in Ladder Networks with Nonlinear Feedback", Proc. 8th Colloquium on Microwave Communications MICROCOLL, pp.506-507, Budapest 1986.

297. D. Mitra : "Criteria for determining if a high-order digital filter using saturation arithmetic is free of overflow oscillations". *Bell System Technical Journal*, **56**, pp.1679-1699, 1977.

298. D. Mitra : " Large amplitude, self-sustained oscillations in difference equations which describe digital filter sections using saturation arithmetic". *IEEE Trans. Acoustics, Speech, Signal Processing*, **ASSP-25**, pp.134-143, 1977.

299. D. Mitra : "A bound on limit cycles in digital filters which exploits a particular structural property of the quantization".*IEEE Trans. Circuits Systems*, **CAS-24**, pp.581-589, 1977.

300. D. Mitra :"The absolute stability of high-order discrete-time systems utilizing the saturation nonlinearity". *IEEE Trans. Circuits Systems*, **CAS-25**, pp.365-371, 1978.

301. F. Moon : *Chaotic Vibrations*, J.Wiley & Sons, New York 1987.

302. O. Morgul, E. Solak : "On the Synchronization of Chaotic Systems by Using State Observers", preprint, 1996.

303. E. Morse, G. Hedlund : "Symbolic Dynamics", *Amer. Journ. Math.*, • Part I - **60**, pp.815-866, 1938, • Part II - **62**, pp.1-42, 1940.

304. F. Mossayebi, H.K. Qammar and T.T. Hartley : "Adaptive Estimation and Synchronisation of Chaotic Systems", *Phys. Letters A*, **161**, No.3, pp.255-262, 1991.

305. G.J. Mpitsos, R,M. Burton Jr. : "Convergence and divergence in neural networks: Processing of chaos and biological analogy", *Neural Networks*, **5**, pp.605-625, 1992.

306. C.A. Murphy and M.P. Kennedy, "Chaos controller for autonomous circuits", in *Proc. NDES'95*, Dublin, 28–29 July 1995, pp. 225–228.

307. S. Nasri : "Contribution à l'Analyse et à la synthèse des redresseurs boucles à thyristors", Thèse de Doctorat de 3ème cycle, Institut National des Sciences Appliqués de Toulouse, Juin 1985.

308. Ju.I. Nejmark : *Point mapping method in the theory of nonlinear oscillations*, (in Russian) Nauka, Moscow 1972.

309. Ju.I. Nejmark : "Symbolic dynamics governing homoclinic structures", *Differencyjalnyje Urownienija*, (in Russian), **12**, No.2, pp.156-262, 1976.

310. V.I.Nekorkin, L.O. Chua, "Spatial disorder and wavefronts in a chain of coupled Chua's circuits". (preprint 1993).

311. R.W. Newcomb, N. El-Leithy : "Chaos Generation Using Binary Hysteresis", *Circuits, Systems and Signal Processing*, **5**, No.3, pp.321-341, 1986.

312. R.W. Newcomb, S. Sethyan : "An RC Op - Amp Chaos Generator", *IEEE Trans. Circuits Syst.*, **CAS-30**, No.1, pp.54-56, 1983.

313. S. Newhouse, J. Palis : "Cycles and Bifurcation Theory", Société Mathématique de France, *Astérisque*,**31**. pp.43-140, 1976.

314. J.S. Nicolis : "Chaotic dynamics in biological information processing: A heuristic outline", In *Chaos in Biological Systems* (ed. H. Degn, A.V. Holden & L.F. Olsen), pp.221-232, New York, Plenum Press, 1987.

315. Y. Nishio, S. Mori : "Chaotic Phenomena in a Four- Dimensional Circuit with Two Ideal Diodes", in *Proc 1989 European Conference on Circuit Theory and Design*, Brighton . IEE Publication 308, pp.157-161, 1989.

316. Y. Nishio, N. Inaba, S. Mori, T. Saito : "Rigorous Analysis of Windows in a Symmetric Circuit", in *Proc. 1989 International Symposium on Circuits and Systems*, **3**, pp.2151-2154, Portland-Oregon 1989.

317. K. Niwa, T. Araseki, T. Nishitani : "Digital Signal Processing for Video", *IEEE Circuits and Devices Magazine*, **6**, pp.27-33, 1990.

318. H.E. Nusse, J.A. Yorke : *DYNAMICS: Numerical Explorations*, Springer Verlag, Applied Mathematical Sciences Vol.101, New York 1994.

319. M.J. Ogorzałek : "'Latch-up' Property in Second Order RC Active Oscillators", in *Proc. 1985 IEEE International Symposium on Circuits Syst.*, **2**, pp.587-590, Kyoto 1985.

320. M.J. Ogorzałek : "Some Observations on Chaotic Motion in Single Loop Feedback Systems", in *Proc. 1986 IEEE Conference on Decision and Control*, **1**, pp. 588-589, Athens, 1986.

321. M.J. Ogorzałek : *Analysis of RC oscillatory circuits with active nonlinear elements*, doctoral thesis, AGH Kraków 1987.

322. M.J. Ogorzałek : "Investigations of Chaotic Motion in RC Ladder Networks with Nonlinear Feedback" *Electronics and Telecommunications Letters*, **2**, No.4, pp.13-19, 1987.

323. M.J. Ogorzałek : "Chaotic Regions from Double Scroll", *IEEE Trans. Circuits Systems*, **CAS-34**, No.2, pp.201-203, 1987.

324. M.J. Ogorzałek : "Some Properties of a Chaos Generator with RC Ladder Network", *Proc. ECCTD'87*, **2**, pp.519-524, Paris 1987.

325. M.J. Ogorzałek : "Chaotic Motion in a Parabolic System with Nonlinear Boundary Feedback", in *Proc. IMACS/IFAC Symposium on Distributed Parameter Systems*, pp.171-177, Hiroshima 1987.

326. M.J. Ogorzałek : "Chaotic Phenomena in Simple Autonomous Electronic Circuits: Two Steps Towards Unified Analysis", in *Proc. IMACS World Congress on Scientific Computation*, **4**, pp.445-447, Paris, 1988.

327. M.J. Ogorzałek : "Order and Chaos in a Third Order RC Ladder Network with Nonlinear Feedback", *IEEE Trans. Circuits and Systems*, **CAS.36**, No.9, pp. 1221-1230, 1989.

328. M.J. Ogorzałek : "Towards Unified Analysis of Autonomous Chaotic Electronic Circuits", in *Numerical and Applied Mathematics*, Ed.: W.F.Ames , pp.303-309, J.C.Baltzer AG, Basel 1989.

329. M.J. Ogorzałek : "Remerging Bifurcation Sequences in an Autonomous Electronic Circuit", in *Proc. Conference on Circuit Theory*

and Design, Brighton . IEE Publication308, pp.142-146, 1989.

330. M.J. Ogorzałek : "Mechanisms of Chaos Generation in autonomous electronic systems with feedback" Archiwum Automatyki, (in Polish) No. 1-2, 1989, pp.

331. M.J. Ogorzałek : "The Squeezed Spiral Map", in *Proc. 1990 IEEE International Symposium on Circuits Syst.*, **1**, pp.595-598, New Orleans 1990.

332. M.J. Ogorzałek : A One-Dimensional Model of Dynamics for a Class of Third Order Systems. Int. J. Circuit Theory and Appl. **18**, No.6, pp.595-624, 1990.

333. M.J.Ogorzałek : "Complex behavior in digital filters", *International Journal of Bifurcation and Chaos*, **2**, no.1, pp.11-29, 1992.

334. M.J. Ogorzałek, Z. Galias : "Arnold Tongues and Devil's Staircase in a Digital Filter Employing Saturation Arithmetic". *Proceedings IEEE ISCAS*, vol.1, pp.384-387, 1991, Singapore 1991.

335. M.J. Ogorzałek, Z. Galias : "Limit Sets of Trajectories in a Nonlinear Digital System". in *Proc. 1991 International Symposium on Mathematical Theory of Networks and Systems*, Kobe, June 17-21, 1991.

336. M.J. Ogorzałek, P. Grabowski : "On the Design of RC Oscillators with Single Nonlinear Active Element", in *Proc. 1985 European Conference on Circuit Theory and Design*, **1**, pp.287-290, Praga 1985.

337. M.J. Ogorzałek, E. Mosekilde : "Noise Induced Effects in an Autonomous Chaotic Circuit", in *Proc. 1989 IEEE International Symposium on Circuits and Systems*, **1**, pp.578-581, Portland-Oregon 1989.

338. M.J. Ogorzałek, E. Mosekilde : "Behavior of Autonomous Chaotic Feedback Systems in the Presence of Noise", in *Proc. MTNS'89 - Mathematical Theory of Networks and Systems*, Amsterdam 1989.

339. M.J. Ogorzałek, E. Mosekilde : "Multiple Periodic Orbits ans Structure of Basin Boundaries in a Digital Filter Employing Saturation Arithmetic", in *Proceedings of The 1991 European Conference on Circuit Theory and Design*, Copenhagen 1991.

340. M.J. Ogorzałek : "Controlling chaos in Chua's circuit", In *Proc. 1993 Int. Symp. Nonlinear Theory and Applications*, Honolulu, pp.47-52, 1993.

341. M.J. Ogorzałek : "Taming Chaos: Part I - Synchronization", *IEEE Trans. Circuits Syst.*, **CAS-40**, pp.693-699, 1993.

342. M.J. Ogorzałek : "Taming Chaos: Part II - Control", *IEEE Trans. Circuits Syst.*, **CAS-40**, pp.700-706, 1993.

343. M.J. Ogorzałek : "Chaos control techniques: A study using Chua's circuit", In *Nonlinear Dynamics of Electronic Systems*, ed. A.C. Davies and W. Schwarz, World Scientific, pp.89-101, 1994.

344. M.J. Ogorzałek :"Chaos control: How to avoid chaos or take advantage of it", *J. Franklin Inst.*, **331B**, No.6, pp681-704, 1994.

345. M.J. Ogorzałek : "Controlling chaos in electronic circuits", *Phil. Trans. Roy. Soc. London*, **353A**, No.1701, pp.127-136, 1995.

346. M.J. Ogorzałek " Design considerations for Electronic Chaos Con-

trollers", em Chaos, Solitons and Fractals (in press) 1997

347. M.J. Ogorzałek, Z. Galias : "Characterisation of chaos in Chua's oscillator in terms of unstable periodic orbits", *J. Circuits Syst. and Computers*, **3**, pp.411-429, 1993.

348. M.J. Ogorzałek, Z. Galias, L.O. Chua : "Exploring chaos in Chua's circuit via unstable periodic orbits", In *Proc. 1993 IEEE International Symposium on Circuits Syst.*, Chicago, **4**, pp.2608-2611, 1994.

349. Y. Ohmori, N. Nakagawa, T. Saito : "Mutual Coupling of Oscillators with Chaos and Period Doubling Bifurcation", in *Proc. 1985 IEEE ISCAS*, pp.61-64, 1985.

350. A.H. Nayfeh, B. Balachandran : *Applied Nonlinear Dynamics. Analytical, Computational and Experimental Methods*, Wiley Series in Nonlinear Science, John Wiley & Sons 1995.

351. A.V. Oppenheim : *Applications of Digital Signal Processing (Prentice-Hall, Englewood Cliffs, New Jersey), 1978.*

352. *A.V. Oppenheim, R.W. Schafer :* Digital Signal Processing (Prentice-Hall, Englewood Cliffs, New Jersey), 1975.

353. A.V. Oppenheim, G.W. Wornell, S.H. Isabelle and K.M. Cuomo, "Signal Processing in the Context of Chaotic Signals", Proc. IEEE ICASSP, San Francisco, CA, vol.4, pp.IV-117—IV-120, 1992.

354. E. Ott, C. Grebogi, J.A. Yorke : "Controlling Chaos", *Phys. Rev. Letters*, **64**, pp.1196-1199, 1990.

355. E. Ott, C. Grebogi, J.A. Yorke : "Controlling Chaotic Dynamical Systems", In *Chaos: Soviet-American Perspectives on Nonlinear Science*, ed. D.K.Campbell, American Institute of Physics, New York 1990.

356. E. Ott, T. Sauer, J.A. Yorke Eds. : *Coping with chaos: Analysis of Chaotic Data and the Exploitation of Chaotic Systems*, Wiley Series in Nonlinear Science, John Wiley 1994.

357. J. Palmore, C. Herring : "Computer arithmetic, chaos and fractals". *Physica*, **D42**, pp.99-110, 1990.

358. T.S. Parker, L.O. Chua : *Practical Numerical Algorithms for Chaotic Systems*, Springer Verlag, New York 1989,

359. T.S. Parker, L.O. Chua : "A Computer Assisted Study of Forced Relaxation Oscillations", *IEEE Trans. Circuits Systems*, **CAS-30**, No.8, pp.518-533, 1983.

360. T.S. Parker, L.O. Chua : "Steady State Behavior of Nonlinear Circuits Syst.", Lecture Notes - Special Lectures ECCTD'87. pp.69-118, Paris 1987.

361. T.S. Parker, L.O. Chua : "Chaos : A Tutorial for Engineers", *Proc. IEEE*, **75**, No.8, pp.982-1008, 1987.

362. T.S. Parker, L.O. Chua : "INSITE - A Software Toolkit for the Analysis of Nonlinear Dynamical Systems", *Proc. IEEE*, **75**, No.8, pp.1081-1089, 1987.

363. T.S. Parker, L.O. Chua : "The Dual Double Scroll Equation", *IEEE Trans. Circuits Syst.*, **CAS-34**, No.9, pp.1059-1073, 1987.

364. U.Parlitz, L.O.Chua, L.Kocarev, K.S.Halle and A.Shang : "Transmis-

sion of Digital Signals by Chaotic Synchronisation", *Int. J. Bifurcation and Chaos*, **2**, No.4, pp.973-977, 1992.

365. L.M. Pecora and T.L. Carroll : "Synchronization in Chaotic Systems", *Phys. Rev. Letters*, bf 64, No.8, pp.821-824, 1991.

366. L.M. Pecora and T.L. Carrol : "Driving Systems with Chaotic Signals", *Phys. Rev. A*, **44**, No.4, pp.2374-2383, 1991.

367. L.M. Pecora and T.L. Carroll : "Synchronized Chaotic Signals and Systems", Proc. IEEE ICASSP, San Francisco, CA, vol.4, pp.IV-137-IV-140, 1992.

368. L.M. Pecora, T.L. Carroll : "Driving systems with chaotic signals", *Physical Review A*, **44**, pp.2374-2383, 1991.

369. L.M. Pecora, T.L. Carroll : "Pseudoperiodic driving: Eliminating multiple domains of attraction using chaos", *Phys. Rev. Letters*, **67**, pp.945-948, 1991.

370. L.-Q. Pei, F. Guo, S.-X. Wu, L.O. Chua : "Period Adding Route to Chaos in a Two-Transistor Nonlinear Circuit", Proc. 1986 IEEE ISCAS. San Jose, pp.69, 1986.

371. L.-Q. Pei, F. Guo, S.-X. Wu, L.O. Chua : "Experimental Confirmation of the Period-Adding Route to Chaos in a Nonlinear Circuit", *IEEE Trans. Circuits Syst.*, **CAS-33**, No.4, pp.438-442, 1986.

372. V.Perez–Menuzuri, V. Perez–Villar, L.O. Chua : "Propagation failure in linear arrays of Chua's circuits". *Int. J. Bifurcation and Chaos*, **2**, No.2, pp.403–406, 1992.

373. V.Perez–Menuzuri, V. Perez–Villar, L.O. Chua : "Autowaves for image processing an a two-dimensional CNN array of Chua's circuits: Flat and wrinkled labyrinths". Memorandum No. UCB/ERL M92/75, 1 June 1992.

374. I. Peterson : "Ribbon of Chaos: Researchers develop a Lab Technique for snatching Order out of Chaos", *Science News*, **139**, pp.60-61, 1991.

375. C.L. Philips, H.T. Nagle jr. : *Digital Control Systems. Analysis and Design* (Prentice-Hall, Englewood Cliffs, New Jersey), 1984.

376. J. Piesin : "Characteristic Lyapunov Exponents and Smooth Ergodic Theory", *Russ. Math. Surveys*, **32**, No.4, pp. 55-114, 1977.

377. A. Prieto, A. Lloris : "Design of Active Circuits with Nonlinear Transfer Characteristics", *Int.J. Electronics*, **54**, No.6, pp.813-824, 1983.

378. R. Pool, "Is it healthy to be chaotic?". Science, vol.243,

379. Pyragas, K. 1992 Continuous control of chaos by self-controlling feedback. *Physics Letters A*, **A170**, 421-428.

380. L.R. Rabiner, B. Gold : *Theory and Applications of Digital Signal Processing.* (Prenticae Hall, Englewood Cliffs, N.J.), 1975.

381. J.H.F. Ritzerfeld : "A condition for the overflow stability of second-order digital filters that is satisfied by all scaled ststé-space structures using saturation" *IEEE Trans. Circuits Systems*, **CAS-36**, pp.1049-1057, 1989.

382. C. Robinson : *Dynamical Systems. Stability, Symbolic Dynamics, and Chaos*, CRC Press, Boca Raton-Ann Arbor-London-Tokyo 1995.

383. A.B. Rodriguez-Vazquez, J.L. Huertas, L.O. Chua : "Chaos in a Switched - Capacitor Circuit", em IEEE Trans. Circuits Systems, **CAS-32**, No.10, pp.1083-1085, 1985.

384. M.T. Rosenstein, J.J. Collins, C.J. De Luca : "Reconstruction expansion as a geometry-based framework for choosing proper delay times", *Physica D*, **D73**, pp.82-98, 1994.

385. O.E. Rössler : "Continuous Chaos - Four Prototype Equations", *Annals of The New York Academy of Sciences*, **316**, pp.376-392, 1979.

386. R. Roy, T.W. Murphy Jr., T.D. Maier, Z. Gills, E.R. Hunt : "Dynamical Control of a Chaotic Laser: Experimental Stabilization of a Globally Coupled System", *Phys. Rev. Letters*, **68**, pp.1259-1262, 1990.

387. D. Ruelle : "Ergodic Theory of Differentiable Dynamical Systems", *Publications IHES* No.50, pp.27-58, 1979.

388. R.F. Rubio, J. Aracil, E.F. Camacho : "Chaotic Motion in an Adaptive Control System", *Int. J. Control*, **42**, No.2, pp.353-360, 1985.

389. N.F. Rul'kov, A.R. Volkovskii, A. Rodriguez-Lozano, E. del Rio and M.G.Velarde, "Mutual Synchronization of Chaotic Self-Oscillators with Dissipative Coupling", *Int. J. Bifurcation and Chaos*, **2**, No.3, pp.669-676, 1992.

390. T. Saito : "A Chaos Generator Based on a Quasi-Harmonic Oscillator", *IEEE Trans. Circuits Systems*, **CAS-32**, No.4, pp.320-331, 1985.

391. T. Saito : "Chaotic Phenomena in a Coupled Oscillators", *Proc. EC-CTD'87*, **1**, pp.275-280, Paris 1987.

392. T. Saito : "The Diode Chaos Generator", Sagami Institute of Technology, Fujisawa 1988 (preprint).

393. T. Saito : "The Hysteresis Chaos Generator Family", *Proc. 1988 IEEE International Symposium on Circuits Syst.*, **1**, pp.15-18, Helsinki 1988.

394. F.M.A. Salam : "The Melnikov Method for Highly Dissipative Systems", *SIAM J. Appl. Math.*, **47**, pp.232-243, 1987.

395. F.M.A. Salam, S.S. Sastry : "Dynamics of the Forced Josephson Junction: The Regions of Chaos", *IEEE Trans. Circuits and Systems*, **CAS-32**, No.8, pp.784-796, 1985.

396. H. Samueli, A.N. Willson : "Almost period P sequences and the analysis of forced overflow oscillations in digital filters". *IEEE Trans. Circuits Systems*, **CAS-29**, pp.510-515, 1982.

397. H. Samueli, A.N. Willson : "Nonperiodic forced overflow oscillations in digital filters". *IEEE Trans. Circuits Systems*, **CAS-30**, pp.709-722, 1980.

398. T. Sauer, J.A. Yorke, M. Casdagli : "Embedology", *J. Stat. Physics*, **65**, pp.579-, 1991.

399. W.M. Schaffer, S. Truty : *Dynamical Software. Users Manual and Introduction to Chaotic Systems*, Ver.2.1, Dynamical Software Inc., Tucson 1987.

400. M. Schell, S. Fraser, R. Kapral : "Diffusive Dynamics in Systems with

Translational Symmetry : A One-Dimensional-Map Model", *Physical Review A*, **26**, No.1, pp.504-521, 1972.

401. M. Schell, S. Fraser, R. Kapral : "Subharmonic Bifurcation in the Sine-Map: An Infinite Hierarchy of Cusp Bistabilities", *Physical Review A*, **28**, No.1, pp.373-278, 1983.

402. S.J. Schiff, K. Jeger, D.H. Duong, T. Chang, M.L. Spano, W.L. Ditto : "Controlling chaos in the brain", *Nature*, **370**, pp.615-620, 1994

403. I.B. Schwartz, I. Triandaf : "Tracking unstable orbits in experiments. *Phys. Rev. A*, **46**, 7439-7444.

404. R. Seydel : *Practical Bifurcation and Stability Theory. From Equilibrium to Chaos*, Springer Verlag, New York 1994 (Second Edition).

405. L.P. Shil'nikov : "Some cases of birth of periodic solutions in n-dimesional space" *Dok/lady AN SSSR*, (in russian), **143**, No.2, pp.289-292, 1962.

406. L.P. Shil'nikov : "On some cases of birth of periodic solutions from singular trajectories", *Matiematiczeskij Sbornik*, (in Russian), **61**, No.4, pp.443-466, 1963.

407. L.P. Shil'nikov : "On a case of existence of a countable set of periodic motions", *Dok/lady AN SSSR*, (in Russian), **160**, No.3, pp.558-561, 1965.

408. L.P. Shil'nikov : "On birth of periodic solution from a trajectory going from a saddle-saddle fixed point into itself", *Dok/lady AN SSSR*, (in Russian), **170**, No.1, pp.49-52, 1966.

409. L.P. Shil'nikov : "On existence of a countable number of periodic motions in a four-dimensional space in an extended neighborhood of a saddle-focus fixed point", *Dok/lady AN SSSR*, (in Russian), **172**, No.1, pp.54-57, 1967.

410. L.P. Shil'nikov : "On existence of a countable number of periodic solutions in a neighborhood of a homoclinic curve", *Dok/lady AN SSSR*, (in Russian), **172**, No.2, pp.298-301, 1967.

411. L.P. Shil'nikov : "On a Poincaré-Birkhoff problem", *Matiematiczeskij Sbornik*, (in Russian) **74**, No.2, pp.378-397, 1967.

412. L.P. Shil'nikov : "on the problem of structure of neighborhood of a homoclinic tube of an invariant torus", *Dok/lady AN SSSR*, (in Russian), **180**, No.2, pp.286-289, 1968.

413. L.P. Shil'nikov : "On the birth of periodic motions from a trajectory doubly-asymptotic to a fixed point of the saddle type", *Matiematiczeskij Sbornik*, (in Russian), **77**, No.3, pp. 461-472, 1968.

414. L.P. Shil'nikov : "On a new type of bifurcation of multidimensional dynamical systems", *Dok/lady AN SSSR*, (in Russian), **189**, No.1, pp.59-60, 1969.

415. L.P. Shil'nikov : "On the question of structure of the structure of an extended neighborhood of an equilibrium point of saddle-focus type", *Matiematiczeskij Sbornik*, (in Russian), **81**, No.1, pp.92-103, 1970.

416. T. Shinbrot : "Chaos: Unpredictable yet controllable", *Nonlinear Science Today*, **3**, 1-8.

417. T. Shinbrot, W. Ditto, C. Grebogi, E. Ott, M. Spano, J.A. Yorke: "Using sensitive dependence of chaos (the "Butterfly effect") to direct trajectories in an experimental chaotic system", *Phys. Rev. Letters*, **68**, pp.2863-2866, 1992

418. M. Shinriki, M. Yamamoto, S. Mori : "Multimode Oscillations in a Modified Van der Pol Oscillator Containing a Positive Nonlinear Conductance", *Proc. of the IEEE*, **69**, No.3, pp.394-395, 1981.

419. C.-P.Silva, L.O. Chua : "The overdamped double scroll family ; Part I: Piecewise-linear geometry and normal form" *Int.J. Circuit Theory and Applications*, **16**, pp.233-302, 1988.

420. S. Sinha : "An Efficient Control Algorithm for Nonlinear Systems", *Physics Letters A*, **156**, pp.475-478, 1991.

421. S. Smale : "A mathematical model of two cells via Turing's equation", *A. Math. Soc. Proc. Pure Appl. Math.*, **6**, pp.15-26, 1974.

422. P. So, E. Ott and W.P. Dayawansa : "Observing Chaos: deducing and tracking the state of a chaotic system from limited observation", *Phys. Lett. A*, **A176**, pp.421, 1993.

423. P. So, E. Ott and W.P. Dayawansa : "Observing Chaos", *Phys. Rev. E*, **49**, pp.2650-2660, 1994.

424. M.M. Sondhi, D.A. Berkley : " Silencing echos on the telephone network", *Proc. of IEEE*, **68**, pp.948-963, 1980.

425. C.T. Sparrow : "Chaos in a Three-Dimensional Single Loop Feedback System with a Piecewise Linear Feedback Function", *J. Math. Analysis and Appl.*, **83**, pp.275-291, 1981.

426. M. de Sousa Vieira, P. Khoury, A.J. Lichtenberg, M.A. Lieberman, W. Wonchoba, J. Gullicksen, J.Y. Huang, R. Sherman and M. Steinberg, "Numerical and Experimental Studies of Self-Synchronization and Synchronized Chaos", *Int. J. Bifurcation and Chaos*, **2**, No.3, pp.645-657, 1992.

427. T. Stojanovski, L. Kocarev, U. Parlitz, R. Harris : "How to Achieve Synchronized Motion with a Finite Number of Samples", in *Proc. NOLTA'96*, Kochi-Japan, pp.83-86.

428. W. Szlenk :*An Introduction to the Theory of Smooth Dynamical Systems*, PWN/John Wiley and Son, Warszawa/Chichester, 1984.

429. Y.S. Tang, A.I. Mees, L.O. Chua : "Synchronization and Chaos", *IEEE Trans. Circuits and Systems*, **CAS-30**, No.9, pp.620-626, 1983.

430. A. Tesi, A. De Angeli, R. Genesio " On the system decomposition for synchronising chaos". Technical Report, RT 12/93, Universita di Firenze.

431. J. Testa, G.A. Held : "Study of One-Dimensional Map with Multiple Basins", *Physical Review A*, **28**, No.5, pp.3085-3089, 1983.

432. J.M.T. Thompson, H.B. Stewart : em Nonlinear Dynamics and Chaos, John Wiley & Sons, New York 1986.

433. R. Tokunaga, M. Komuro, T. Matsumoto, L.O. Chua : "Lorenz Attractor from an Electrical Circuit with Uncoupled Continuous Piecewise-Linear Resistor", *Int. J. Circuit Theory and Applications*,

17, pp.71-85, 1989.

434. R. Tokunaga, L.O. Chua, T. Matsumoto : "Bifurcation Analysis of a Cusp-Constrained Piecewise-Linear Circuit", *Int. J. Circuit Theory and Applications*, **17**, pp.283-346, 1989.

435. R. Tokunaga, T. Matsumoto, L.O. Chua, K. Miya, A. Hotta, R. Fujimoto : "Homoclinic Linkage : A New Bifurcation Mechanism", *Proc. 1989 IEEE International Symposium on Circuits and Systems*, **2**, pp.826-829, Portland-Oregon 1989.

436. C. Tresser : "Homoclinic Oorbits for Flows in R^n", *J.Physique*, **45**, pp.837-841, Mai 1984.

437. C. Tresser : "About Some Theorems by L.P.Šilnikov", *Ann.Inst. Henri Poincaré*, **40**, No.4, pp.441-461, 1984.

438. A.M. Turing : "The chemical basis of morphogenesis", *Phil. Tran. Roy. Soc.*, pp.37-72, 1953.

439. Y. Ueda : "Randomly Transitional Phenomena in the System Governed by Duffing's Equation". *J. Stat. Physics*, **20**, No.2, pp.181-196, 1979.

440. Y. Ueda : "Random Phenomena Resulting from Nonlineartity in the System Described by Duffing's Equation", *Int.J.Non-Linear Mech.*, **20**, No.5/6, pp.481-491, 1985.

441. Y. Ueda : "Survey of Strange Attractors and Chaotically Transitional Phenomena in the System Governed by Duffing's Equation", in *Complex and Distributed Systems : Analysis, Simulation and Control*, Eds.: S.G.Tzafestas, P.Borne, Elsevier, pp.173-180, 1986.

442. Y. Ueda : *The Road to Chaos*, Aerial Press, Santa Cruz 1993.

443. Y. Ueda, N. Akamatsu : "Chaotically Transitional Phenomena in the Forced Negative - Resistance Oscillator", Research Report IPPJ-482, Oct. 1980, Nagoya University. also *IEEE Trans. Circuits Systems*, **CAS-28**, No.3, pp.217-224, 1981.

444. Y. Ueda, C. Hayashi, N. Akamatsu : "Computer Simulation of Nonlinear Ordinary Differential Equations and Nonperiodic Oscillations", *Electronics and Telecommunications in Japan*, **56A**, No.4, pp.27-34, 1973.

445. T. Uehara, N. Inaba, S. Mori : "On Chaotic Phenomena in an Auto Gain Controlled Oscillator Using Pulse Width Controller". *Proc. EC-CTD'87*, **2**, pp.525-530, Paris 1987.

446. B. Uehleke, O.E. Rössler : "Analytical Results on a Chaotic Piecewise - Linear O.D.E.". *Z.Naturforsch.*, **39a**, pp.342-348, 1983.

447. B. Uehleke, O.E. Rössler : "Complicated Poincaré Half-Maps in a Linear System". *Z. Naturforsch.*, **39a**, pp.1107-1113, 1983.

448. A. Ünal : "An Algebraic Criterion for the Onset of Chaos in Nonlinear Dynamical Systems". *Proc. IEEE International Symposium on Circuits Syst.*, **1**, pp.19-22, Helsinki 1988.

449. E. Walter, L.Pronzato : "Qualitative and Quantitative Experiment Design for Phenomenological Models - A Survey", *Automatica*, **26**,

No. 2, pp. 195-213, 1990

450. E. Walter, L.Pronzato : "Identifiabilités et non linéarités" in *Systèmes non linéaires. 1. Modélisation - estimation* A.J. Fossard and D. Normand-Cyrot Eds., Masson, Paris- Milan- Barcelone- Bonn 1993.

451. H. Wang, E.H. Abed : "Bifurcation Control of Chaotic Dynamical Systems". In *Proc. 2nd IFAC Symposium NOLCOS*, 57-62, Bordeaux-France.

452. C. Vanneste, C.C. Chi : "Accrochage de fréquences et apparition du chaos dans une jonction Josephson soumise à un champ microonde". *Annales des Telecommunications*, **42**, No.5-6, pp.274-284, 1987.

453. G.S.Vernam "Cipher printing telegraph systems for secret wire and telegraphic communications". *J. Amar. Inst. Elec. Eng.*, **55**, pp.109-115, 1926.

454. S. Wiggins : *Global Bifurcations and Chaos. Analytical Methods.* Springer Verlag, New York 1988.

455. R.S. Williams : The "Structure of Lorenz Attractors". *Publications IHES*, **50**, pp.321-345, 1980.

456. A.N. Willson : "Limit cycles due to adder overflow in digital filters". *IEEE Trans. Circuit Theory*, **CT-19**, pp.342-346, 1972.

457. A.N. Willson : "Computation of the periods of forced overflow oscillations in digital filters". *IEEE Trans. Acoustics, Speech and Signal Processing*, **ASSP-24**, pp.89-97, 1976.

458. A. Wolf, J.B. Swift : "Universal Power Spectra for the Reverse Bifurcation Sequence", *Phys. Letters A*, **83A**, pp.184-187, 1981

459. A. Wolf, J.B. Swift, H.L. Swinney, J.A. Vastano : "Determining Lyapunov Exponents from a Time Series'. *Physica D*, **16D**. pp.285-317, 1985.

460. C-W. Wu, L.O. Chua "A unified framework for Synchronisation and Control of Dynamical Systems" *Int. J. Bifurcation and Chaos,*

461. C-W. Wu, L.O. Chua "Synchronising Nonautonomous Chaotic Systems without Phase-locking" *J. Circuits, Systems and Computers*

462. C.-W. Wu, N.F. Rulkov : "Studying Chaos via 1-D maps — A tutorial", *IEEE Trans. Circuits Syst.*, **CAS-40**, No.10, pp.707-721, 1993.

463. S. Wu : "Chua's Circuit Family". *Proc. IEEE*, **75**, No.8, pp.1022-1032, 1987.

464. S. Wu, L. Pei, F. Guo : "U-Sequence and Period Tripling Phenomena in a Forced Nonlinear Oscillator", *Chinese Phys. Letters*, **2**, No.5, pp.213-216, 1985.

465. J.A. Yorke : DYNAMICS. Department of Mathematics, University of Maryland, 1988.

466. L.S. Young : "Entropy, Lyapunov exponents, and Hausdorff Dimension in Differentiable Dynamical Systems". *IEEE Trans. Circuits Syst.*, **CAS-30**, No.8, pp.599-607, 1983.

467. Yang L., Liao Y. : Self-Similar Bifurcation Structures from Chua's Circuit . Int.J. Circuit Theory and Applications. **15**, pp.189-192, 1987.

468. G.-Q. Zhong, F. Ayrom : "Periodocity and Chaos in Chua's Circuit". *IEEE Trans. Circuits and Systems*, **CAS-32**, No.5, pp.501-503, 1985

469. G.-Q. Zhong, F. Ayrom : "Experimental Confirmation of Chaos from Chua's Circuit". *Int.J.Circuit Circuit Theory and Applications*, **13**, pp.93-98, 1985.

470. Proceedings of the IEEE, Special Issue on Cryptography, May 1988.

471. Special Issue on Digital Signal Processing , *Proc IEEE*, **63**, April 1975.

472. Special Issue on "Engineering Chaos", *Trans. Inst. Electronics, Information and Communication Eng. IEICE Japan*, vol.E73, No.6, June 1990.

473. Special Issues on "Chaos in Nonlinear Electronic Circuits", Part A - *IEEE Trans. Circuits Syst. Part I: Fundamental Theory Appl.*, **CAS-40**, No.10, October 1993; Part B - *IEEE Trans. Circuits Syst. Part I: Fundamental Theory Appl.*, **CAS-40**, No.11, November 1993; Part C - *IEEE Trans. Circuits Syst. Part II: Analog and Digital Signal Processing*, **CAS-40**, No.10, October 1993;

474. Special Issue on "Chaos and Nonlinear Dynamics", *J. Franklin Institute*, vol.331B, No.6, November 1994.

475. Special Issue on "Nonlinear waves, Patterns and Spatio-Temporal Chaos in Dynamic Arrays", *IEEE Trans. Circuits Syst. Part I: Fundamental Theory Appl.*, **CAS-42**, No.10, October 1995.

476. Theme Issue on "Chaotic behaviour in electronic circuits" J.M.T. Thompson and L.O. Chua Eds. *Philosophical Trans. Royal Society London*, Series A, vol.353, No.701, October 1995.

477. Part Special Issue on "Nonlinear Dynamics of Electronic Systems", *Int. J. Electronics*, Vol.79, No.6, December 1995.

INDEX